# Protein Synthesis and Targeting in Yeast

# NATO ASI Series

## Advanced Science Institutes Series

*A series presenting the results of activities sponsored by the NATO Science Committee, which aims at the dissemination of advanced scientific and technological knowledge, with a view to strengthening links between scientific communities.*

The Series is published by an international board of publishers in conjunction with the NATO Scientific Affairs Division

| | |
|---|---|
| A Life Sciences<br>B Physics | Plenum Publishing Corporation<br>London and New York |
| C Mathematical and Physical Sciences<br>D Behavioural and Social Sciences<br>E Applied Sciences | Kluwer Academic Publishers<br>Dordrecht, Boston and London |
| F Computer and Systems Sciences<br>G Ecological Sciences<br>H Cell Biology<br>I Global Environmental Change | Springer-Verlag<br>Berlin Heidelberg New York<br>London Paris Tokyo Hong Kong<br>Barcelona Budapest |

## NATO-PCO DATABASE

The electronic index to the NATO ASI Series provides full bibliographical references (with keywords and/or abstracts) to more than 30000 contributions from international scientists published in all sections of the NATO ASI Series. Access to the NATO-PCO DATABASE compiled by the NATO Publication Coordination Office is possible in two ways:

- via online FILE 128 (NATO-PCO DATABASE) hosted by ESRIN, Via Galileo Galilei, I-00044 Frascati, Italy.
- via CD-ROM "NATO Science & Technology Disk" with user-friendly retrieval software in English, French and German (© WTV GmbH and DATAWARE Technologies Inc. 1992).

The CD-ROM can be ordered through any member of the Board of Publishers or through NATO-PCO, Overijse, Belgium.

Series H: Cell Biology, Vol. 71

# Protein Synthesis and Targeting in Yeast

Edited by

Alistair J. P. Brown

Department of Molecular & Cell Biology
University of Aberdeen
Marischal College
Aberdeen, AB9 1AS, U.K.

Mick F. Tuite

Biological Laboratory
The University
Canterbury, Kent, CT2 7NJ, U.K.

John E. G. McCarthy

Department of Gene Expression
GBF
Mascheroder Weg 1
3300 Braunschweig, Germany

Springer-Verlag Berlin Heidelberg GmbH

Proceedings of the NATO Advanced Research Workshop on Protein Synthesis and Targeting in Yeast, held at Canterbury, Kent, U.K., July 4–8, 1992

ISBN 978-3-642-84923-7 ISBN 978-3-642-84921-3 (eBook)
DOI 10.1007/978-3-642-84921-3


Originally published by Springer-Verlag Berlin Heidelberg New York London Paris Tokyo in 1993
Softcover reprint of the hardcover 1st edition 1993

Typesetting: Camera ready by author
31/3145 - 5 4 3 2 1 0 - Printed on acid-free paper

# INTRODUCTION

Translation of mRNA in eukaryotic cells is achieved by a battery of translation factors which orchestrate the binding of the ribosome to the mRNA template, the location of the beginning of the coding region and to translate the genetic code into a string of amino acids that constitute the polypeptide. Once synthesis of the polypeptide chain has been completed it either remains in the cell's cytoplasm or is targeted to one of a number of alternative intracellular compartments or to the outside of the cell. How protein synthesis and protein targeting are carried out and coordinated with such efficiency and accuracy represents one of the major focuses of current research in eukaryotic cell and molecular biology.

To analyse these processes in detail the researcher must have at her or his fingertips a barrage of genetic and biochemical tools with which to probe the chosen cell or organism. The arrival of recombinant DNA-based 'reverse genetics' has provided additional and powerful tools with which to probe deeper into the mechanism of protein synthesis and targeting in both simple and complex eukaryotes, but it is the availability of a complementary 'classical' genetic approach that has turned researchers to the use of single-celled yeasts as models for such studies. At the forefront in this respect is the budding yeast Saccharomyces cerevisiae. Over the past decade this remarkably 'simple' unicellular eukaryote has provided us with new, unanticipated insights into eukaryotic gene structure, expression and regulation. It was the realisation that a major worldwide effort was under way to further exploit S. cerevisiae and its closely related species to probe the mechanism of protein synthesis and targeting that persuaded us that this would be an opportune time to gather together all those with active research interests in these two areas.

As far back as the late 1960's (i.e. before the development of gene cloning and splicing technology) a collection of conditional-lethal (temperature-sensitive) mutants had been isolated by Cal McLaughlin and Lee Hartwell among which were identified a number of protein synthesis mutants. Surprisingly it was the lack of effective biochemical systems for dissecting these mutants that delayed further progress in defining the process of protein synthesis in yeast. This technical block was released with the development of an efficient mRNA-dependent translation system was finally achieved again by Cal McLaughlin in collaboration this time with Kivie Moldave. With these biochemical tools finally in place and with the concomitant development of plasmid-based transformation systems and their associated genetic tricks, we now have the means of addressing three key questions: Does yeast use the same translation factors as those biochemically defined in mammalian cells? Can new genetic screens be developed (and exploited) that will identify previously undefined translation factors? and, How is protein synthesis controlled at both the global and mRNA-specific levels?

Studies of protein targeting have also benefited from this surge of interest in S.cerevisiae and, as with the studies on protein synthesis, work in this area began with the isolation of a large collection of conditional-lethal secretion (sec) mutants by Randy Sheckman and his co-workers. These mutants have provided a wealth of information on the process of secretion in S. cerevisiae, from how the secretory proteins are translocated across the membrane of the endoplasmic reticulum into the secretory pathway, to their subsequent migration either to the vacuole, the periplasmic space or to the outside of the cell. By exploiting novel mutant screens that utilise both classical genetics and molecular genetic components the number of such targeting mutants has been further increased. These screening procedures have also revealed, at least at the genetic level, key components of the machinery which targets polypeptides to the two other major organelles of the yeast cell, namely the mitochondrion and the nucleus.

The broad range of topics that could be covered in a Workshop aiming to discuss both protein synthesis and targeting, together with the restrictions in size imposed by the NATO-ASI format, confronted us with a very difficult task in choosing only 50 representatives from a much larger number of leading scientists in these fields even though we were focusing primarily (but not exclusively) on studies carried out on a single organism, S. cerevisiae. It was, we believe, a testimony to the timeliness of holding this meeting on protein synthesis and targeting that we had a very positive response from those we invited to attend. Judging by the response of the participants, both during and after the Workshop, the effort in organising the meeting had been worthwhile. Throughout the four days the level of presentation was very high, the discussions lively and all participants felt they had learnt something new. Furthermore the meeting brought together two usually disparate groups of researchers under a single roof united by their common interest in exploiting the same model cell to address their particular questions.

The present volume contains contributions from over 90% of the speakers at the Canterbury workshop and therefore provides a representative and up-to-date cross section of the activities of the leading groups in their respective fields of research. The contributions are in general review-orientated, and thus offer the reader a series of overviews covering most of the key aspects of protein synthesis and targeting in yeasts.

We are very grateful to the NATO Scientific Affairs Division for providing the major financial support for this Workshop. In addition smaller, but no less valuable financial contributions are gratefully acknowledged from The Wellcome Trust, Glaxo Group Research, ICI Pharmaceuticals and Pfizer UK. We would also like to thank members of our respective laboratories and the staff of Darwin College (University of Kent) for helping to make the Workshop both an enjoyable and a productive experience.

September 1992

Alistair J P Brown, Aberdeen, UK
Mick F Tuite, Canterbury, UK
John E G McCarthy, Braunschweig, FRG

# CONTENTS


mRNA STABILITY AND TRANSLATION

Chapter 1: **Identification of the *cis*-acting sequences and *trans*-acting factors involved in nonsense-mediated mRNA decay** . . . . . . . . . . . . . . 1
S W Peltz, C Trotta, H Feng, A Brown, J Donahue, E Welch and A Jacobson

Chapter 2: **RNA14 and RNA15, two proteins regulating mRNA stability in *Saccharomyces cerevisiae*** . . . . . . . . . . . . . . . . . . . . . . . . . . . . . . . 11
A Petitjean, L Minvielle-Sebastia, E Mandart, N Bonneaud and F Lacroute

Chapter 3: **Translational control in *Saccharomyces cerevisiae* studies *in vivo* and *in vitro*** . . . . . . . . . . . . . . . . . . . . . . . . . . . . . . . . . . 19
J E G McCarthy, B Gerstel, M R Vega Laso, D. Zhu, M F Tuite, F Sagliocco and A J P Brown

THE RIBOSOME

Chapter 4: **Thoughts on the regulation of ribosome synthesis in *Saccharomyces cerevisiae*.** . . . . . . . . . . . . . . . . . . . . . . . . . . . . . . . 29
J R Warner

Chapter 5: **Regulation of ribosomal protein synthesis in yeasts (*Saccharomyces and Kluyveromyces*)** . . . . . . . . . . . . . . . . . . . . . . . . . 39
R J Planta, R Hoekstra, G K Bergkamp-Steffens, P M Goncalves and W H Mager

Chapter 6: **The evolution of ribosomal proteins and yeast** . . . . . . . . . . . . . . . . . . 53
E Otaka, T Hashimoto, K Mizuta and K Suzuki

Chapter 7: **The acidic ribosomal proteins and the control of protein synthesis in yeast.** . . . . . . . . . . . . . . . . . . . . . . . . . . . . . . . . . . . . . . 67
J P G Ballesta, M Remacha, T Naranda, C Santos, B Bermejo, A Jimenez-Diaz and B Ortiz-Reyes

TRANSLATIONAL FIDELITY

Chapter 8: ***SUP35* and *SUP45* genes code for ribosome-bound proteins involved in the control of translational fidelity in yeast** . . . . . . . . . . . . 81
M D Ter-Avanesyan, S A Didichenko, V V Kushnirov and A R Dagkesamanskaya

Chapter 9: **Mutations affecting translational accuracy in yeast** . . . . . . . . . . . . . . . 91
S W Liebman, A Vincent, J M Song and J All-Robyn

Chapter 10: **Dosage-dependent modifiers of *psi*-dependent omnipotent suppression in yeast.** . . . . . . . . . . . . . . . . . . . . . . . . . . . . . . . . . . . . 101
Y O Chernoff and B Ono

Chapter 11: **Aberrant mRNA decoding by the dimorphic yeast *Candida albicans*** . . . . . . . . . . . . . . . . . . . . . . . . . . . . . . . . . . . . . . 111
M A Santos, G Keith and M F Tuite

TRANSLATION FACTORS

Chapter 12: **Effect of initiation factor eIF-5A depletion on cell proliferation and protein synthesis** . . . . . . . . . . . . . . . . . . . . . . . . . . . 123
H A Kang, H G Schwelberger and J W B Hershey

Chapter 13: **Analysis of the genes encoding eIF-4A from yeast** . . . . . . . . . . . . . . . 131
S R Schmid, P Buser, R Coppolecchia, A Fischli and P Linder

Chapter 14: **Uncharged tRNA and derepression of the general amino acid control: autoregulation of yeast lysyl-tRNA synthetase** . . . . . . . . . . . . 143
S Lanker, J L Bushman, A G Hinnebusch, H Trachsel and P P Meuller

TRANSLATION IN THE MITOCHONDRION

Chapter 15: **Positive control of translation in organellar genetic systems** . . . . . . . . 157
T D Fox and Z Shen

Chapter 16: **Translation in yeast mitochondria** . . . . . . . . . . . . . . . . . . . . . . . . . . 167
L A Grivell, P J T Dekker and H J Pel

Chapter 17: **Yeast mitochondrial translation: nuclear genes involved in the expression of the mitochondrial genome** . . . . . . . . . . . . . . . . . . . . 177
T Rinaldi, M Valens and M Bolotin-Fukuhara

MITOCHONDRIAL IMPORT AND SORTING

Chapter 18: **Early events in protein import into mitochondria** . . . . . . . . . . . . . . . 193
K Kassenbrock, M Douglas and D Cyr

Chapter 19: **Targeting pathways to the mitochondrial inner membrane** . . . . . . . . . 203
M G Cumsky, B R Miller, A S Gaikwad and L A Jung

Chapter 20: **Intra-mitochondrial sorting of precursor proteins** . . . . . . . . . . . . . . . 213
R A Stuart, H Koll and W Neupert

Chapter 21: **Genetic characterization of the intermembrane space sorting domains of yeast cytochrome $b_2$** . . . . . . . . . . . . . . . . . . . . . . . . . . . 223
E M Beasley

Chapter 22: **Mitochondrial import of cytochrome C** . . . . . . . . . . . . . . . . . . . . . . . 235
M E Dumont

NUCLEAR TRANSPORT

Chapter 23: **Using yeast to study exchange of macromolecules between the cytoplasm and the nucleus** . . . . . . . . . . . . . . . . . . . . . . . . . . . . . . 245
P Silver

Chapter 24: **Nucleocytoplasmic transport in ribosome biogenesis** . . . . . . . . . . . . . 257
H M Fried

Chapter 25: **Approaches towards a genetic analysis of the nuclear pore complex in yeast** . . . . . . . . . . . . . . . . . . . . . . . . . . . . . . . . . . . . . 269
C Wimmer, V Doye, U Nehrbass, N Schlaich and E C Hurt

TARGETING TO THE ENDOPLASMIC RETICULUM AND BEYOND

Chapter 26: **The SRP-dependent protein targeting pathway in *Saccharomyces cerevisiae*.** . . . . . 283
P Walter, S C Ogg and B C Hann

Chapter 27: **Similarities between *S.cerevisiae* Sec61p and *E. coli* SecY suggest a common origin for protein translocases of the eukaryotic ER and the bacterial plasma membrane** . . . . . 293
C J Stirling

Chapter 28: **ATP transport into yeast ER is a prerequisite for preprotein transfer across the ER membrane** . . . . . 307
Peter Mayinger and David I Meyer

Chapter 29: **Cross-linking signal sequences to components of yeast microsomes.** . . . . . 315
O Westwood and B Austen

Chapter 30: **The nucleotide cycle of SEC4 is important for its function in vesicular transport** . . . . . 325
P Novick, P Brennwald, M D Garrett, M Moya, D Roberts and R Bowser

Chapter 31: **Vacuolar protein sorting in yeast** . . . . . 329
B F Horazdovsky, T R Graham and S D Emr

Chapter 32: **Eukaryotic MDR1/P-glycoprotein homologues: unconventional secretion processes mediated by a growing family of ATP-dependent membrane translocators** . . . . . 339
M Goransson and J Thorner

INTRACELLULAR PROTEIN FOLDING

Chapter 33: **The role of protein disulfide isomerase in yeast** . . . . . 349
T Mizunaga, H Tachikawa and T Miura

Chapter 34: **Novel chaperone-like pro-sequences allow secretion of recombinant human insulin-like growth factor-1 from yeast** . . . . . 359
B Chaudhuri and A Hinnen

Chapter 35: **mRNA translation and protein folding *in vivo*** . . . . . 369
A J P Brown and T Crombie

PROTEIN SPLICING AND DEGRADATION

Chapter 36: **Protein splicing of yeast TFP1: evidence for a new class of mobile genetic elements** . . . . . 379
A A Cooper and T H Stevens

Chapter 37: **Ubiquitin-dependent protein degradation** . . . . . 389
S Jentsch, H-P Hauser, R Heinlein, J Jungmann, S Schlenker, W Seufert, T Sommer and S Springer

Chapter 38: **The *TIF463A* and *TIF463B* genes encode the large subunits of the yeast cap binding protein complex** . . . . . 395
C Goyer, H S Lee, A Blanc, M Altmann, H Trachsel and N Sonenberg

SUBJECT INDEX . . . . . 405

PARTICIPANTS AND CONTRIBUTORS

Austen, Dr. B.M. Department of Surgery, St. George's Hospital Medical School, Cranmer Terrace, Tooting, London, SW17 DCPT.

Ballesta, Dr. J.P.G. Centre de Biologia Molecular, CSIC-UAM, Canto Blanco, 28049 Madrid, Spain.

Beasley, Dr. E. Biozentrum, University of Basel, Klingelbergstr. 70, CH-4056 Basel, Switzerland.

Bolotin-Fukuhara, Dr. M. Lab. de Genetique Moleculaire, Bat. 400, Universite Paris-Sud, Inst. de Genetique et Microbiologie, Centre d'Orsay, 91 405 Orsay, Cedex, France.

Brown, Dr. A.J.P. Department of Molecular & Cell Biology, University of Aberdeen, Marischal College, Aberdeen, AB9 1AS.

Chaudhuri, Dr. B. Ciba-Geigy AG, Biotechnology Department K681.146, CH-4002 Basel, Switzerland.

Cumsky, Dr. M. Dept. Molecular Biology & Biochemistry, University of California, Irvine, California 92717, U.S.A.

Donahue, Dr. T.F. Department of Biology, Indiana University, Jordan Hall, Room A305, Bloomington, Indiana 47405, U.S.A.

Dumont, Dr. M. Department of Biochemistry, University of Rochester, P.O. Box 607, Rochester, NY 14642, U.S.A.

Egerton, Dr. M. ICI Pharmaceuticals, Biotechnology Department, Mereside, Alderley Park, Macclesfield, Cheshire, SK10 4TG.

Emr, Dr. S.D. Div. of Cellular & Molecular Medicine, Howard Hughes Medical Institute, University of California, San Diego, La Jolla, CA 92093-0668, U.S.A.

Featherstone, Dr. C. Elsevier Trends Journals, 68 Hills Road, Cambridge, CB2 1LA.

Fox, Dr. T. Section of Genetics & Development, Cornell University, Biotechnology Building, Ithaca, NY 14853-2730, U.S.A.

Fried, Dr. H.M. Department of Biochemistry, University of North Carolina, CB 7260 335 Fac. Lab. Off. Blds., Chapel Hill, North Carolina 27599, U.S.A.

Görranson, Dr. M. Division of Biochemistry & Molecular Biology, Department of Molecular & Cell Biology, Room 401, Barker Hall, University of California, Berkeley, CA 94720, U.S.A.

Grivell, Dr. L.A. Section for Molecular Biology, University of Amsterdam, Kruislaan 318, Amsterdam 1098SM, The Netherlands.

Hershey, Dr. J.W. Department of Biological Chemistry, University of California, Davis, CA 95616, U.S.A.

Hinnebusch, Dr. A. National Institute of Health, Building 6, Room 320, Bethesda, MD 20892, U.S.A.

Hurt, Dr. E.C. EMBL, Meyerhofstr. 1, W-6900 Heidelberg, FRG.

Jacobson, Dr. A. Dept. of Molecular Genetics & Microbiology, University of Massachusetts Medical School, 55 Lane Avenue North, Worcester, MA 01655, U.S.A.

Jentsch, Dr. S. Friedrich-Miescher-Laboratorium der Max-Planck-Gesellschaft, Spemannstr. 37 - 39, W-7400 Tübingen, F.R.G.

Kassenbrock, Dr. K. Dept. of Biochemistry & Biophysics, CB 7260 405 FLOB, University of North Carolina at Chapel Hill, Chapel Hill, NC 27599, U.S.A.

Lacroute, Dr. F. Centre Genetic Moleculaire, Du CNRS, Gif sur Yvette, 91190, France.

Lanker, Dr. S. Inst. für Biochemie und Molekularbiologie, Universität Bern, Bühlstr. 28, 3012 Bern, Switzerland.

Liebman, Dr. S.W. Laboratory for Molecular Biology, Department of Biology, University of Illinois, Box 4348, Chicago, IL 6060, U.S.A.

Linder, Dr. P. Department of Microbiology, Biozentrum, Klingelbergstr. 70, CH-4056 Basel, Switzerland.

McCarthy, Dr. J.E.G. Department of Gene Expression, GBF, Mascheroder Weg 1, W-3300, Braunschweig, F.R.G.

Meyer, Dr. D.I. Department of Biological Chemistry, UCLA School of Medicine, Los Angeles, CA 90024, U.S.A.

Mizunaga, Dr. T. Department of Agricultural Chemistry, Faculty of Agriculture, University of Tokyo, Bunkyo-Ku, Tokyo 113, Japan.

Novick, Dr. P. Department of Cell Biology, Yale University School of Medicine, P.O. Box 3333, New Haven, Connecticut, U.S.A.

Olievera, Miss C. Department of Gene Expression, GBF, Mascheroder Weg 1, W-3300, Braunschweig, F.R.G.

Ono, Dr. B. Laboratory of Environmental Hygiene Chemistry, Faculty of Pharm. Sciences, Tshushima Naka 1-1-1-, Okyama 700, Japan.

Otaka, Dr. E. Department of Biochemistry & Biophysics, Research Institute for Nucl. Medicine & Biology, Hiroshima University, Kasumi 1-2-3, Minami-Ku, Hiroshima 734, Japan.

Paddon, Dr. C.J. Glaxo Group Research Ltd., Greenford Road, Greenford, Middlesex, UB6 0HE.

Pelham, Dr. H.R.B. Laboratory of Molecular Biology, Hills Road, Cambridge, CB2 2QH.

Planta, Professor R. Department of Biochemistry & Molecular Biology, Free University, de boelelaan 1083, 1081 Amsterdam, The Netherlands.

Santos, Mr. M.A. Biological Laboratory, The University, Canterbury, Kent, CT2 7NJ, U.K.

Sherman, Dr. F. Department of Biochemistry, University of Rochester, 601 Elmwood Avenue, Rochester, NY 14642, U.S.A.

Silver, Dr. P. Molecular Biology Department, Princeton University, Princeton, NJ 08544, U.S.A.

Smith, Dr. R. Department of Molecular & Cell Biology, University of Aberdeen, Marischal College, Aberdeen, AB9 1AS, Scotland, U.K.

Sonenberg, Dr. N. Department of Biochemistry, McGill University, 3655 Drummond Str., Montreal P.Q., Canada H3G 1Y6.

Stansfield, Dr. I. Biological Laboratory, The University, Canterbury, Kent, CT2 7NJ, U.K.

Stevens, Dr. T.H. Institute of Molecular Biology, University of Oregon, Eugen, OR 97403, U.S.A.

Stirling, Dr. C.J. Department of Biochemistry & Molecular Biology, University of Manchester, Oxford Road, Manchester, M13 9PT, U.K.

Stuart, Dr. R. Inst. für Physiologische Chemie der Universität München, Goethestr. 33, 8000 München 2, F.R.G.

Ter-Avanasyen, Dr. M.D. Institute of Experimental Cardiology, USSR Cardiology Research, 3rd Cherepkovskaya Street 15 A, 121552 Moscow, U.S.S.R.

Tuite, Dr. M.F. Biological Laboratory, The University, Canterbury, Kent, CT2 7NJ, U.K.

Varshavsky, Dr. A. Division of Biology, California Institute of Technology, Pasadena, CA 91125, U.S.A.

Walter, Dr. P. Department of Biochemistry, University of California, San Francisco, San Francisco, CA 94143-0448, U.S.A.

Warner, Dr. J.R. Department of Cell Biology & Biochemistry, Albert Einstein College of Medicine, 1300 Morris Park Avenue, Bronx, NY 10461, U.S.A.

Wickstead, Mr. B. Department of Molecular & Cell Biology, University of Aberdeen, Marischal College, Aberdeen, AB9 1AS, Scotland, U.K.

Zanchin, Mr. N. Department of Gene Expression, GBF, Mascheroder Weg 1, W-3300, Braunschweig, F.R.G.

---

PARTICIPANTS AT THE NATO-SPONSORED WORKSHOP:
CANTERBURY, JULY 1992

PROTEIN SYNTHESIS AND TARGETING IN YEAST

1, Al Brown; 2, Mick Tuite; 3, Bun-ichiro Ono; 4, Susan Liebman; 5, Fred Sherman; 6, Eiko Otaka; 7, Carol Featherstone; 8, Carla Oliveira; 9, Nilsson Zanchin; 10, Mark Dumont; 11, Takemitsu Mitzunaga; 12, Alex Varshavsky; 13, Michael Ter-Avanesyan; 14, Mark Egerton; 15, Chris Paddon; 16, Manuel Santos; 17, Barton Wickstead; 18, Ellen Beasley; 19, Ken Kassenbrock; 20, Peter Novick; 21, Monique Bolotin-Fukuhara; 22, Rudi Planta; 23, Francois Lacounte; 24, Les Grivell; 25; Tom Stevens; 26, Pam Silver; 27, Stefan Lanker; 28, Patrick Linder; 29, Albert Hinnen; 30, Ian Stansfield; 31, Rachel Smith; 32, Peter Walter; 33, Brian Austen; 34, Scott Emr; 35, Colin Stirling; 36, Jon Warner; 37, Stefan Jentsch; 38, Ed Hurt; 39, Allan Jacobson; 40, Tom Fox; 41, Hugh Pelham; 42, Howard Fried; 43, John McCarthy; 44, David Meyer; 45, Mikael Goransson.

Missing: John Hershey, Tom Donahue, Nahum Sonenberg, Michael Cumsky, Rosemary Stuart, Juan Ballesta.

# Identification of the cis-acting sequences and trans-acting factors involved in nonsense-mediated mRNA decay

Stuart W. Peltz, Christopher Trotta, He Feng, Agneta Brown, Janet Donahue, Ellen Welch and Allan Jacobson
Department of Molecular Genetics and Microbiology
University of Massachusetts Medical School
Worcester, Massachusetts 01655

**Introduction.** In both prokaryotes and eukaryotes nonsense mutations in a gene can enhance the decay rate or reduce the abundance of the mRNA transcribed from that gene (see Peltz and Jacobson, 1992, for a review). We use the term nonsense-mediated mRNA decay to describe this phenomenon. To understand the mechanism of nonsense-mediated mRNA decay, we have inserted amber mutations at six different positions in the coding region of *PGK1*, a yeast gene which encodes a transcript that is normally stable and abundant. We find that: a) a subset of the nonsense mutations have a significant destabilizing effect on PGK1 mRNA, b) in addition to a nonsense codon, downstream sequences are required for mRNA destabilzation, c) a nonsense codon and a downstream sequence are not sufficient for nonsense-mediated mRNA decay because the pathway is regulated by an independent, stabilizing, sequence element, d) the product of the *UPF1* gene (Upf1p) is essential for full activity of the nonsense-mediated mRNA decay pathway and e) Upf1p is localized to polysomes. The experiments that lead us to these conclusions are discussed below.

**Destabilization of the PGK1 transcript is dependent on the position of a nonsense codon within the PGK1 coding region.** To investigate the relationship between the location of a nonsense mutation in the *PGK1* protein coding region and its effect on mRNA half-life, a linker harboring an amber codon was inserted (in separate constructs) into six convenient restriction sites of the *PGK1* gene. The *PGK1* allele used contains a DNA tag inserted into its 3'-UTR, so that the mRNA decay rates of wild-type and mutant *PGK1* alleles could be monitored with a radioactive probe specific for only the tag sequence. The mutant *PGK1* alleles on centromere plasmids were transformed into yeast cells harboring a temperature-sensitive allele of RNA polymerase II (*rpb1-1*; Nonet et al, 1987) and the abundance and decay rates of the wild-type and mutant PGK1 mRNAs were determined (Parker et al., 1991). The results of these experiments indicate that 5'-proximal nonsense mutations accelerate the PGK1 mRNA decay rate more than 3'-proximal mutations, although the relationship is nonlinear (Fig. 1; see UPF1(+) column). Nonsense

NATO ASI Series, Vol. H 71
Protein Synthesis and Targeting in Yeast
Edited by A. J. P. Brown, M. F. Tuite, and J. E. G. McCarthy

mutations that terminate translation after 55% or less of the PGK1 coding sequence accelerate the PGK1 mRNA decay rate approximately twelve-fold. A nonsense mutation that allows translation of 67% of the protein coding region still decreases the PGK1 mRNA half-life four-fold while nonsense mutations inserted in the last quarter of the PGK1 transcript had no effect (Fig. 1). The decay rate of an unrelated, stable mRNA encoded by the *CYH2* gene was essentially identical in all cells ($t_{1/2}$ = 40-44 min), regardless of the nature of their respective *PGK1* alleles (Fig. 1). Furthermore, the abundance of the *PGK1* nonsense-containing mRNAs is directly related to their respective decay rates (Fig. 1).

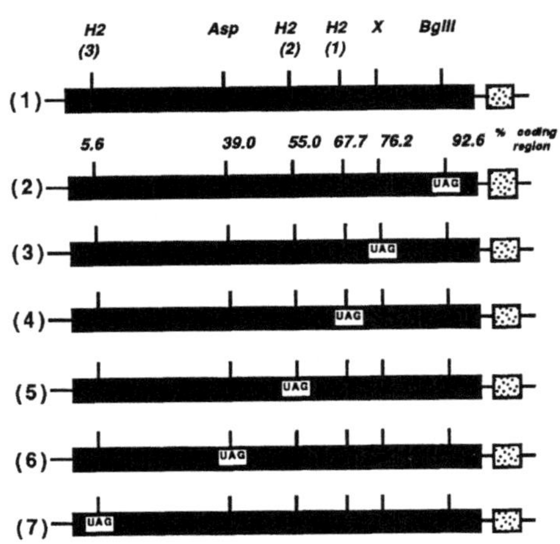


| UPF(+) | | | UPF(-) |
|---|---|---|---|
| PGK1 | | CYH2 | PGK1 |
| $t_{1/2}$ (min) | mRNA Abundance | $t_{1/2}$ (min) | $t_{1/2}$ (min) |
| 64 | 1.0 | 43 | 62 |
| 60 | 0.94 | 41 | 61 |
| 62 | 1.1 | 40 | 63 |
| 15 | 0.23 | 44 | 64 |
| 5 | 0.10 | 44 | 63 |
| 5 | 0.1 | 40 | 62 |
| ≤3 | 0.04 | 43 | 63 |

**Figure 1. mRNA decay rates for *PGK1* and its nonsense alleles in *UPF1*$^+$ and *upf1*$^-$ strains.** mRNA decay rates were measured as described by Parker et al. (1991). mRNA abundance was measured by RNA blot analysis of equal amounts of RNA from the time zero points of each of the mRNA decay measurements, normalizing to the abundance of the CYH2 transcript. The decay rate of the CYH2 transcript was determined in the *UPF1*$^+$ strain. UPF(+): wild-type for the *UPF1* gene; UPF(-); a deletion of the *UPF1* gene.

**The *UPF1* gene encodes a trans-acting factor involved in nonsense-mediated mRNA decay.** The product of the yeast *UPF1* gene is required for the rapid turnover of mRNAs containing a premature translational termination codon (Leeds et al, 1991; reviewed Peltz and Jacobson, 1992). Either the *upf1-2* mutation, or deletion of the *UPF1* gene from the yeast genome (*upf1Δ*), will selectively stabilize mRNAs containing early nonsense mutations while the decay rates of most other mRNAs are unaffected (Leeds et al, 1991; Peltz and Jacobson, 1992). The consequences of deleting the *UPF1* gene from the yeast chromosome on the mRNA decay rates of the six *PGK1* nonsense alleles were determined. In a *upf1Δ* strain, all mRNAs were as stable as the wild-type PGK1 mRNA, with half-lives on the order of 60 min, regardless of the location of the nonsense codon (see UPF(-), Fig. 1).

The *UPF1* loss of function mutations provide valuable tools for the study of the nonsense-mediated mRNA decay pathway. We can now define a cis-acting element that promotes nonsense-mediated mRNA decay as a sequence that accelerates mRNA decay in wild-type cells, but which is inactivated in strains deleted for the *UPF1* gene.

The *UPF1* gene product may regulate the decay rates of transcripts with upstream open reading frames (reviewed in Peltz and Jacobson, 1992) or control the abundance of inefficiently spliced pre-mRNAs or pre-mRNAs whose splicing is regulated. Inefficently spliced pre-mRNAs encoded by the *CYH2* or *RP51B* genes, or the unspliced transcript from the *MER2* gene (Engebrecht et al. 1991), are stabilized five- to ten-fold in a *upf1*$^-$ strain. Since the intron sequences in both transcripts are 5′-proximal and contain in-frame nonsense codons, it is not surprising that the unspliced pre-mRNAs are unstable in *UPF1*$^+$ cells and stabilized in a strain lacking *UPF1* function.

**Nonsense-mediated decay of the PGK1 mRNA requires both a nonsense codon and sequences downstream of the nonsense codon.** The discontinuous relationship between the position of a nonsense codon and its effect on mRNA turnover (Fig. 1) suggested that sequences downstream of the nonsense codon may play a role in nonsense-mediated decay, either because they act as sites accessible to nuclease attack or because sequences in addition to a nonsense codon are required to trigger mRNA decay. To determine whether sequences 3′ of a nonsense mutation are required for nonsense-mediated mRNA decay, we deleted various amounts of *PGK1* protein coding sequence downstream of a nonsense mutation located at 5% of the *PGK1* coding region and the mRNA half-lives of these *PGK1* alleles were determined (Fig. 2). Deletions of the PGK1 protein coding

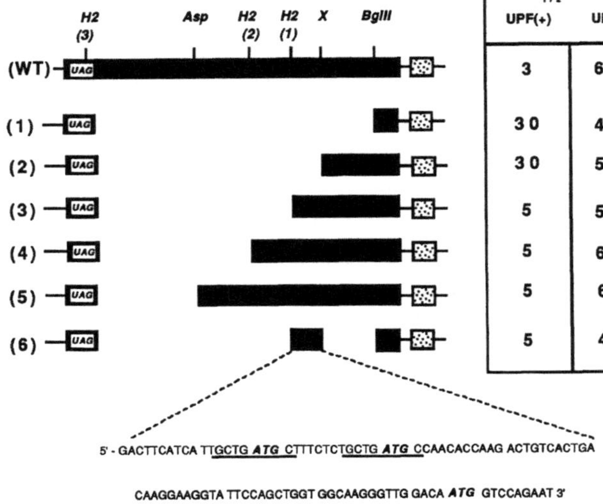


| | $t_{1/2}$(min.) | |
|---|---|---|
| | UPF(+) | UPF(-) |
| (WT) | 3 | 63 |
| (1) | 30 | 48 |
| (2) | 30 | 57 |
| (3) | 5 | 50 |
| (4) | 5 | 63 |
| (5) | 5 | 60 |
| (6) | 5 | 49 |

**Figure 2. Decay rates for PGK1 mRNAs containing deletions downstream of a nonsense codon.** *PGK1* nonsense alleles containing a 5'-proximal nonsense mutation (located in the H2(3) site) and deletions of various amounts of the coding sequence downstream of the nonsense mutation are depicted and their mRNA decay rates in *UPF1*$^{+}$ and *upf1*Δ strains are summarized. The *PGK1* coding sequence is represented by the thick bar and the sequence that was deleted is represented by the absence of the bar. mRNA decay rates were determined as described by Parker et al. (1991). The deletions comprise the sequences between: (1) 5.6 and 92%, (2) 5.6 and 76.2%, (3) 5.6 and 67.7%, (4) 5.6 and 55%, and (5) 5.6 and 39% of the *PGK1* coding region. The construct shown in (6) contains the 106 nt downstream element inserted 3' to the amber mutation in the H2(3) site.

region that removed sequences between 5 and 67% (Fig. 2, [3]), 5 and 55% (Fig. 2 , [4]), and 5 and 39% of the *PGK1* protein sequence [Fig. 2, [5]) did not inactivate the nonsense-mediated mRNA decay pathway. The mRNAs encoded by these constructs all had half-lives of 5 min. However, mRNA decay rates for two *PGK1* nonsense alleles that deleted either between 5 and 92% or 5 and 76% of the *PGK1* protein coding region downstream of the amber mutation were stabilized approximately ten-fold.(Fig. 2, [1] and [2]). Experiments that inserted small regions of the deleted DNA back into a *PGK1* nonsense allele in which most of the protein coding region was deleted (*PGK1* mini-gene) demonstrated that a 106 nt sequence element, when positioned downstream of the nonsense codon, can promote rapid decay of its mRNA (Fig. 2, [6]). Although the 106 nt element is specific, it is not unique. Deletion of the 106 nt fragment from an otherwise intact *PGK1* gene containing an early nonsense mutation did not stabilize the resultant transcript, indicating that there are redundant 3′ cis-acting element(s) in the *PGK1* gene that can promote nonsense-mediated mRNA decay (data not shown).

The 5′ and 3′ boundaries of the 106 nt downstream element in the *PGK1* mini-gene were identified by deletion analysis. Deleting 34 nt or more from the 5′ end of the downstream element stabilized the transcript, from a half-life of 5 min or less to approximately 30 min. A 19 nt deletion from the 5′ end of the downstream element resulted in an intermediate mRNA decay rate ($t_{1/2}=15$ min). The results from the 3′ deletion series demonstrated that removing 25 nt from the 3′ end of the downstream element had no effect on the decay rate of the mini-PGK1 transcript while deleting either 50 or 75 nt resulted in intermediate decay rates ($t_{1/2}=15$ min); larger 3′ deletions stabilized the transcripts such that they had the same decay rates as the PGK1 mRNA lacking the downstream element.

**What is the role of the downstream element?** Three observations suggest that the role of the downstream element in the nonsense-mediated mRNA decay pathway may be to promote translational reinitiation: a) three ATG codons lie within the 106 nt sequence that serves as a functional downstream element; b) insertion of a stem-loop structure, which inhibits both translational initiation and reinitiation, immediately downstream of a nonsense codon stabilizes an otherwise unstable PGK1 transcript (S.W.P, A.H.Brown and A.J., unpublished observations), and c) 3-aminotriazole, an inhibitor of amino acid biosynthesis that has been shown to reduce the capacity of cells to reinitiate translation at downstream start codons (as in the case of the GCN4 transcript; Hinnebusch and Liebman, 1991) also stabilizes mRNAs with nonsense mutations without affecting the decay of wild-type transcripts (S.W.P, A.H.B, and A.J., unpub. exps.). Based on these observations we determined whether mutating the potential translational reinitiation sites (the ATG codons) in the downstream element sequence would affect the ability of the downstream sequence elements to promote nonsense-mediated mRNA decay. We reasoned that the ATG codons would be the most likely (but not the only) potential translational reinitiation sites. Deletion analysis demonstrated that the third ATG triplet in the downstream element is not required for nonsense-mediated mRNA decay. The two other ATG triplets reside in the first 34 nt of the element shown to be essential for rapid mRNA decay. These ATG triplets are 12 nt apart and are each contained within identical nine nt sequences. Consistent with the possibility that these sequences are involved in translational reinitiation or ribosome pausing, this nine nucleotide element is complementary to sequences in the 18S rRNA (see below). To assess the importance of these ATG codons in the promotion of nonsense-mediated mRNA decay, the G nucleotide in each was mutated to other bases. The mutant 106 nt downstream elements were inserted downstream of the 5′-proximal nonsense mutation in the mini-*PGK1* gene (i.e., the *PGK1* gene that lacked sequences from 5 to 92% of the PGK1 protein coding region) and mRNA decay rates were determined. The results from these experiments indicate that certain combinations of mutations that altered the G nucleotide of the first two ATG triplets (ATG-1 and ATG-2) decreased the ability of the downstream element to promote nonsense-mediated mRNA decay while other mutation combinations had no effect. In two constructs in which the downstream element contained a mutation of ATG-1 to an ATA codon the mRNAs were stabilized 6- to 9-fold. Mutations in the G nucleotide in ATG-2 had only modest effects when an ATA codon was substituted for ATG-1. Changing the G nucleotide of ATG-1 to either C or T in combination with various altered triplets in ATG-2 did not affect the nonsense-mediated mRNA decay pathway significantly. Taken together with the deletion analyses, these results indicate that ATG-1 and ATG-2 are important components of the downstream element.

**The position-dependent effects of nonsense mutations on mRNA turnover are a consequence of modulations in the activity of the nonsense-mediated mRNA decay pathway.** When between 67 and 76.6% of the PGK1 protein coding sequence has been translated the PGK1 transcript becomes insensitive to the nonsense-mediated decay pathway (Fig. 1). This transition indicates that either that 3′-proximal amber mutations lack the necessary downstream element or that factor(s) required for nonsense-mediated mRNA decay are lost progressively from the ribosome as it translocates down the mRNA. To test these possibilities, sequences 3′ of the nonsense mutation inserted at 92% of the *PGK1* protein coding region were replaced with protein coding regions and a 3′-UTR sequence from a *PGK1* gene harboring downstream elements capable of promoting nonsense-mediated mRNA decay. The mRNA decay rates of these *PGK1* alleles were unaffected by these insertions ($t_{1/2}$=60-62 min). These results indicate that 3′-proximal nonsense mutations are resistant to the nonsense-mediated mRNA decay pathway for reasons other than a lack of a specific downstream element, and that there must be sequences in the PGK1 mRNA that, when translated, neutralize the destabilizing effects of any downstream nonsense mutations. Recent experiments indicate that the PGK1 mRNA contains such a stabilizer element. This element may promote resistance to the nonsense-mediated decay pathway by promoting the loss of a ribosome-associated factor required for nonsense-mediated mRNA decay. We consider Upf1p, the protein encoded by the *UPF1* gene, to be an excellent candidate for such a ribosome-associated, stabilizer-regulated, factor.

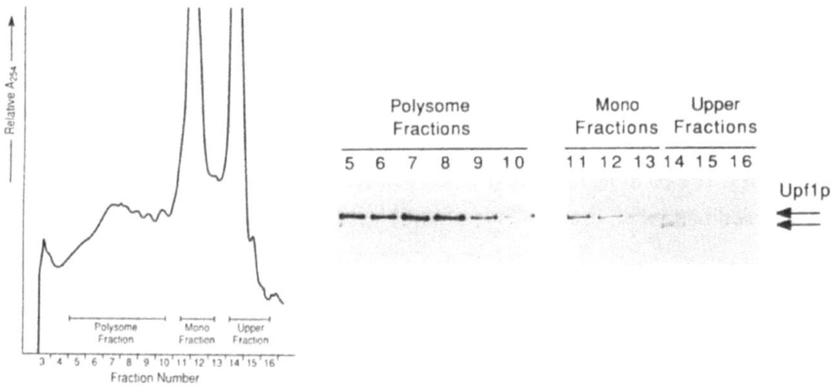


**Figure 3. Localization of the FLAG-Upf1p protein in yeast extracts.** A yeast extract isolated from a upf1Δ strain harboring the FLAG-UPF1 allele inserted into a high copy plasmid was fractionated on a sucrose gradient. The absorbance profile is shown on the left. Aliquots from the polysome fractions, the monosome fractions, and fractions from the upper part of the sucrose gradient were concentrated and analyzed by western blotting, probing with a FLAG-specific monoclonal antibody.

**The Upf1 protein is associated with polysomes.** The hypothesis described above suggests that the Upf1 protein (Upf1p) is associated with ribosomes in order for it to be active. In order to determine the subcellular localization of Upf1p, an epitope tag (FLAG sequence: Met Asp Tyr Lys Asp Asp Asp Asp Lys; Hopp et al., 1988) was inserted at the amino terminal end of the *UPF1* gene resulting in the FLAG-*UPF1* allele. An antibody against this nanopeptide is commercially available. The FLAG-*UPF1* gene was subcloned into both centromere and high copy yeast plasmids and transformed into a *upf1*Δ strain. Both biochemical and genetic assays demonstrated that the activity of Upf1p was unaffected by the FLAG sequence. The decay of the his4-38 mRNA, encoded by a *his4* frameshift allele that prematurely terminates translation, was degraded rapidly in both *UPF1*$^+$ and FLAG-*UPF1*$^+$ strains, but was stabilized five-fold in a *upf1*$^-$ strain. Furthermore, the FLAG-*UPF1* allele was able to abrogate the omnipotent suppression phenotype for the *leu2-2* allele observed in a *upf1*$^-$ strain. To determine whether Upf1p is ribosome-associated, post-mitochondrial extracts prepared from cells harboring the FLAG-*UPF1* gene inserted into a high copy plasmid were fractionated by centrifugation through sucrose gradients. The presence of the FLAG-Upf1 protein in the gradient fractions was determined by western blotting, probing the blots with a monoclonal antibody directed against the FLAG epitope (Fig. 3). Most of the FLAG-Upf1p is polysome-associated, although a small amount of the protein is also contained in the monosome and upper fractions (Fig. 3). Upf1p associated with the polysome fraction consists predominantly of a 116 kD polypeptide and a smaller amount of a 108kD molecular weight polypeptide. This result suggests that Upf1p may be modified when it associates with ribosomes.

**Identifying factors that interact with Upf1p.** In order to identify factors that interact with the *UPF1* protein, we have utilized a novel genetic system to detect protein-protein interactions *in vivo* (Fields and Song, 1989 ,Chien et al, 1991). The system is based on the fact that the yeast *GAL4* transcriptional activator protein contains separable DNA-binding and transcription activation domains. These two domains, even when present on separate polypeptides, are capable of activating transcription if they are brought into close proximity by a protein-protein interaction. A hybrid gene consisting of the *GAL4* DNA binding domain and the complete *UPF1* protein coding region was cotransformed with a library of genomic sequences fused to the *GAL4* activation domain into a strain harboring a *GAL1/lacZ* reporter gene. By screening for yeast cells expressing lacZ activity when both plasmids are present in the same cell, at least six genes whose products putatively interact with Upf1p have been identified. One of these genes, *DBP2*, encodes a yeast homolog of a putative human RNA helicase (Iggo et al, 1991). Another gene isolated, *MNS1*, encodes a yeast α-mannosidase, an enzyme involved in post-translational modifications of proteins (Camirand et al, 1991). The relevence of this protein in mRNA decay is currently being investigated.

Although the function of these genes in mRNA turnover has not been established, the identification of a putative RNA helicase is a striking finding. Many of the known translational initation factors are also RNA helicases (Yoon et al, 1992). As described above, we hypothesized that translational reinitiation may be part of the nonsense-mediated mRNA decay pathway. Perhaps these factors are necessary for this process. Alternatively, RNA helicases may be required to unwind the RNA in order for it to be a substrate for the turnover machinery. Further work is required to establish their roles in both nonsense-mediated mRNA decay and in the turnover of inherently unstable transcripts.

**Translation and turnover are intimately linked.** Experiments reported previously have demonstrated that ongoing translation is important for promoting rapid mRNA turnover or its regulation (reviewed in Peltz et al, 1991; Peltz and Jacobson, 1992). Consistent with this linkage of translation and turnover, we find that premature translational termination destabilizes mRNAs, and that translational elongation inhibitors (i.e., cycloheximide) or mutations that inhibit protein synthesis stabilize PGK1 mRNAs containing nonsense mutations. These observations are consistent with an essential role for ribosome translocation in mRNA decay.

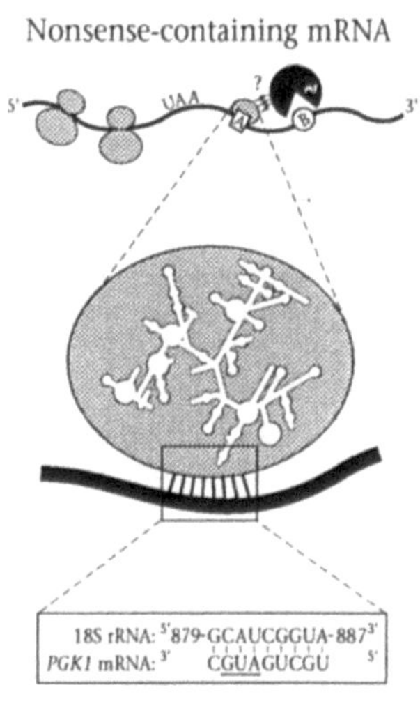


**Figure 4. Two-site model for the decay of mRNAs containing premature translational termination codons.** The "A" site of the element is postulated to be a position of ribosome pausing, possibly as a consequence of mRNA:rRNA interactions. Examples of possible base-pairing are shown for a segment of yeast 18S rRNA of the sequences flanking a downstream AUG codon required for destabilization of a PGK1 mRNA containing a premature translational termination codon. The "B" site indicates a position of endonucleolytic cleavage. Staggered lines between the 40S subunit and the putative nuclease are intended to suggest a possible interaction. Numbers flanking rRNA sequences are the nucleotide positions within the primary sequence.

Nonsense-mediated mRNA decay requires a sequence downstream of a nonsense codon that may serve as a site of translational re-initiation and/or mRNA:rRNA interaction (Fig. 4). A likely consequence of either event may be ribosome pausing. We consider it possible that a ribosome paused at a specific site may expose downstream nuclease recognition sites which could then be cleaved by either a soluble or a ribosome-bound or ribosome-activated nuclease. A two-site model, in which the first site potentiates the cleavage mechanism and the second site is the actual position of the nucleolytic attack (see Fig. 4), is consistent with the deletion data of Fig. 2. Moreover, the dependence on a ribosome-bound or ribosome-activated nuclease is consistent with the available data for both the coding region stabilizer element and the *UPF1* gene product.

**Acknowledgments.** This work was supported by a grant (GM27757) to AJ from the NIH and by a postdoctoral fellowship to SWP from the American Cancer Society.

**References**

**Camirand, A., A. Heysen, B. Grondin, and A. Herscovics.** 1991. Glycoprotein biosynthesis in Saccharomyces cerevisiae. Isolation and characterization of the gene encoding a specific processing α-mannosidase. J. Biol. Chem. **266**:15120-15127.

**Chien, C.-T., P.L. Bartel, R. Sternglanz, and S. Fields.** 1991. The two hybrid system: a method to identify and clone genes for proteins that interact with a protein of interest. Proc. Natl. Acad. Sci. USA **88**:9578-9582.

**Engebracht, J., K. Voelkel-Meiman, and G.S. Roeder.** 1991. Meiosis-specific RNA splicing in yeast. Cell **66**:1257-1268.

**Fields, S. and O.-K. Song.** 1989. A novel genetic system to detect protein-protein interactions. Nature **340**:245-246.

**Hinnebusch, A.G. and S.W. Liebman.** 1991. Protein synthesis and translational control in Saccharomyces cerevisiae. In The Molecular and Cellular Biology of the Yeast Saccharomyces: Volume I. Genome Dynamics, Protein Synthesis, and Energetics (J.R. Broach, J.R. Pringle, and E.W. Jones, Eds.). Cold Spring Harbor Laboratory Press, pp. 627-735.

**Hopp, T.P., K.S. Prickett, V.L. Price, R.T. Libby, C.J. March, P. Ceretti, D.L. Urdal, and P.J. Conlon.** 1988. A short polypeptide marker sequence useful for recombinant protein identification and purification. Biotechnology **6**:1204-1210.

**Iggo, R.D., D.J. Jamieson, S.A. MacNeill, J. Southgate, J. McPheat, and D.P. Lane.** 1991. p68 RNAhelicase:identification of a nucleolar form and cloning of related genes containing a conserved intron in yeasts. Mol. Cell. Biol. **11**:1326-1333.

**Leeds, P., S.W. Peltz, A. Jacobson, and M.R. Culbertson.** 1991. The product of the yeast UPF1 gene is required for rapid turnover of mRNAs containing a premature translational termination codon. Genes and Devel. **5**:2303-2314.

**Nonet, M., C. Scafe, J. Sexton, and R. Young.** 1987. Eucaryotic RNA polymerase conditional mutant that rapidly ceases mRNA synthesis. Mol. Cell. Biol. **7**:1602-1611.

**Parker, R., D. Herrick, S.W. Peltz, and A. Jacobson.** 1991. Measurement of mRNA decay rates in Saccharomyces cerevisiae. In Methods in Enzymology: Molecular Biology of Saccharomyces cerevisiae. (Eds. C. Guthrie, G. Fink), Academic Press, pp. 415-423.

**Peltz, S.W., G. Brewer, P. Bernstein, and J. Ross.** 1991. Regulation of mRNA turnover in eukaryotic cells. Crit. Rev. Euk. Gene Exp. **1**:99-126.

**Peltz, S.W. and Jacobson.** 1992. mRNA Turnover in Saccharomyces cerevisiae. In: Control of mRNA Stability (Eds: G. Brawerman & J.Belasco) Academic Press, 1992.

**Yoon, H. and T.F. Donahue.** 1992. Control of translation initiation in Saccharomyces cerevisiae. Molec. Microbiol. **6**:1413-1419.

# RNA14 AND RNA15, TWO PROTEINS REGULATING mRNA STABILITY IN *SACCHAROMYCES CEREVISIAE*

A. Petitjean, L. Minvielle-Sebastia, E. Mandart, N. Bonneaud, and F. Lacroute
Centre de Génétique Moléculaire
Laboratoire propre du CNRS associé à l'Université P. et M. Curie
91198 Gif-sur-Yvette Cedex
France

## Introduction

In *S. cerevisiae*, thermosensitive mutations in the unlinked genes *RNA14* and *RNA15* cause very similar phenotypes once the cells are shifted to the non permissive growth temperature. The length of the mRNA poly(A) tails is shortened and the mRNA stability is strongly decreased whilst the polymerase II transcription rate is only slightly modified in an allelic-dependent manner (Bloch *et al.*, 1978, Minvielle-Sebastia *et al.*, 1991). Moreover there is in both mutants a significant increase of monosomes concomitant with a strong reduction in the polysome population (A. Petitjean, unpublished results). This could either reflect the mRNA instability or a more specific impairment of these mutants in translation initiation (Petersen and Mc Laughlin, 1974).

## Sequence determination and functional analysis of *RNA14* and *RNA15*

Both the *RNA14* and *RNA15* genes are essential for yeast viability (Minvielle-Sebastia *et al.*, 1991). *RNA14*, encodes a protein of 636 amino acids which does not display any known consensus sequence. Very recently, it has been observed that the protein encoded by the *Drosophila melanogaster* gene suppressor of forked (*su(f)*), is significantly homologous to RNA14 (26% identity, 47% homology, M. Simonelig, personal communication). Mutations in the gene *su(f)* suppress the morphological defect caused by a dysfunction in the forked gene which is due to the insertion of a gypsy element. If the suppression of the forked mutation occurs *via* a modification in mRNA processing this could mean that *RNA14* and *su(f)* are functionally related.

*RNA15*, encodes a protein of 296 amino acids which is not homologous to RNA14 and which has several consensus sequences. Its N-terminal part contains an RNA-binding domain followed by an OPA-like sequence (Minvielle-Sebastia *et al.*, 1991). Recently, it has been

NATO ASI Series, Vol. H 71
Protein Synthesis and Targeting in Yeast
Edited by A. J. P. Brown, M. F. Tuite, and J. E. G. McCarthy

observed that the C-terminus of the RNA15 protein displays homology with the polymerase module defined by Argos (Argos, 1988) which is also found in eucaryotic poly(A) polymerases (Raabe *et al.*, 1991, Lingner *et al.*, 1991) and also some similarities with the DNA primase of *Escherichia coli* (L. Minvielle-Sebastia, unpublished observations). Moreover, *in vitro* synthesised RNA15 protein binds with homopolymers of poly(U) or poly(T) with high affinity .

Subcellular localisation of the RNA14 and the RNA15 proteins

Rabbit antibodies have been raised against the RNA14 protein and the C-terminal half of the RNA15 protein. They have been used to localise these proteins in the different compartments of the yeast cells. The RNA14 protein is localised primarily in the nucleus but could also be present in the yeast cytosol. The full length RNA15 protein is found in the nucleus and a shorter protein is recognised by anti-RNA15 antibodies in the cytosolic polysomes. (N. Bonneaud, manuscript in preparation). It is not yet clear whether this short protein results from a post-translational modification of the RNA15 protein or is a different polypeptide sharing common epitopes with RNA15.

Isolation of suppressors of the *rna14* and *rna15* mutations

Besides causing the same defects when mutated, it is very likely that the RNA14 and RNA15 proteins interact closely with each other. Indeed, the thermosensitive growth defect of the *rna14* mutants cells is suppressed when they contain multiple copies of the wild-type allele of the *RNA15* gene and *vice versa* (Minvielle-Sebastia *et al.*, 1991). In order to better define the function of RNA14 and RNA15, we have undertaken a systematic search for genes acting in the same pathway. To identify these genes, we have looked for suppressors that could relieve, even partially, the thermosensitive growth defect of *rna14* and/or *rna15* mutants. Two kinds of suppressors have been selected: (i) classical secondary mutations that result in the reversal of the mutant thermosensitive phenotype, (ii) genes whose wild-type allele, when present in high copy number, compensate partially the ts defect.

At present, only one suppressor mutation has been studied in more detail. It arises at a very high frequency, and corrects almost totally the thermosensitivity of all the *rna14* mutants but not that of the *rna15* mutants. This mutation has been named *ssm4* (suppressor of mRNA stability mutant).( E. Mandart, personnal communication). The other approach, which is now frequently used in yeast and has allowed to identify closely interacting proteins in many case ( see above the suppression of *rna14* by *RNA15* and *vice versa*) has led us to isolate three genes from a yeast genomic library constructed in 2μ vectors. Two of them, *SSM1* and *SSM2* are specific suppressors of the *rna14* mutants (A. Petitjean, manuscript in preparation) while the third one, *SSM3* suppresses both mutants. (E. Mandart, unpublished results)

Characterisation of *ssm4*.

The fact that the *ssm4* mutation appears at a very high frequency, associated with its ability to suppress the thermosensitivity of two different *rna14* alleles and its recessivity, suggest that this mutation causes a loss of function. The recessivity of the mutation should have therefore allowed us to easily clone the wild-type allele of *SSM4* by transformation of a double mutant strain *rna14 ssm4* with a yeast genomic DNA library. Normally, the transformants bearing a wild-type allele of *SSM4* would have been detected by replica plating and incubation at the restrictive temperature since they should have recovered the *rna14* mutation ts phenotype. Despite many independent transformations of *rna14 ssm4* strains with different yeast genomic libraries, constructed either in 2μ or ARS-CEN vectors, we failed to isolate the gene *SSM4* by this method. Nevertheless, the Ty transposition technique (Garfinkel *et al.*, 1991) has recently allowed us to isolate DNA from the *SSM4* locus. All attempts to isolate (by hybridisation in yeast genomic DNA libraries) or to reconstruct (by cloning in *E. coli*) a complete *SSM4* gene have failed suggesting that this gene is lethal when expressed in the bacteria. The gene *SSM4* is currently being sequenced.

Characterisation of the *SSM3* suppressor

Multiple copies of the wild-type allele of *SSM3* suppress, but only partially, the thermosensitive growth defect of both *rna14* and *rna15* mutants. Indeed, it allows the growth of the mutants cells up to 33°C while the restrictive temperature of the isogenic wild-type strain FL100 is superior to 37°C. The determination of the nucleotide sequence of *SSM3* has revealed that it is identical to that of *CDC48* (Fröhlich *et al.*, 1991), a gene originally identified by recessive mutations causing cold sensitive cell division cycle arrest and the accumulation of yeast cells with large buds and whose specific function remains unknown (Moir *et al.*,1982, Fröhlich *et al.*,1991). Nevertheless the CDC48 protein could be member of a new protein family comprising VCP, NSF, SEC18, PAS1 and TBP-1 (see Fröhlich *et al.*, 1991, Erdmann *et al.*, 1991 and references therein) which has been very recently enlarged by the cloning of two new genes *SUG1* (Swaffield *et al.*, 1992) and *MSS1* (Shibuya *et al.*, 1992) both encoding proteins involved in transcription activation respectively in yeast and human. All these proteins share one or two copies of a large domain of homology (≈ 220 amino-acids). This domain contains a motif formed by a consensus ATPase site positioned 88 amino-acids from another highly conserved region of homology (SRH) of unknown function. A functional linkage between all these proteins remains to be established. In view of the suppressor effect of *CDC48* on the ts phenotype of *rna14* and *rna15* it would be interesting to test whether this gene has any role in transcription like its homologues *SUG1* and *MSS1*.

Characterisation of *SSM1*

When the wild-type allele of the gene *SSM1* is present in high copy number in the *rna14* mutant cells, it suppresses the ts growth defect up to the semi-restrictive temperature 33°C. It has no effect on the ts phenotype of the *rna15* mutant. The gene *SSM1* contains an ORF encoding a basic polypeptide of 217 amino-acids (pI:10.4). This ORF has a codon bias characteristic of highly expressed genes. Northern blotting experiments have confirmed that *SSM1* is highly transcribed since its mRNA level is comparable to that of the actin mRNA which is very abundant in yeast.

The *SSM1* encoded protein displays a significant homology to the archae- and eubacterial L1 protein (Itoh, 1988, Post *et al.*, 1979) which is associated in these organisms with the large ribosomal subunit. Antibodies raised against the SSM1 protein have allowed us to show that in yeast SSM1 is also a ribosomal protein from the large 60S subunit. This is the first eucaryotic L1-like protein found. Southern analysis has revealed the existence of another copy of the gene *SSM1*a in the yeast genome which was cloned and sequenced. This gene, *SSM1*b, contains an ORF which is 98.5 % homologous with the *SSM1*a ORF but encodes a strictly identical protein. Either the chromosomal deletion of *SSM1*a, or that of *SSM1*b allows the vegetative growth of the haploid yeast cells but the absence of both *SSM1* genes is lethal.

Both 5' and 3' flanking regions of these ORF are widely divergent and this allowed us to design specific probes to measure the steady-state level of *SSM1*a and *SSM1*b mRNA by Northern analysis. This showed that in rapidly growing wild-type cells (YPD medium) the level of *SSM1*b mRNA is three times higher than *SSM1*a. This also revealed that in a haploid strain growing in the same conditions, the deletion of the gene *SSM1*a does not induce any change of the *SSM1*b mRNA level, but the deletion of the *SSM1*b gene results in a three fold increase of the *SSM1*a mRNA level. Despite this dosage compensation effect, the strain deleted for *SSM1*b has a strongly reduced growth rate compared to wild-type while the *SSM1*a deletion does not result in decrease in growth. Polysome profile analysis of these strains has shown that both deletions induce the appearance of half-mers to a different extent, a phenotype which is reminiscent of a defect in translation initiation and which could be due to a shortage in 60S subunit. The translational defect is more acute in the *SSM1*b deleted strain which has a strong reduction in the number of large size polysomes.

Unexpectedly, the gene *SSM1*b has no suppressor effect on the thermosensitive growth defect of the *rna14* mutant despite the fact that its mRNA level is twice higher than that of *SSM1*a in yeast cells containing these genes cloned in a 2µ vector. There is no obvious explanation for this since both genes are overexpressed when present in multiple copies and encode identical proteins. One possibility could be that there is a post-transcriptional mechanism which involves the SSM1 protein itself and regulates more strictly the translational efficiency of the gene *SSM1*b so impeding the creation of a free pool of SSM1b protein but not of SSM1a.

Characterisation of *SSM2*.

The gene *SSM2* has the same suppressor effect as *SSM1*. It contains an ORF encoding a very basic protein (pI:11.6) of 78 amino-acids which displays a high codon bias suggesting a high level of expression. This protein shares 50 % of identity with the rat ribosomal protein L38 (Kuwano *et al.*., 1991), a protein which seems typically eucaryotic since it has no prokaryotic homologue. In view of this homology and the general primary sequence conservation of ribosomal proteins in eukaryotes, it is tempting to hypothesise that, like SSM1, SSM2 is a protein from the large 60S ribosomal subunit. The chromosomal deletion of the gene *SSM2* does not induce any modification of the growth rate. This can be easily explained by the existence in the yeast genome of a gene which hybridises to the *SSM2* ORF, even in stringent conditions, and which probably encodes a highly homologous protein.

Physiological studies of the suppressor effect.

To try to understand at which level the *ssm* genes identified exert their suppressor effect, we have started to analyse whether they suppress other known phenotypes of *rna14* and *rna15* mutants. We first measured the ratio of poly($A^+$) mRNA versus total RNA in *rna14* or/and *rna15* cells in the presence of the *ssm* suppressors and in conditions of suppression of their ts phenotype. Indeed, one of the defects engendered by the ts *rna14* and *rna15* mutations is a decrease of the amount of poly($A^+$) mRNA, newly synthesised at the restrictive growth temperature (Bloch *et al.*, 1978, Minvielle-Sebastia *et al.*, 1991). The *ssm4-1* mutation raises significantly the poly($A^+$) mRNA level in the *rna14-1* mutant cells, up to the restrictive temperature 37°C (Table 1). On the contrary, the genes *SSM1*, *SSM2* and *SSM3* have no effect on the amount of poly($A^+$) mRNA in mutant cells even at the semi-restrictive temperature 33°C.

Table 1 Effect of the *ssm4* suppressor on the Poly($A^+$) mRNAs synthesised in *rna14* mutant cells

| Relevant Genotype | poly($A^+$)/ poly($A^+$)+ poly($A^-$) ratio in % | |
|---|---|---|
| | Temperature | |
| | 22°C | 37°C |
| *RNA14 SSM4* | 23.8 | 41.8 |
| *rna14-1 SSM4* | 23.3 | 8.8 |
| *rna14-1 ssm4* | 23.8 | 20.3 |

The ratio of poly($A^+$)/ poly($A^+$)+ poly($A^-$) has been measured in yeast cells labelled for 5 minutes with $^3$H-adenine either at 22°C or after a shift of 10 minutes to 37°C.

Since the shift of *rna14* and *rna15* mutants to a non permissive growth temperature induces the disappearance of a large fraction of the polysomes and the concomitant accumulation of monosomes, we have begun to study the effect of the different suppressors on this phenotype. We have compared the polysome profile of the wild-type strain FL100, the *rna14-1* mutant and this of the same mutant containing the gene *SSM1*a on a 2µ vector grown at the permissive temperature 22°C and after a one hour shift at the restrictive temperature 33°C. At 22°C, the ratio of polysomes over 80S monosomes is similar in all three strains while after the shift it is respectively 3.37, 1.15 and 1.63. This shows that *SSM1*a increases the amount of polysomes and eliminates partially the accumulation of monosomes in the *rna14* mutant cells. Since it does not increase the amount of poly($A^+$) mRNA this could mean that it exerts its suppressor effect at the translational level of these mRNAs.

## Discussion.

The new data presented in this paper pertain mainly to the cellular localisation of the *RNA14* and *RNA15* encoded proteins and to the characterisation of different suppressors of the ts phenotype caused by mutations in these genes. The first evidence coming out is that the RNA14 and RNA15 proteins are very likely part of a complex, at least transitory, in which they interact with each other. The second is that they appear to have a role both in the nucleus and the cytoplasm, perhaps associated with the polysomes.

The two arguments supporting their nuclear function are their clear immunological detection in nuclear protein extracts and the allele-dependent modification of the polymerase II transcription rate observed in ts mutant cells. Several observations suggest that these proteins interact with the polysomes. A possible cytoplasmic localisation for RNA14 and a probable RNA15 immunological detection in purified polysomes. Mutated RNA14 and RNA15 proteins have strong repercusions on poly($A^+$) mRNA stability, even long after they have been synthesised and when they are mainly located in polysomes (Bloch *et al.*, 1978). Last but not least, two suppressors of the *rna14* mutation are ribosomal proteins of the large 60S subunit.

If one admits these conclusions, one can ask what the role of these proteins in mRNA metabolism could be? Remembering the binding specificity of the RNA15 protein to poly(U), and the sequence homologies found in its C-terminal part, with the so called polymerase module, it is tempting to hypothesise that the RNA14-RNA15 tandem is involved, both in the nucleus and the cytoplasm, in a secondary synthesis of poly(A) at the mRNA 3' end. The fact that RNA15 binds preferentially to poly(U) would be required either if this protein has an affinity for the poly(U) rich sequences often found both in the 5' and the 3' ends of yeast mRNAs or if RNA15 uses a small poly(U) rich RNA as a template for poly(A) elongation. In this model, the RNA14 protein would not have any catalytic function but would rather regulate the RNA15 activity, depending of the balance existing in the yeast cell between the availability of mRNAs and that of the translation system components.

Acknowledgments

We are grateful to Alexi Harrington and Michel Werner for critical reading of this manuscript. A. P. has been supported during a part of this work by a long term EMBO post-doctoral fellowship. L. M. S. has been supported by a fellowship of the french "Ministère de la Recherche et de la Technologie" and of the "Société de Secours des Amis des Sciences".

References

Argos, P. (1988). A sequence motif in many polymerases. Nucleic Acids Res. *16*, 9909-9916

Bloch, J. C., Perrin, F. and Lacroute, F. (1978). Yeast temperature-sensitive mutants impaired in processing of poly(A)-containing RNAs. Mol. Gen. Genet. *165*, 123-127

Erdmann, R., Wiebel, F. F., Flessau, A., Rytka, J., Beyer, A., Frölich, K. and Kunau, W. (1991). *PAS1*, a yeast gene required for peroxisome biogenesis, encodes a member of a novel family of putative ATPases. Cell *64*, 499-510

Frölich, K., Fries, H., Rüdiger, M., Erdmann, R., Botstein, D. and Mecke, D. (1991). Yeast cell cycle protein CDC48p shows full-length homology to the mammalian protein VCP and is a member of a protein family involved in secretion, peroxisome formation, and gene expression. J. Cell. Biol. *114*, 443-453

Garfinkel, D. J., Mastrangelo, M. F., Sanders, N. J., Shafer, B. K., and Strathern, J. N. (1988). Transposon tagging using Ty elements in yeast. Genetics *120*, 95-108

Itoh, T. (1988). Complete nucleotide sequence of the ribosomal 'A' protein operon from the archaebacterium, *Halobium halobium*. Eur. J. Biochem. *176*, 297-303

Kuwano, Y., Olvera, J. and Wool, I. G. (1991). The primary structure of rat ribosomal protein L38. Biochem. Biophys. Res. Commun. *175*, 551-555

Lingner, J., Kellerman, J. and Keller, W. (1991). Cloning and expression of the essential gene for poly(A) polymerase from *S. cerevisiae*. Nature *354*, 496-498

Minvielle-Sebastia-Sebastia, L., Winsor, B., Bonneaud, N. and Lacroute,F. (1991). Mutations in the yeast *RNA14* and *RNA15* genes result in an abnormal mRNA decay rate; sequence analysis reveals an RNA-binding domain in the RNA15 protein. Mol. Cell. Biol. *11*, 3075-3087

Moir, D., Steward, S. E., Osmond, B. C. and Botstein, D. (1982). Cold-sensitive cell-division-cycle mutants of yeast: isolation, properties, and pseudoreversion studies. Genetics *100*, 547-563

Petersen, N. S. and Mc Laughlin, C. S. (1974). Polysome metabolism in protein synthesis mutants of yeast. Mol. Gen. Genet. *129*, 189-200

Post, L. E., Strycharz, G. D., Nomura, M., Lewis, H. and Denis, P. P. (1979). Nucleotide sequence of the ribosomal protein gene cluster adjacent to the gene for RNA polymerase subunit beta in *Escherichia coli*. Proc. Natl. Acad. Sci. U. S. A. *76*, 1697-1701

Raabe, T., Bollum, F. J. and Manley, J. L. (1991). Primary structure and expression of bovine poly(A) polymerase. Nature *353*, 229-239

Shibuya, H., Irie, K., Ninomiya-Tsuji, J., Goebl, M., Taniguchi, T. and Matsumoto, K. (1992). New human gene encoding a positive modulator of HIV Tat-mediated transactivation. Nature *357*, 700-702

Swaffield, J. C., Bromberg, J. F. and Johnston, S. A. (1992). Alterations in a yeast protein resembling HIV Tat-binding protein relieve requirement for an acidic activation domain in GAL4. Nature *357*, 698-700

# TRANSLATIONAL CONTROL IN *SACCHAROMYCES CEREVISIAE* STUDIES IN VIVO AND IN VITRO

**J.E.G. McCarthy, B. Gerstel, M.R. Vega Laso, Dept. of Gene Expression, GBF, W-3300 Braunschweig, FRG**

**and**

**D. Zhu, M.F. Tuite, Biological Laboratory, University of Kent, Kent CT2 7NJ, UK**

**and**

**F. Sagliocco, A.J.P. Brown, Dept. of Molecular and Cell Biology, University of Aberdeen, UK**

There is still relatively little known about the control of translational efficiency in eukaryotes. The focus of control and of regulatory circuits is generally the initiation process, whereby ribosomal subunits locate the translational start codon and begin synthesis of the polypeptide chain. The most generally applied working hypothesis of eukaryotic translational initiation, the scanning model (Kozak, 1989), states that eukaryotic ribosomes, unlike their prokaryotic counterparts, are unable to bind to or recognize directly the initiation codon. Instead, their "search" for a start codon is reduced to a one-dimensional, ATP-dependent migration along the mRNA leader which follows an initial binding at or near the 5' end of the message. This binding is usually only efficient in the presence of a 5' cap structure. The translational start codon of eukaryotes is almost exclusively AUG, and the start codon selected is usually the first AUG encountered by ribosomes (preinitiation complexes) scanning from the cap. Experiments with *S. cerevisiae* have demonstrated that recognition of the start codon is mediated by the initiator $tRNA^{met}$ in a process influenced by initiation factor 2 [eIF-2 (Cigan et al., 1988; Donahue et al., 1988)]. Moreover, there is evidence that the nucleotide sequence context of the start-codon represents a further factor influencing start-site selection. Interestingly, the so-called consensus context sequences, which indicate the nucleotides occurring most frequently at the positions adjacent to the start codon, differ greatly between the different groups of eukaryotes. Thus whereas the consensus sequence for 2595 vertebrate mRNAs is cgccgccA/GcCAUGgc (bases in lower case are 'preferred', bases in upper case are 'consensus'), it is aaaaaaaAaaAUGuc for yeast (Cavener and Ray, 1991), whereby the presence of a purine at position -3 is perhaps the most important component (Kozak, 1989). This could at least theoretically relate to intrinsic differences in the AUG selection pathways/mechanisms. A 'poor' context may allow eukaryotic ribosomes to overlook a potential start codon and proceed further, possibly to recognize an AUG later in the sequence.

Apparent deviations from the basic scanning model of translational initiation have

NATO ASI Series, Vol. H 71
Protein Synthesis and Targeting in Yeast
Edited by A. J. P. Brown, M. F. Tuite, and J. E. G. McCarthy

been described. For example, there are some mRNAs in mammals and yeast that seem to allow reinitiation to occur (Hinnebusch et al., 1990; Kozak, 1992). On the other hand, experimental data from several laboratories are at least consistent with the occurrence of internal translational initiation on certain viral and cellular mRNAs, independent of the 5' end of the mRNA (Jackson et al., 1990; Macejak and Sarnow, 1991). It is believed that internal ribosomal entry sites (IRES) allow direct interaction of initiation complexes with internal start sites. There is as yet no evidence of internal initiation in yeast.

Little is known about the range of translational initiation rates that occurs in eukaryotic cells, and how the efficiency of translation of each individual mRNA is controlled. Potential control points are the interaction of the 40S ribosomal subunit-factor complex with the 5'end (cap) of the mRNA, migration along the leader (scanning), and initiation at the start codon. Any of these steps could, at least theoretically, be influenced by the sequence of the leader, the start codon context, and/or secondary structure in the early part of the mRNA. Stem-loop structures in leader regions have been shown to inhibit translation in both higher eukaryotes (Pelletier and Sonenberg, 1988; Kozak, 1986, 1989) and in yeast (Baim and Sherman, 1988; Bettany et al., 1989; Cigan et al., 1988). The strength of the inhibitory effect is apparently linked to its thermodynamic stability. The stabilities of stem-loop structures that have been shown to strongly inhibit translation in yeast are lower than those reported to be necessary for comparative effects in higher eukaryotes (Baim and Sherman, 1988; Cigan et al., 1988). However, no direct comparisons using identical messages in systems derived from both a lower and a higher eukaryote have been made up to now. The position of a stem-loop structure has also been shown to modulate the observed inhibitory effect in rabbit reticulocyte lysates (Kozak, 1989), but the generality of this effect in eukaryotes remains unclear.

The present short article focusses on how the sequence and structure of the leader, and the state of modification of mRNA influence translational control in yeast.

**The roles of 5'-capping and 3'-polyadenylation in vitro**

A direct demonstration of the roles of 5'-capping and of 3'-polyadenylation on specific mRNAs in yeast has awaited the use of cell-free translation extracts (Gerstel et al., 1992). In this recent work it was shown that these modifications not only enhance translation but also stabilize the mRNA in yeast cell-free extracts. The stabilization effects observed were to some extent dependent upon the nature of the transcript. On the other hand, a striking observation was that the cap-dependence of translation was markedly reduced in the presence of the *HSP26* (heat shock protein 26 gene) leader before the reading-frame (whether this was the *HSP26* gene itself or the *cat* reading-frame). This might indicate the operation of a form of posttranscriptional regulation in yeast cells under stress conditions. In extracts from Ehrlich cells, the interaction of the cap-binding-protein complex with capped mRNA is impaired under heat-shock conditions (Lamphear and Panniers, 1991). Thus by analogy, the observed (partial) independence of initiation from the 5' cap (Gerstel et al., 1992) on the *HSP26* leader might allow heat-shock mRNAs to be selectively translated under stress conditions. This possibility needs to be investigated in more detail in vivo.

**A new modular expression system for the study of translational control in vivo and in vitro**

In further work, we have performed a systematic study of the inhibition of translational initiation in yeast as a function of the stability and position of stem-loop structures in the mRNA leader. For this purpose, we constructed a new modular expression system that facilitates the manipulation of all essential components (Fig. 1).

The central element of the expression system (Fig. 1) is a modular region of DNA bearing *cat* as reporter gene flanked by a 5' leader and a 3' trailer. This combination of gene and flanking sequences can be inserted into an in vitro transcription vector [pHST7, derived from pHSTO (Jobling et al., 1988)] and an in vivo expression vector (a derivative of YCp50). The *cat* gene was modified in such a way as to allow the exchange of the first 70 base pairs and the last 53 base pairs of the reading frame using unique restriction endonuclease sites (*NdeI/StuI* and *EagI/XbaI*, respectively). The introduction of the *NdeI* restriction site at the translational start codon facilitated the exchange of leader sequences. The *NdeI* restriction site (CATATG) has the advantages of not intruding into the reading frame beyond the start codon and of being A/T-rich (McCarthy, 1991). Transcription from the T7 promoter is initiated at the G in the *BglII* restriction site (AGATCT). Immediately following this site is the ACCACC sequence shown in Fig. 2.

A series of vectors for in vivo expression was designed to facilitate the expression of the same set of gene/flanking region combinations transcribed in vitro. The *cat* gene plus flanking regions could be excised as a whole on a *BglII-BamHI* fragment and inserted into the *BamHI* site of an expression vector based on YCp50 (Fig. 1). The constitutively active *TEF1* promoter was present as a *ClaI-BamHI* fragment, whereby the *BamHI* site is located at the major transcriptional start site(s) of this promoter, and is used as the cloning site for the *BglI-BamHI* fragments derived from the in vitro vector system. The *BamHI-NruI* fragment downstream of the promoter region is synthetic DNA bearing the *CYC1* terminator region. The less-well defined *HSP26* terminator region was inserted as a *NotI-ClaI* module upstream of the *TEF1* promoter in order to interrupt any transcription originating 5' of the planned transcription unit.

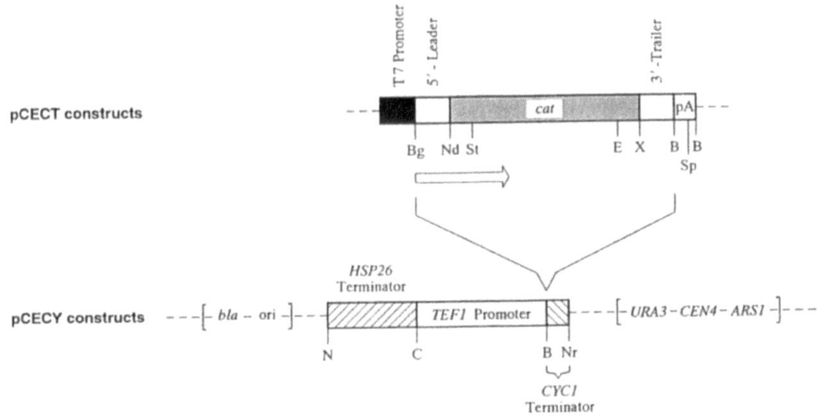


**Fig. 1** The modular system for studying gene expression in vivo and in vitro in *S. cerevisiae*. All gene flanking region combinations were constructed initially in the pCECT system. The *cat* gene was borne on an *NdeI-XbaI* fragment flanked by synthetic leader (*BglII-NdeI*) and trailer (*XbaI-BamHI*) sequences. The reading frame contained newly engineered *StuI* and *EagI* restriction sites designed to allow rapid exchange of the N-terminal and C-terminal regions of the gene. The synthetic polyA tail cassette was inserted as a *BamHI* fragment behind the trailer. Cleavage at the *SpeI* site in this module allowed transcription from the T7 promoter to run from the *BglII* site through the reading frame and trailer, to end with a polyA sequence of 50 bases. The gene/flanking regions could be excised as a *BglII-BamHI* fragment and inserted into the YCp50-derivative vectors, thus creating the parallel series of pCECY plasmids. Transcription of the gene/flanking regions in the pCECY constructs was initiated by the *TEF1* promoter and terminated in the *CYC1* terminator. *URA3* was the selection marker used to ensure maintenance of the single-copy vectors in an appropriate yeast mutant strain.
B=*BamHI*; Bg=*BglII;* C=*ClaI*; E=*EagI*; N=*NotI*; Nd=*NdeI*; *Nr=NruI*; Sp=*SpeI*; St=*StuI*; *X=XbaI.*

The described expression vector system allowed parallel investigations of nearly identical transcripts to be performed in vivo and in vitro. A range of synthetic flanking sequences was used to create secondary structures at various positions (Fig. 2). The stem-loops of pCEC143 and pCEC130 are of very similar sequence, structure and stability, but

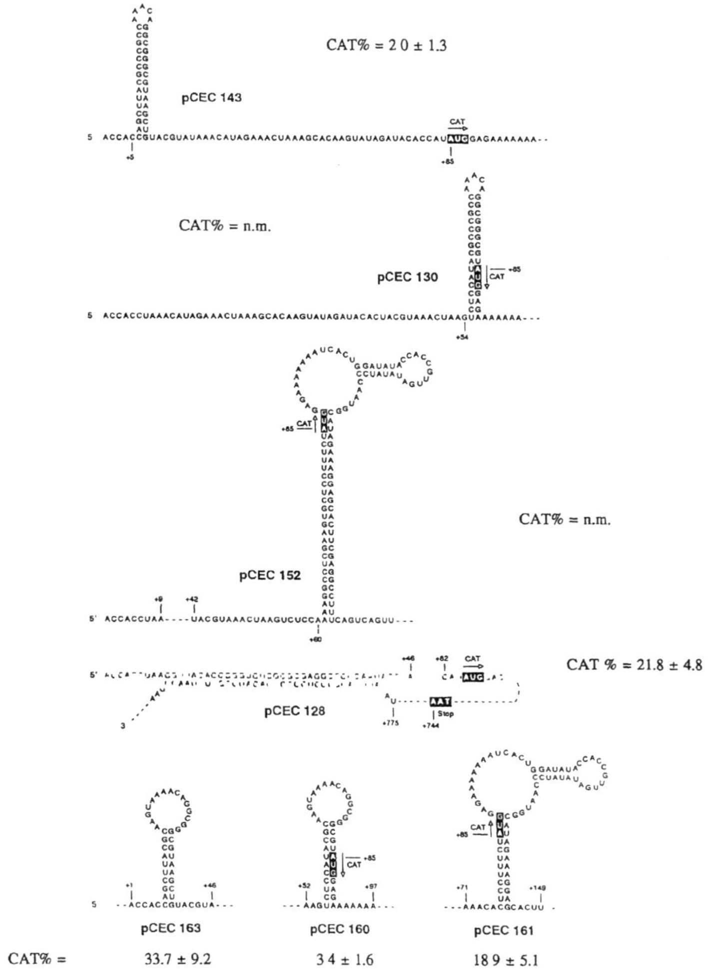


**Fig. 2** Stem-loop structures inserted into the leader region and/or N-terminal region of the *cat* reading frame. This figure shows a number of the hairpin-loops designed for the experiments described in this article. They were inserted at effectively four positions: position 1, close to the 5' end (pCEC143 and pCEC163); position 2, with the start codon in the lower part of the right arm (pCEC 130 and pCEC 160); position 3, with the start codon in the left arm immediately adjacent to the loop (pCEC152 and pCEC161); position 4, where the two arms of the potential stem are located in the leader and trailer regions, respectively. Leader length was in all cases maintained at 84 bases. The start codon of the cat gene is boxed. In the case of pCEC128, there was continuous complementarity of 32 base pairs between the leader and trailer regions. The predicted stabilities of the stem-loop structures, calculated at 28 °C, were as follows: pCEC143, -28.3 kcal $mol^{-1}$; pCEC130, -29.9 kcal $mol^{-1}$; pCEC152, -49.4 kcal $mol^{-1}$; pCEC128, -59.8 kcal $mol^{-1}$; pCEC163, -12.6 kcal $mol^{-1}$; pCEC160, -14.2 kcal $mol^{-1}$; pCEC161, -19.5 kcal $mol^{-1}$. The relative in vivo CAT activities (CAT %) are presented as percentages of the control level (100 %) obtained using pCEC127, which bears the leader of the *HSP26* gene. n.m.= too small to be accurately measurable.

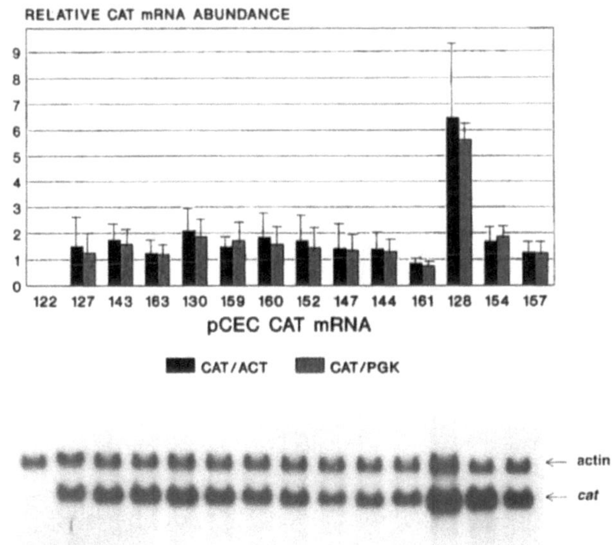


**Fig. 3** Relative amounts of cellular *cat* mRNAs. Northern blots were prepared using total cellular RNA derived from the respective yeast transformants. Hybridization was performed using probes specific for *cat* and for actin mRNA (as internal standard). Quantitation of the radioactivity in the respective bands yielded average amounts that could be expressed as ratios to either of the internal standards used (actin or *PGK* mRNA). The columns in the graph represent average ratios from at least three determinations, normalized against the ratio observed in cells bearing pCEC126. The thin bars indicate the standard deviations for each of the averages. The autoradiograph is from one typical experiment, so that the intensities of the respective bands do not correspond exactly to the average values shown in the graph.

inserted at two different positions (1 and 2), relative to the start-codon and 5' end of the mRNA. The large hairpin in pCEC152 involves 27 bases of the N-terminal part of the *cat* reading frame (position 3), whereby codon-usage changes were necessary to allow formation of the hairpin. Further constructs were prepared in which smaller stem-loops were inserted at positions 1 (pCEC163), 2 (e.g. pCEC160) and 3 (pCEC161). The stem-loops in these clones were derived from the more stable ones already described by eliminating base-pairs in the respective stems. An alternative type of construct was designed to allow base-pairing between the trailer and leader (pCEC128). Finally, a range of control constructs (not shown) were made which allowed us to distinguish between the effects of specific sequences, or of codon-usage changes, and the influence of stem-loop structures on translation.

**Inhibition of translational initiation is determined by the stability and position of stem-loops in the leader.**

In order to be able to draw unambiguous conclusions about the influence of stem-loop structures upon translational efficiency, we performed analyses of both the relative quantities of mRNA in vivo and of the structures assumed by the transcripts in vitro. Northern blots prepared using extracts from the transformants (Fig. 3) revealed the presence of *cat* mRNA species in the expected range of 840 - 900 nucleotides. The quantities of the respective mRNAs were assessed relative to two internal controls: the *PGK* and actin mRNAs. The majority of the strains showed similar steady-state *cat*-specific mRNA levels (falling e.g. within the range ± 30% of the average amount recorded for all constructs except pCEC128). However, the strain bearing pCEC128Y fell well outside this range, containing at least three times as much *cat*-specific mRNA as the average value of all the constructs examined.

Structural analysis was performed using two methods. First, primer-extension analysis of total RNA derived from the various strains yielded major termination bands that corresponded to the expected positions of the various stem-loops predicted to form in the leader region. Second, single-strand-specific and double-strand-specific ribonucleases were used to analyze in vitro transcripts. Both methods confirmed the existence of the stem-loops shown in Fig. 2 (data not shown).

Transformants of *S. cerevisiae* bearing the various pCECY plasmids expressed the *cat* gene at rates that were strongly influenced by the nature of the flanking regions inserted into the expression vector (Fig. 2). Measurement of CAT activities in the cell extracts revealed that the introduction of the described stem-loop structures into the leader reduced to various extents the expression of the *cat* gene in vivo. That the respective CAT activities reflected the various amounts of CAT protein present in the cell could be confirmed by means of Western blotting (data not shown). Given the limited variation in mRNA abundancy (Fig. 3), the relative CAT activities summarized in Fig. 2 can be assumed to reflect differences in translational efficiency attributable to the inhibitory effects of secondary structure upon the initiation step.

In general, the translational rates of the pCECT transcripts in vitro followed the same trend as the in vivo expression data. The inhibitory effects of stem-loop structures, especially of the more stable ones, were more striking in vivo than in the yeast in vitro extract (Fig. 4A). On the other hand, comparative experiments revealed that the reticulocyte lysate system showed generally less sensitivity to secondary structure than the yeast in vitro and in vivo systems (Fig. 4B).

**Conclusions**

Investigations of the translation of in vitro transcribed mRNAs in yeast cell-free extracts confirm that capping and polyadenylation are important for optimal gene expression in yeast. Both modifications enhance the translation and stability of mRNA in vitro. Thus at least in this respect, yeast gene expression is influenced in a similar way by these modifications as is gene expression in higher systems.

The significance of other aspects of mRNA sequence and structure may differ between higher and lower eukaryotes. One example of this is illustrated by the distinctive "consensus" sequences for translational initiation in vertebrates and yeast (see e.g. Cavener and Ray, 1991). Start-site selection in yeast is evidently less dependent upon specific base patterns near the start codon. Beyond that, it is important to compare how the efficiency of translational initiation is controlled by the primary and higher-order structure of mRNA in yeast and mammals. Earlier studies revealed that yeast translation shows increased sensitivity to inhibition by stem-loop structures in the leader compared to translation in higher eukaryotes. In the present study we have been able to confirm this, and also to analyze the role of secondary structure in more detail. We have observed that the inhibitory effect of a given stem-loop is modulated by its position, so that it is more effective when positioned close to the start codon than when near the 5' end of the leader. This is the opposite "polarity" in terms of inhibitory action to that reported previously for mammalian systems (Kozak, 1989).

The described differences in behaviour of the translational apparatuses of higher and lower eukaryotes may reflect the existence of (perhaps subtle) differences in the respective-pathways of translational initiation.

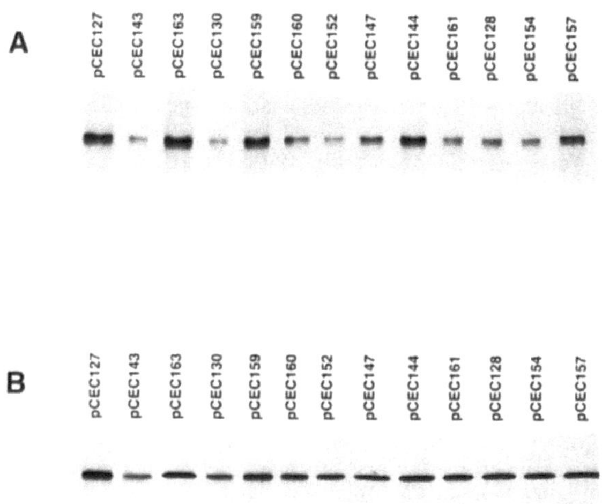


**Fig. 4.** Translation of in vitro synthesized *cat* mRNAs in cell-free extracts. The fluorographs reflect the relative levels of radioactive incorporation into CAT specified by the respective mRNAs obtained using yeast cell-free extracts (A) and rabbit reticulocyte lysates (B). The results clearly illustrate the greater sensitivity of the yeast translational apparatus to secondary structure in the leader.

**Acknowledgements**

This work was supported by grant SCI*0193-C(SMA) of the Commission of the European Communities and by grants from the Deutscher Akademischer Austauschdienst (DAAD) and the British Council.

**References**

Baim, S.B. and F. Sherman. 1988. mRNA structures influencing translation in the yeast *S. cerevisiae*. Mol. Cell. Biol. 8: 1591 - 1601.

Bettany, A.J.E., P.A. Moore, R. Cafferkey, L.D. Bell, A.R. Goodey, B.L.A. Carter and A.J.P. Brown. 1989. 5'-secondary structure formation, in contrast to a short string of non-preferred codons, inhibits the translation of the pyruvate kinase mRNA in yeast. Yeast 5: 187 - 198.

Cavener, D.R. and S.C. Ray 1991. Eukaryotic start and stop translation sites. Nucl. Acids Res. 19: 3185 - 3192.

Cigan, A.M., L. Feng and T.F. Donahue 1988. tRNAi$^{met}$ functions in directing the scanning ribosome to the start site of translation. Science 242: 93 -97.

Cigan, A.M., E.K. Pabich and T.F. Donahue. 1988. Mutational analysis of the *HIS4* translational initiatior region in *Saccharomyces cerevisiae*. Mol. Cell. Biol. 8: 2964 - 2975.
Donahue, T.F., A.M. Cigan, E.K. Pabich and B.C. Valavicius 1988. Mutations at a Zn(II) finger motif in the yeast eIF- 2ß gene alter ribosomal start-site selection during the scanning process. Cell 54: 621 - 632.
Gerstel, B., M.F. Tuite and J.E.G. McCarthy. 1992. The effects of 5'-capping, 3'-polyadenylation and leader composition upon the translation and stability of mRNA in a cell-free extract derived from the yeast *Saccharomyces cerevisae*. Molec. Microbiol. 6: 2339-2348.
Hinnebusch, A.G., J.-P. Abastado, E.M. Hannig, B.M. Jackson, P.F. Miller, M. Ramirez, R.C. Wek and N.P. Williams 1990. Translational control of the transcriptional activator *GCN4* involves upstream open reading frames, a general initiation factor and a protein kinase, p. 325 - 335. In J.E.G. McCarthy and M.F. Tuite (ed.), Post-transcriptional control of gene expression. Springer-Verlag, Berlin, Heidelberg.
Jackson, R.J., M.T. Howell and A. Kaminski. 1990. The novel mechanism of initiation of picornavirus RNA translation. Trends Biochem. Sci. 15: 477 - 483.
Jobling, S.A., C.M. Cuthbert, S.G. Ropgers, R.T. Fraley and L. Gehrke. 1988. In vitro transcription and translational efficiency of chimeric SP6 messenger RNAs devoid of 5'vector nucleotides. Nucl. Acids Res. 16: 4483 - 4498.
Kozak, M. 1992. A consideration of alternative models for the initiation of translation in eukaryotes. Crit. Revs. in Biochem. and Molec. Biol. in press.
Kozak, M. 1986. Influences of mRNA secondary structure on initiation by eukaryotic ribosomes. Proc. Natl. Acad. Sci. U.S.A. 83: 2850 - 2854.
Kozak, M. 1989. Circumstances and mechanisms of inhibition of translation by secondary structure in eukaryotic mRNAs. Mol. Cell. Biol. 9: 5134 - 5142.
Lamphear, B.J. and Panniers, R., 1991. Heat shock impairs the interaction of cap-binding protein complex with 5' mRNA cap. J. Biol. Chem. 266: 2789-2794.
Macejak, D.G. and P. Sarnow. 1991. Internal initiation of translation mediated by the 5' leader of a cellular mRNA. Nature 353: 90 - 94.
McCarthy, J.E.G.. 1991. Optimizing post-transcriptional steps of gene expression in *Escherichia coli*. In "Advances in Gene Technology II" (P.J. Greenaway, ed.) JAI Press, London, pp. 145 - 175.
Pelletier, J. and N. Sonenberg. 1988. Internal initiation of translation of eukaryotic mRNA directed by a sequence derived from poliovirus RNA. Nature 334: 320 - 325.

# THOUGHTS ON THE REGULATION OF RIBOSOME SYNTHESIS IN *SACCHAROMYCES CEREVISIAE*

J.R. Warner
Department of Cell Biology
Albert Einstein College of Medicine
1300 Morris Park Avenue
Bronx, New York 10461
USA

For a meeting on Protein Synthesis in Yeast, it seems fitting to focus on the chief agent of protein synthesis, the ribosome, and particularly on the control of the synthesis of this agent. Ultimately, the control of ribosome synthesis is the control of protein synthesis, and the control of protein synthesis is the control of growth. Let us examine what little we know about the control of ribosome synthesis in yeast, and attempt to pose the questions that will occupy us for the duration of this millenium.

In recent years, the biosynthesis of ribosomes in *Saccharomyces cerevisiae* has been reviewed perhaps more often than warranted by the progress that has been made (Mager, 1988; Mager, Planta, 1990; Mager, Planta, 1991; Planta, Raué, 1988; Raué, Planta, 1991; Warner, 1989; Warner et al.1990; Woolford, 1991; Woolford, Warner, 1991). The reader is referred to these reviews for the experimental observations. I will state only briefly the major elements of ribosome biosynthesis that have been established, as a basis for further discussion.

**Transcription of Ribosomal RNA:** All 140 copies of the ribosomal RNA genes are arranged in a single tandem array on chromosome XII. The transcription units, of 6580 nucleotides, are separated by a 'non-transcribed spacer' of about 2410 nucleotides, that itself carries the 5S rRNA gene, transcribed in the opposite direction. The promoter elements involved in the initiation of rRNA transcription have been established both *in vivo* and *in vitro* (Butlin, Quincey, 1991; Choe et al. 1992; Kulkens et al. 1991; Musters et al. 1989). There are core regions, from approximately -76 to +8, as well as an upstream element, from -146 to -91. Although one expects that a number of DNA binding proteins and auxiliary transcription factors are involved in the initiation of transcription, none has yet been identified. However, the recent establishment of efficient *in vitro* transcription (Riggs, Nomura, 1990; Schultz et al. 1991) and the introduction of a powerful genetic approach (Nogi et al. 1991a; Nogi et al. 1991b), lead to hope for rapid progress. In addition there is an unusual enhancer of rRNA

NATO ASI Series, Vol. H 71
Protein Synthesis and Targeting in Yeast
Edited by A. J. P. Brown, M. F. Tuite, and J. E. G. McCarthy

transcription, that lies 2.2 kb upstream from the initiation site (Elion, Warner, 1984; Elion, Warner, 1986; Johnson, Warner, 1989) and is at least partially included in the upstream transcription unit (Johnson, Warner, 1991; Kempers-Veenstra et al. 1986; van der Sande et al. 1989).

**Transcription of Ribosomal Protein Genes:** The genes for ribosomal proteins are unusual in at least three respects. Frequently they are duplicated; most have an intron; their transcription is highly coordinated. The basis for this coordination may lie in the fact that most have a UAS consisting of two sites for the binding of Rap1p, a T-rich element, and a TATA box. Some, however, have Abf1p sites instead of Rap1p sites. Both Abf1p and Rap1p bind upstream of the genes of numerous 'housekeeping' proteins and glycolytic enzymes, play a role in mating type silencing, and perform other functions in chromosome replication and maintenance.

**Regulation of Ribosome Biosynthesis:** The ribosome is composed of four RNA molecules and about 75 proteins. Each RNA has about 140 genes; many of the proteins have two genes. With a few possible exceptions (Baronas-Lowell, Warner, 1990; Sachs, Davis, 1990), each component is needed for a functional ribosome. Under normal growing conditions there is a balance among the syntheses of the components of the ribosome. Ribosomal precursor RNA is processed into mature species with nearly 100% efficiency (Udem, Warner, 1972). Most of the ribosomal proteins seem also to be utilized with high efficiency, although precise data is more difficult to obtain. The importance of this careful regulation is emphasized by the fact that in a rapidly growing cell rRNA synthesis accounts for nearly 60% of total transcription, while ribosomal protein synthesis accounts for about 15% of total translation. How, then, does the cell coordinate the transcription of nearly three hundred genes to produce equimolar amounts of nearly eighty macromolecules? It is a problem of cordination that has few rivals in biology.

Not only does this coordination take place during log-phase growth, but it is also responsive to an abrupt improvement in carbon source (Kief, Warner, 1981), to the deprivation of amino acids (Moehle, Hinnebusch, 1991; Warner, Gorenstein, 1978), to the imposition of conditions inducing sporulation (Kraig et al. 1982; Pearson, Haber, 1980), to the gradual deterioration of growth conditions that occurs during late log phase of growth (Q. Ju & JRW, unpublished) and undoubtedly to a great many other environmental insults. As an overall process, the synthesis of ribosomes is exquisitely sensitive to the culture conditions, sensing in some subtle way the first or second derivative of the future growth rate. What, then, are the genetic and molecular elements that lead to this coordination?

**Regulation of Ribosomal RNA Transcription:** Although there is substantial information about the key sequences of the rRNA promoter, regulatory elements have yet to be identified. Indeed, the regulation of rRNA transcription, at least in response to the carbon source, is due not to the promoter but to the enhancer (BE Morrow, SP Johnson, & JRW, submitted). The rRNA enhancer has a number of features that are unusual, if not unique for yeast enhancers: it works from both upstream and downstream positions; it can work from a distance of > 5 kb; a single enhancer can affect transcription of more than one gene, and even can affect transcription of a distal gene; the enhancer, itself, is part of the upstream transcription unit (Johnson, Warner, 1991; Kempers-Veenstra et al. 1986; van der Sande et al. 1989). Do one or more of these features account for the coupling of the transcription of all 140 rRNA genes? or for the sensitivity of the transcription to the environmental conditions? It has recently been shown that a single TATA binding protein, when accompanied by the appropriate associated proteins, can function in transcription by PolI, by PolII, and by PolIII (Comai et al.1992; Cormack, Struhl, 1992; Schultz et al. 1992). Although the enhancer has sites for two abundant DNA-binding proteins, Reb1p and Abf1p, as well as a T-rich region that has been implicated in transcription by both PolI and PolII (Chasman et al. 1990; Lorch et al. 1990; Struhl, 1985), none of these features are essential either for the basal activity of the enhancer in a poor carbon source, or for the stimulation of activity on addition of a good carbon source. Conclusion: the transcription of rRNA is carefully regulated, in at least certain instances through the enhancer, but we are entirely ignorant either about the molecules that first signal a change, possibly small molecules analogous to ppGpp that is used by *E. coli*, or about the macromolecules that transduce that signal to influence the transcription of ribosomal RNA.

The observation that the transcription of rRNA continues into the enhancer element has stimulated suggestions that there may be a functional and/or physical association of the end of one transcription unit with the beginning of another (Elion, Warner, 1987; Johnson, Warner, 1989; Kempers-Veenstra et al. 1986), suggestions reinforced by the unusual sensitivity to depletion of topoisomerase exhibited by rRNA transcription (Brill et al. 1987). The models that have been proposed are esthetically appealing but experimentally appalling, since the predictions they make involve as yet inaccessible aspects of chromatin structure. The presence of two binding sites for the protein Reb1p, one in the enhancer and one just upstream from the promoter (Morrow et al. 1990), led to the hope that interaction between these proteins, with or without the assistance of other proteins, might be the factor necessary to couple initiation with termination. Such a hope has been at least partially dashed by the finding that deletion of either or both sites has little if any effect on the trancription of an rRNA minigene carried on a plasmid (Kulkens et al. 1989). The possibility remains, however, that in the context of the 1.2 megabase rDNA locus on chromosome XII, the effect of Reb1p binding might be more influential. Conclusion: The location of the enhancer, and the likelihood that termination and initiation are somehow linked, suggests that the three dimensional organization of the rDNA plays a fundamental role in the transcription of rRNA and its regulation. Current efforts to locate the rDNA in the nucleolus of higher eukaryotes (Thiry, 1992) may lead us to consider how the structure of the nucleolus is involved in rRNA transcription.

**Regulation of Ribosomal Protein Synthesis:** There is evidence that the Rap1p binding sites are responsible for the elevation of the level of ribosomal protein mRNAs in response to the addition of glucose to a culture growing on a non-fermentable carbon source (Donovan, Pearson, 1986; Herruer et al. 1987), as well as for the decrease of those mRNAs on depletion of an amino acid (Moehle, Hinnebusch, 1991). Yet, Rap1p can hardly be the entire story. In neither case referred to above has an alteration in the level of Rap1p binding activity been observed. Many other genes with a Rap1p site have at best a qualitatively simlar response (Capieaux et al. 1989). Moreover, in cells carrying a ts mutant of <u>RAP1</u> the level of mRNA for ribosomal proteins is unaffected for several hours after the cells are shifted to the non-permissive temperature (D. Shore, pers. comm.). Finally, a sizeable fraction of the ribosomal protein genes have Abf1p sites instead of Rap1p sites. Abf1p sites are also found upstream of a number of genes encoding proteins involved in general metabolism, such as transcription, translation, and glycolysis. (Della-Seta et al. 1990) Conclusion: Rap1p and Abf1p clearly play a role in coordinating the expression of the ribosomal protein genes, but they are likely to be subordinant to another factor(s). One of these may be Gcr1p, a non-essential protein that appears to facilitate the transcription of many genes (Santangelo, Tornow, 1990).

The one feature in common between the genes for ribosomal proteins and those for ribosomal RNA is the T-rich sequence. Is it possible that this element is responsible for the coordination of the two sets of genes? This hypothesis seems unlikely because such an element is present 5' of almost every gene.

Many of the ribosomal proteins are represented by two genes, both of which are active. Others have a single gene. In the few cases examined, the level of mRNA for both types of proteins is the same (Kim, Warner, 1983). Thus the enhancer/promoter region of the duplicated genes have evolved to compensate for the duplication. It is interesting that during this evolution toward less efficient transcription, none of the genes has lost either the T-rich region or one of the Rap1p binding sites.

To assess the types of controls available for ribosomal protein synthesis, a number of groups have introduced into cells multicopy plasmids with a ribosomal protein gene. The qualitative result has been that transcription proceeds *ad libitum*, i.e., the transcription is proportional to the number of genes (Abovich et al. 1985; el Baradi et al. 1986; Pearson et al. 1982; Tsay et al. 1988). In general, then, there is no feedback repression at the level of transcription. The implication is that some global regulator controls the synthesis of the ribosomal proteins as a group. On the other hand, the level of mRNA is seldom fully proportional to the copy number of the gene, suggesting that there may be some modulation of either the transcription or the degradation of the mRNA. Furthermore, in at least one instance the disruption of one of a pair of ribosomal protein genes leads to the increased transcription of the other, suggesting that feedback control of transcription can occur (A. Paulovich & J. Woolford, pers. comm.).

In the case of one ribosomal protein, L32, there is now strong evidence that the protein itself is responsible for feedback regulation at the level of splicing, in connection with a specific structure largely within the 5' exon (Dabeva et al. 1986; Eng, Warner, 1991). Ribosomal protein genes are unusual in *S. cerevisiae*, in that most have an intron, whereas introns are generally rare. The finding that L32 uses the regulation of splicing as a method to maintain feedback control appeared to explain the maintenance of introns in the ribosomal protein genes of an organism that seems to be in the process of losing most of its introns (Fink, 1987). However, a search for other genes that use the same type of regulation has been fruitless. Is there something special or aberrant about L32? If so, it is biologically relevant, because the structure has been conserved during the evolutionary divergence of *S. cerevisiae* and *Kluyveromyces lactis* (Eng, Warner, 1991).

In the final resort, the cell has mechanisms that rapidly degrade any ribosomal proteins that are overexpressed, with half lives of as little as 30 seconds (elBaradi et al. 1986; Gorenstein, Warner, 1977; Maicas et al. 1988; Pearson et al. 1982; Tsay et al. 1988). There is no doubt that cells view with alarm the accumulation of a rougue protein that binds strongly to RNA, and have devised methods to deal with such a menace. But how can a cell determine, in less than 30 seconds, that a protein synthesized in the cytoplasm is not needed in the nucleolus?

**Global Regulation:** Thus far we have dealt separately with the regulation of transcription of rRNA and of ribosomal protein genes. Yet these two processes must be coupled! A key question that remains is whether a single regulatory signal influences, in parallel, the two sets of transcription reactions, or whether the signal influences one, e.g. the transcription of rRNA, and the regulation of that process in some way leads to the regulation of the other. There is little information on this question.

A serious experimental impediment to the understanding of such regulation is the fact that the synthesis of ribosomal components is so exquisitely sensitive to the environmental conditions. To distinguish mutations in the key regulatory gene(s) from mutations that interrupt some element of metabolism with a secondary effect on ribosome synthesis will be difficult, indeed. Conclusion: although we recognize that the regulation of ribosome synthesis is a key element in the regulation of growth, not only of yeast but of mammalian cells as well, and although we recognize many of the features involved in the transcription of ribosomal RNA and protein genes, we are indeed far from understanding the integration of these features into the metabolism and control of the cell.

**Acknowledgement**

Research in this laboratory was supported by Public Health Service grants GM25532 and CA13330 from the National Institutes of Health and by American Cancer Society grant NP-793.

**References**

Abovich N, Gritz L, Tung L, Rosbash M (1985) Effect of RP51 gene dosage alterations on ribosome synthesis in *Saccharomyces cerevisiae*. Mol Cell Biol 5:3429-3437

Baronas Lowell DM, Warner JR (1990) Ribosomal protein L30 is dispensable in the yeast Saccharomyces cerevisiae. Mol Cell Biol 10:5235-5243

Brill SJ, DiNardo S, Voelkel-Meiman K, Sternglanz R (1987) Need for DNA topoisomerase activity as a swivel for DNA replication and for transcription of ribosomal RNA. Nature 326:414-416

Butlin M, Quincey R (1991) Activity of promoter mutants of the yeast ribosomal RNA gene with and without the enhancer. Yeast 7:679-690

Capieaux E, Vignais M-L, Sentenac A, Goffeau A (1989) The yeast $H^+$-ATPase gene is controlled by the promoter binding factor TUF. J Biol Chem 264:7437-7448

Chasman DI, Lue NF, Buchman AR, LaPointe JW, Lorch Y, Kornberg RD (1990) A yeast protein that influences the chromatin structure of $UAS_G$ and functions as a powerful auxiliary gene activator. Genes and Develop 4:503-514

Choe SY, Schultz MC, Reeder RH (1992) In vitro definition of the yeast RNA polymerase I promoter. Nucleic Acids Res 20:279-285

Comai L, Tanese N, Tjian R (1992) The TATA-binding protein and associated factors are integral components of the RNA polymerase I transcription factor, SL1. Cell 68:965-976

Cormack BP, Struhl K (1992) The TATA-binding protein is required for transcription by all three nuclear RNA polymerases in yeast cells. Cell 69:685-696

Dabeva MD, Post Beittenmiller MA, Warner JR (1986) Autogenous regulation of splicing of the transcript of a yeast ribosomal protein gene. Proc Natl Acad Sci U S A 83:5854-5857

Della-Seta F, Ciafre S-A, Marck C, Santoro B, Presutti C, Sentenac A, Bozzoni I (1990) The ABF1 factor is the transcriptional activator of the L2 ribosomal protein genes in *Saccharomyces cerevisiae*. Mol Cell Biol 10:2437-2441

Donovan DM, Pearson NJ (1986) Transcriptional regulation of ribosomal proteins during a nutritional upshift in Saccharomyces cerevisiae. Mol Cell Biol 6:2429-2435

elBaradi TT, van der Sande CA, Mager WH, Raué HA, Planta RJ (1986) The cellular level of yeast ribosomal protein L25 is controlled principally by rapid degradation of excess protein. Curr Genet 10:733-739

Elion EA, Warner JR (1984) The major promoter element of rRNA transcription in yeast lies 2 kb upstream. Cell 39:663-673

Elion EA, Warner JR (1986) An RNA polymerase I enhancer in Saccharomyces cerevisiae. Mol Cell Biol 6:2089-2097

Elion EA, Warner JR (1987) Characterization of a yeast RNA polymerase I enhancer. In: Granner DK, Rosenfeld G, Chang S (eds)Transcriptional Control Mechanisms. Alan R. Liss, Inc., New York, pp 21-29

Eng FJ, Warner JR (1991) Structural basis for the regulation of splicing of a yeast messenger RNA. Cell 65:797-804

Fink GR (1987) Pseudogenes in yeast? Cell 49:5-6

Gorenstein C, Warner JR (1977) Synthesis and turnover of ribosomal proteins in the absence of 60S subunit assembly in *Saccharomyces cerevisiae*. Mol Gen Genet 157:327-332

Herruer MH, Mager WH, Woudt LP, Nieuwint RT, Wassenaar GM, Groeneveld P, Planta RJ (1987) Transcriptional control of yeast ribosomal protein synthesis during carbon-source upshift. Nucleic Acids Res 15:10133-10144

Johnson SP, Warner JR (1989) Unusual enhancer function in yeast rRNA transcription. Mol Cell Biol 9:4986-4993

Johnson SP, Warner JR (1991) Termination of transcription of ribosomal RNA in Saccharomyces cerevisiae. Mol Cell Biochem 104:163-168

Kempers-Veenstra AE, Oliemans J, Offenberg H, Dekker AF, Piper PW, Planta RJ, Klootwijk J (1986) 3'-End formation of transcripts from the yeast rRNA operon. EMBO J 5:2703-2710

Kief DR, Warner JR (1981) Coordinate control of syntheses of ribosomal ribonucleic acid and ribosomal proteins during nutritional shift-up in Saccharomyces cerevisiae. Mol Cell Biol 1:1007-1015

Kim CH, Warner JR (1983) Messenger RNA for ribosomal proteins in yeast. J Mol Biol 165:79-89

Kraig E, Haber JE, Rosbash M (1982) Sporulation and rna2 lower ribosomal protein mRNA levels by different mechanisms in *Saccharomyces cerevisiae*. Mol Cell Biol 2:1199-1204

Kulkens T, Van Heerikhuizen H, Klootwijk J, Oliemans J, Planta RJ (1989) A yeast ribosomal DNA-binding protein that binds to the rDNA enhancer and also close to the site of Pol I transcription initiation is not important for enhancer functioning. Curr Genet 16:351-359

Kulkens T, Riggs DL, Heck JD, Planta RJ, Nomura M (1991) The yeast RNA polymerase I promoter: ribosomal DNA sequences involved in transcription initiation and complex formation in vitro. Nucleic Acids Res 19:5363-5370

Lorch Y, Lue NF, Kornberg RD (1990) Interchangeable RNA polymerase I and II enhancers. Proc Natl Acad Sci 87:8202-8206

Mager WH (1988) Control of ribosomal protein gene expression. Biochim et Biophy Acta 949:1-15

Mager WH, Planta RJ (1990) Multifunctional DNA-binding proteins mediate concerted transcription activation of yeast ribosomal protein genes. Biochim Biophys Acta 1050:351-355

Mager WH, Planta RJ (1991) Coordinate expression of ribosomal protein genes in yeast as a function of cellular growth rate. Mol Cell Biochem 104:181-187

Maicas E, Pluthero FG, Friesen JD (1988) The accumulation of three yeast ribosomal proteins under conditions of excess mRNA is determined primarily by fast protein decay. Mol Cell Biol 8:169-175

Moehle CM, Hinnebusch AG (1991) Association of RAP1 binding sites with stringent control of ribosomal protein gene transcription in Saccharomyces cerevisiae. Mol Cell Biol 11:2723-2735

Morrow BE, Ju Q, Warner JR (1990) Purification and characterization of the yeast rDNA binding protein REB1. J Biol Chem 265:20778-20783

Musters W, Knol J, Maas P, Dekker AF, Van Heerikhuizen H, Planta RJ (1989) Linker scanning of the yeast RNA polymerase I promoter. Nucleic Acids Res 17:9661-9678

Nogi Y, Vu L, Nomura M (1991a) An approach for isolation of mutants defective in 35S ribosomal RNA synthesis in Saccharomyces cerevisiae. Proc Natl Acad Sci U S A 88:7026-7030

Nogi Y, Yano R, Nomura M (1991b) Synthesis of large rRNAs by RNA polymerase II in mutants of Saccharomyces cerevisiae defective in RNA polymerase I. Proc Natl Acad Sci U S A 88:3962-3966

Pearson NJ, Fried HM, Warner JR (1982) Yeast use translational control to compensate for extra copies of a ribosomal protein gene. Cell 29:347-355

Pearson NJ, Haber JE (1980) Changes in regulation of ribosomal protein synthesis during vegetative growth and sporulation of *Saccharomyces cerevisiae*. J Bacteriol 143:1411-1421

Planta RJ, Raué HA (1988) Control of ribosome biogenesis in yeast. Trends Genet 4:64-68

Raué HA, Planta RJ (1991) Ribosome Biogenesis in Yeast. Prog Nucl Acid Res &Molec Biol 41:89-129

Riggs DL, Nomura M (1990) Specific transcription of Saccharomyces cerevisiae 35 S rDNA by RNA polymerase I in vitro. J Biol Chem 265:7596-7603

Sachs AB, Davis RW (1990) Translation initiation and ribosomal biogenesis: involvement of a putative rRNA helicase and RPL46. Science 247:1077-1079

Santangelo GM, Tornow J (1990) Efficient transcription of the glycolytic gene ADH1 and three translational component genes requires the GRC1 product, which can act through TUF/GRF/RAP binding sites. Mol Cell Biol 10:859-862

Schultz MC, Choe SY, Reeder RH (1991) Specific initiation by RNA polymerase I in a whole-cell extract from yeast. Proc Natl Acad Sci U S A 88:1004-1008

Schultz MC, Reeder RH, Hahn S (1992) Variants of the TATA-binding protein can distinguish subsets of RNA polymerase I, II, and III promoters. Cell 69:697-702

Struhl K (1985) Naturally occurring poly (dA-dT) sequences are upstream promoter elements for constitutive transcription in yeast. Proc Natl Acad Sci USA 82:8419-8423

Thiry M (1992) Ultrastructural detection of DNA within the nucleolus by sensitive molecular immunocytochemistry. Exp Cell Res 200:135-144

Tsay YF, Thompson JR, Rotenberg MO, Larkin JC, Woolford JL,Jr. (1988) Ribosomal protein synthesis is not regulated at the translational level in Saccharomyces cerevisiae: balanced accumulation of ribosomal proteins L16 and rp59 is mediated by turnover of excess protein. Genes Dev 2:664-676

Udem SA, Warner JR (1972) Ribosomal RNA synthesis in *Saccharomyces cerevisiae*. J Mol Biol 65:227-242

van der Sande CA, Kulkens T, Kramer AB, de Wijs IJ, Van Heerikhuizen H, Klootwijk J, Planta RJ (1989) Termination of transcription by yeast RNA polymerase I. Nucleic Acids Res 17:9127-9146

Warner JR (1989) Synthesis of ribosomes in Saccharomyces cerevisiae. Microbiol Rev 53:256-271

Warner JR, Baronas-Lowell DM, Eng FJ, et al (1990) Genetic approaches to ribosome biosynthesis in the yeast, *Saccharomyces cerevisiae*. In: Hill WE, Dahlberg A, Garrett RA, et al (eds)The Ribosome: Structure, Function, & Evolution. ASM, Washington,DC, pp 443-451

Warner JR, Gorenstein C (1978) Yeast has a true stringent response. Nature 275:338-339

Woolford JL,Jr. (1991) The structure and biogenesis of yeast ribosomes. Adv Genet 29:63-118

Woolford JL,Jr., Warner JR (1991) The Ribosome and Its Synthesis. In: Broach JR, Pringle JR, Jones EW (eds)The Molecular and Cellular Biology of the Yeast Saccharomyces Genome Dynamics, Protein Synthesis, and Energetics. Cold Spring Harbor Laboratory Press, New York, pp 587-626

# REGULATION OF RIBOSOMAL PROTEIN SYNTHESIS IN YEASTS (*SACCHAROMYCES* AND *KLUYVEROMYCES*)

Rudi J. Planta, Ruurdtje Hoekstra, G. Karin Bergkamp-Steffens, Paula M. Gonçalves and Willem H. Mager
Department of Biochemistry and Molecular Biology, Vrije Universiteit, de Boelelaan 1083, 1081 HV Amsterdam, The Netherlands

## INTRODUCTION

Expression of ribosomal protein genes (rp-genes) in the yeast *Saccharomyces cerevisiae* is accurately adjusted to meet the cellular need for ribosomes, *i.e.* protein synthetic capacity, under different growth conditions. Coordinated transcription of rp-genes ensures the supply of equimolar amounts of all rp-mRNAs. This balance is maintained upon alterations in growth conditions: *e.g.* addition of glucose to ethanol-grown cells triggers an immediate and simultaneous increase in all rp-mRNA levels (Mager and Planta, 1991).

Examination of a large number of rp-gene promoters led to the identification of the *cis*-acting elements responsible for transcription-activation of these genes. In front of most rp-genes a fourteen nucleotides long element, called RPG-box, is present, usually in duplicate, in either orientation and at a distance of 250-500 bp upstream of the ATG codon. These RPG-boxes act as a binding site for an abundant, multifunctional protein RAP1, formerly called TUF (Vignais et al., 1987; Shore and Nasmyth, 1987; Kimmerly et al., 1988). The same protein is also involved in transcription activation of a variety of other yeast genes, including those involved in glycolysis, cellular differentiation, nutrient transport and the cell

NATO ASI Series, Vol. H 71
Protein Synthesis and Targeting in Yeast
Edited by A. J. P. Brown, M. F. Tuite, and J. E. G. McCarthy

division cycle (Mager and Planta, 1991). Furthermore, RAP1 binds to telomeres (Berman et al., 1986; Buchman et al., 1988) and plays a role in anchoring DNA to the nuclear scaffold (Hofmann et al., 1989).

A minority of rp-genes (see Fig. 1) contains, instead of a RPG-box, another conserved regulatory sequence in the upstream region to which a different abundant protein, designated ABF1, binds (Herruer et al., 1989; Dorsman et al., 1989; Della Seta et al., 1990; Kraakman et al., 1991). The ABF1-binding-site is characterized by a dual binding motif (Dorsman et al., 1989) and is found in a single copy in either orientation at a distance of around 160 to 220 bp upstream of the translation start codon (see Fig. 1).

Like RAP1, also ABF1 is a multifunctional protein. In addition to acting

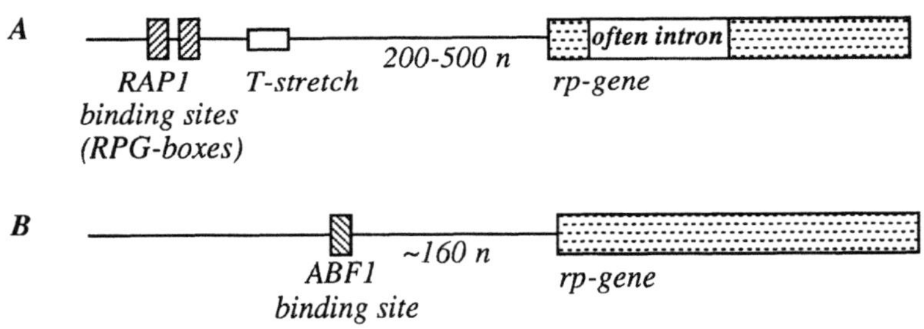


| | | | |
|---|---|---|---|
| A: | S10A, S10B | L4B, L15A, L15B | RP28A, RP28B |
| | S11, S13 | L16A, L16B | RP51A, RP51B |
| | S16aA, S16aB | L17a, L25, L29 | RP59A |
| | S24, S26, S28 | L46 | |

B: S4, S33, L2A, L2B, L3, L4A, L45

Figure 1. Regulation of transcription of ribosomal protein genes in *Saccharomyces cerevisiae*. The *cis*-acting elements and the *trans*-acting factors are indicated. See text for further explanation.

as a transcriptional activator for a large and diverse set of nuclear genes encoding ribosomal proteins, mitochondrial proteins, proteins involved in general metabolism, and components of the transcriptional machinery, it also functions as a transcriptional repressor for the silent mating type loci. Finally, ABF1 binds to an ARS-associated sequence element suggesting that it is involved in DNA replication (reviewed by Mager and Planta, 1991).

The existence of two subgroups of rp-gene promoters in *Saccharomyces* raises the important question of how the coordination of transcription from ABF1- and RAP1-regulated promoters is brought about, in particular in response to changing physiological conditions. Are both factors modified (*e.g.* by phosphorylation) in a similar fashion or do they interact with common additional factors. In order to find some clues in answering this question, we decided to investigate to what extent the regulatory machinery controlling rp-gene transcription in *Saccharomyces* has been conserved in other yeast species, for instance in *Kluyveromyces*.

## ISOLATION AND SEQUENCING OF *KLUYVEROMYCES* RP-GENES

Various restriction digests of both *K. lactis* and *K. marxianus* total DNA were analyzed by Southern blotting with probes derived from the coding region of several *S. cerevisiae* rp-genes. In all cases strong hybridization signals were obtained (results not shown) indicating a high degree of homology between the *Kluyveromyces* and *Saccharomyces* rp-genes. In contrast to the situation in *S. cerevisiae*, where most of the rp-genes are duplicated (Planta and Raué, 1988), *Kluyveromyces* strains appear to contain only a single copy of the rp-genes with the exception of the gene for S24 which is present in two copies.

The rp-genes in *Kluyveromyces*, sequenced so far, have several interesting characteristics. Firstly, the coding regions of the *Kluyveromyces* genes closely resemble those of their *Saccharomyces* counterparts, showing about 90% sequence similarity at the nucleotide and 85-95% sequence similarity at the protein level (see Fig. 2A and 2B for an outline).

A.

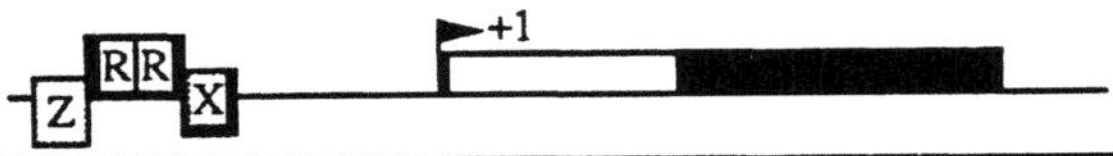


| | |
|---|---|
| **Isolated genes** | *K. lactis* L25-gene and *K. marxianus* S10-, L25-, RP 59-, S24- and L46-gene |
| **Number of copies** | 1 (exception: 2 copies of S24) |
| Length **ORF**<br>Homology | = *S. cerevisiae*<br>± 90% |
| Length **introns** | usually longer than in *S. cerevisiae* |
| **RPGK 's** | -NACAYCCANANANN-; - 250 to - 600 |
| **X-box** | GGGACTGT; - 270 to - 600 |
| **Z-box** | GRARATTTNNRRYYY; - 350 to - 680 |

B.

| | |
|---|---|
| **Isolated genes** | *K. lactis* and *K. marxianus* S33- gene |
| **Number of copies** | 1 |
| Length **ORF**<br>Homology | = *S. cerevisiae*<br>90% |
| **Intron** | not present |
| **ABF-I site** | ATCATTGCATACG; - 215 |

Figure 2 **A.** General structure of some RAP1-regulated ribosomal protein genes in *Kluyveromyces*.
Structure and position of conserved sequences in the upstream region are also indicated. A detailed description of the genes and their flanking regions is given in Bergkamp-Steffens et al. (1992).

**B.** Characteristics of an ABF1-regulated ribosomal protein gene (encoding S33) in *Kluyveromyces*. A detailed description of the gene and its flanking regions is given in Hoekstra et al. (1992).

Furthermore, the lengths of the coding regions of the *Kluyveromyces* genes are the same as in *Saccharomyces*. This is in contrast to the recently isolated *LEU2* gene from *K. marxianus* (Bergkamp et al., 1991), which lacks 5 codons as compared to the corresponding gene in *S. cerevisiae*. Secondly, both the occurrence and the position of the intron has been conserved in the *Kluyveromyces* rp-genes. Also, both the S33-gene and the S24-1 gene from *Kluyveromyces* lack an intron like in *S. cerevisiae*. The intron sequences do not show any conservation in either length or sequence, except for the 5'- and the 3'-splice sites and the TACTAAC branchpoint sequence. The conservation of the splicing signals in *Kluyveromyces* is in agreement with the finding of Deshler et al. (1989) that the actin gene from *K. lactis* is properly spliced in *S. cerevisiae*.

Finally, the genetic organization of the rp-genes in *Kluyveromyces* in terms of physical linkage and orientation seems to be similar to the situation in *Saccharomyces*, at least in the cases studied so far. For instance, in *Kluyveromyces* the S24-1- and the L46-genes are physically linked in a head-to-head arrangement like in *Saccharomyces*, and in addition the RP59-gene is located closely downstream from the S24-1-gene in the opposite orientation (*cf.* Leer et al., 1985, and Larkin et al., 1987).

The 5'- and 3'-flanking regions of the *Kluyveromyces* rp-genes show no significant sequence similarity with the corresponding regions of their respective counterparts from *Saccharomyces*, except for a well-conserved binding site for ABF1 in front of the S33-gene, and one or more copies of a sequence with a high similarity to the *S. cerevisiae* RPG-box. The significance of these sequence elements will be discussed below.

## TRANSCRIPTIONAL REGULATION OF *KLUYVEROMYCES* RP-GENES

By Northern blot analysis it was shown that all the identified *Kluyveromyces* rp-genes are actively transcribed and that usually the transcripts derived from these genes are significantly longer than their *Saccharomyces* counterparts (Hoekstra et al., 1992; Bergkamp-Steffens et al., 1992). In at least one case it has been established that the difference in transcript length is due to a difference in the size of the trailer and/or the poly(A)-tail.

Sequence comparison of the upstream regions of the isolated *Kluyveromyces* rp-genes revealed the presence of three types of a conserved sequence element in front of all the genes, except for the S33-gene (see below). These sequence elements, which may represent potential transcription regulatory elements, are defined in Fig. 2A. One element is very similar, though not identical to the RPG-box found in front of the majority of the rp-genes in *S. cerevisiae* and has been designated the $RPG_K$-box. These $RPG_K$-boxes were found at positions varying from - 603 (S24-gene) to - 232 (L46-gene) with respect to the ATG-start codon, which is on the average about 200-300 nucleotides further upstream than the RPG-box in front of the *Saccharomyces* rp-genes (Leer at al., 1985). The consensus sequence of the $RPG_K$-box in *Kluyveromyces* is: 5'-NACAYCCANANANN-3', in which the extra nucleotides N flanking the actual consensus sequence indicate that the binding site for a *trans*-acting protein might well be longer than the binding site for RAP1 in *Saccharomyces* (as judged from band-shift assays with various oligonucleotides and a *Kluyveromyces* protein extract). The most striking difference between the $RPG_K$-box and the RPG-box in *Saccharomyces* is the loss of the strict conservation of a C-residue at position 11 in the RPG-box, which apparently can be substituted by any other nucleotide in the $RPG_K$-box.

Another conserved element, designated X-box, consists of eight nucleotides with the sequence 5'-GGGACTGT-3', which in most cases occurs in the coding strand. This conserved sequence, which can be located either upstream or downstream of the $RPG_k$-boxes, has not been found in the upstream regions of the rp-genes in *Saccharomyces*.

Finally, a third conserved element, designated Z-box, comprises 15 nucleotides with the sequence 5'-GRARATTTNNRRYYY-3', and is also usually found in the coding strand. The position of this element varies from

- 353 to - 684 with respect to the translation start codon. The Z-box shows a strong similarity to the binding site of the product of the regulatory gene *LAC9* (5'-CGGAAATTTGTGGTCCG-3'; *cf.* Ruzzi et al., 1987; Leonardo et al., 1987). The *LAC9* protein activates transcription of the *LAC4*-gene, involved in the lactose and galactose metabolism, and is homologous to the *GAL4* protein of *S. cerevisiae.*

In the upstream region of the *Kluyveromyces* S33-gene a sequence, (5'-ATCATTGCATACG-3'), is present from position - 215 to - 227 in the non-coding strand which matches very well to the consensus binding site for the ABF1 protein of *S. cerevisiae* (Dorsman et al., 1990). In the upstream region of the *S. cerevisiae* S33-gene the ABF1-binding site is located closer to the gene (positions - 150 to - 162) but also in the non-coding strand (Herruer et al., 1989). Furthermore, in the upstream region of the S33-gene of *K. marxianus* two T-rich sequences are found, which have been implicated in the transcription activation of the S33-gene in *S. cerevisiae* (Herruer et al., 1989).

To determine whether the conserved sequence elements present in the upstream regions of the *Kluyveromyces* rp-genes, are indeed involved in transcription regulation, two different approaches were followed. In one approach, the intergenic region between the S24- and the L46-gene of *Kluyveromyces* (which contains the $RPG_K$-, X- and Z-boxes) flanked by either the S24-gene or the L46-gene was cloned into the multicopy *K. lactis* vector LEpTRP. This vector was derived from the plasmid YIplac204 (Gietz and Sugino, 1988) by insertion of the KARS2 sequence from *K. lactis* (Das and Hollenberg, 1982) which ensures autonomous replication in *K. lactis* cells. In addition, various deletions of the intergenic region between the S24-gene and the L46-gene coupled with either the S24-gene or the L46-gene were also cloned in the vector LEpTRP. The various plasmids were transformed into *K. lactis* MSK110 and the level of the S24- or L46-transcript was determined in duplicate by Northern blot analysis, taking into account the copy number of the plasmid as determined by Southern analysis.

In this way it could be established that the $RPG_K$-box functions as a transcriptional activator, while the X-box appears to act as a transcriptional repressor (Bergkamp-Steffens et al., 1992). The function of the third element, the Z-box, is still unclear.

Another approach was followed for the functional analysis of the ABF1-like binding site in front of the *K. marxianus* S33-gene. In this case the upstream region of the *K. marxianus* S33-gene was fused to the β-glucuronidase gene flanked by a short transcription initiating region (120 bp) and a transcription terminating region (260 bp) derived from the L25-gene from *S. cerevisiae* . This fusion construct, as well as various mutants having deletions in the upstream region of the *K. marxianus* S33-gene were inserted into the multicopy vector LEpTRP, defined above. By measuring the β-glucuronidase activity in the various transformants, it could be unambiguously shown that the ABF1-like binding site acts as a transcriptional activator for the *Kluyveromyces* S33-gene (Hoekstra et al., 1992).

## DOES *KLUYVEROMYCES* CONTAIN FUNCTIONAL RAP1- AND ABF1-HOMOLOGUES?

In order to investigate whether *Kluyveromyces* contains a RAP1-related protein band-shift assays with two different synthetic oligonucleotides were performed. One oligonucleotide is in sequence identical to the consensus RPG-box in *S. cerevisiae*, the other one identical to one of the $RPG_K$-boxes in the S24-L46 intergenic region of *K. marxianus*. Using extracts from *S. cerevisiae* cells we observed complex formation with either probe, and with both probes to the same extent. Competition experiments demonstrated that both complexes could be competed out by adding either the unlabeled RPG-box or unlabeled $RPG_K$-box. This indicates that the protein RAP1 from *S. cerevisiae* can specifically bind to both boxes. On the other hand, using extracts from *Kluyveromyces* cells clear complex formation occurred only with the $RPG_K$-probe and not with the RPG-box. Thus, *K. marxianus* contains a protein that recognizes the $RPG_K$-box but has no, or at best a very low, affinity for the *Saccharomyces* RPG-box. These data indicate that *Kluyveromyces* strains contain a protein that might be functionally homologous to the RAP1 protein in *Saccharomyces*, but that is structurally different. This latter conclusion was supported by the observation that with a DNA fragment carrying the *S. cerevisiae* RAP1-gene as a hybridization probe only a very weak signal was obtained on a Southern blot of either *K. marxianus* or *K. lactis* DNA, or on a Northern blot of *Kluyveromyces* RNA. In spite of this remarkable structural difference several lines of evidence suggest that a RAP1-like protein must be present in *Kluyveromyces*, for instance the finding that a *Kluyveromyces* rp-gene (encoding L25: Rentenaar and Hoekstra, unpublished result; or RP59: Larson and Rossi, 1991) can be expressed in *Saccharomyces* and *vice versa.*

With the putative ABF1-binding site in front of the *K. marxianus* S33-gene as a probe, complex formation was found to occur with extracts of both *S. cerevisiae* and *K. marxianus*. With a *K. lactis* extract only a faint complex with the ABF1-binding sequence was obtained. As judged from the mobility of the DNA-protein complexes formed, the *Kluyveromyces* ABF1-like proteins appear to have a lower molecular mass than *S. cerevisiae* ABF1, in agreement with earlier observations by Dorsman et al., 1990. Furthermore, also the ABF1-like protein from *Kluyveromyces* must be structurally different from the *S. cerevisiae* ABF1, since only weak signals were obtained with a *S. cerevisiae* ABF1-gene as a probe on both a Southern and a Northern blot of *Kluyveromyces* DNA and RNA, respectively.

The finding that the ABF1-like and the RAP1-like proteins in *Kluyveromyces* appear to be structurally rather different from their *Saccharomyces* counterparts, prompted us to set out a strategy for the isolation of the genes encoding ABF1 and RAP1 from *Kluyveromyces*. Comparison of the primary structures of functionally related proteins, which are structurally different, can reveal conserved regions and hence provides the basis for a directed functional analysis.

To this end we constructed a *Saccharomyces* strain conditionally expressing the ABF1-gene under a GAL-promoter. Subsequently, a complementation analysis of this strain with DNA from *K. lactis* was performed. In this way a transformant was obtained in which a *K. lactis* DNA fragment restored growth on glucose. Apparently, this *K. lactis* DNA fragment carries the information for the production of a protein which can take over all the essential ABF1-dependent functions in the cell, including transcription-activation (Gonçalves et al., 1992). Subsequent sequence analysis revealed a gene, encoding a *K. lactis* ABF1-protein which is considerably shorter than *Saccharomyces* ABF1 (482 *vs.* 731 amino acids). In the N-terminal part of *Kl.* ABF1 two areas of high homology with *Sc.* ABF1 can be distinguished. The first one contains a Zn-finger motif and is involved in DNA-binding (Halfter et al., 1989; Diffley and Stillman, 1989). The second homologous domain might contribute to the specificity of the DNA-binding. Contrary to the N-terminal region, the C-terminal domain is globally divergent. The marked acidic character of the *Sc.* ABF1 C-terminal region has led to the suggestion that it might have a function in transcription-activation (Halfter et al., 1989), since the activator regions of well-studied factors have been shown to exhibit this common feature. However, the equivalent region in *Kl.* ABF1 is not very acidic (net charge only - 3). This suggests that transcription-activation by ABF1 probably occurs through a mechanism distinct from the one used by the "acidic" activators.

Close to the C-terminus of both proteins a 20-amino acid motif is present which is remarkably well-conserved (17 identical amino acids). This stretch of conserved residues might be important for interactions with additional, as yet unidentified protein factors. Also in the heat shock factors from *S. cerevisiae* and *K. lactis* a conserved element within an otherwise divergent sequence has been implicated in protein-protein interactions (Jakobsen and Pelham, 1991). Alternatively, the presence of two serine residues in the center of the conserved motif in the ABF1 proteins suggests the possibility of regulation through specific phosphorylation at this site. Interestingly, part of the conserved motif is also present in the C-terminal portion of *Saccharomyces* RAP1 (*cf.* Gonçalves et al., 1992).

In summary, the availability of the *Kluyveromyces* ABF1-gene opens new possibilities for a functional domain analysis of the multifunctional ABF1 protein, and might elucidate the link in the concerted action of ABF1 and RAP1 in the expression of rp-genes in yeast.

## ACKNOWLEDGEMENTS

This work was supported in part by the Netherlands Foundation for Chemical Research (S.O.N.) with financial aid from the Netherlands Organization for Scientific Research (N.W.O.). It was also financially supported by Unilever Research Laboratory, Vlaardingen, and Gist-Brocades, Delft. We are indebted to Wietse Mulder (University of Amsterdam) for providing the *K. lactis* DNA library.

## REFERENCES

Bergkamp RJM, Geerse RH, Verbakel JMA, Musters W, Planta RJ (1992) Cloning and disruption of the *LEU2* gene of *Kluyveromyces marxianus* CBS6556. Yeast 7, 963-970

Bergkamp-Steffens GK, Hoekstra R, Planta RJ (1992) Structural and putative regulatory sequences of *Kluyveromyces* ribosomal protein genes. Yeast, in press

Berman J, Tachinaba CY, Tyė BK (1986) Identification of a telomere-binding activity from yeast. Proc. Natl. Acad. Sci. USA 83, 3713-3717

Buchman AR, Kimmerly WJ, Rine J, Kornberg RD (1988) Two DNA-binding factors recognize specific sequences at silencers, upstream activating sequences, autonomously replicating sequences and telomeres in *Saccharomyces cerevisiae*. Mol. Cell. Biol. 8, 210-225

Das S, Hollenberg CP (1982) A high-frequency transformation system for the yeast *Kluyveromyces lactis*. Curr. Genet. 6, 123-128

Della Seta F, Ciafré SA, Marck C, Santoro B, Presutti C, Sentenac A, Bozzoni I (1990) The ABF1 factor is the transcriptional activator of the L2 ribosomal protein genes in *Saccharomyces cerevisiae*. Mol. Cell. Biol. 10, 2437-2441

Deshler JO, Larson GP, Rossi JJ (1989) *Kluyveromyces lactis* maintains *Saccharomyces cerevisiae* intron-encoded splicing signals. Mol. Cell. Biol. 9, 2208-2213

Diffley JFX, Stillman B (1989) Similarity between the transcriptional silencer binding proteins ABF1 and RAP1. Science 246, 1034-1038

Dorsman JC, Doorenbosch MM, Maurer CTC, De Winde JH, Mager WH, Planta RJ, Grivell, LA (1989) An ARS/silencer binding factor also activates two ribosomal protein genes in yeast. Nucl. Acids Res. 17, 4917-4923

Dorsman JC, Van Heeswijk WC, Grivell LA (1990) Yeast general transcription factor GF1 : sequence requirements for binding to DNA and evolutionary conservation. Nucl. Acids Res. 18, 2769-2776

Gietz RD, Sugino A (1988) New yeast-*Escherichiae coli* shuttle vectors constructed with in vitro mutagenized yeast genes lacking six-base pair restriction sites. Gene 74, 527-534

Gonçalves PM, Maurer K, Mager WH, Planta RJ (1992) *Kluyveromyces* contains a functional ABF1-homologue. Nucl. Acids Res. 20, 2211-2215

Halfter H, Kavety B, Vanderkerkhove J, Kiefer F, Gallwitz D (1989) Sequence, expression and mutational analysis of BAF1, a transcriptional activator and ARS1-binding protein of the yeast *Saccharomyces cerevisiae*. EMBO J 8, 4265-4272

Herruer MH, Mager WH, Doorenbosch TM, Wessels PLM, Wassenaar TM, Planta, RJ (1989) The extended promoter of the gene encoding ribosomal protein S33 in yeast consists of multiple protein binding elements. Nucl. Acids Res. 17, 7427-7439

Hoekstra R, Moradas-Ferreira P, Bootsman TC, Mager WH, Planta RJ (1992) Structure and expression of the ABF1-regulated ribosomal protein S33-gene in *Kluyveromyces*. Yeast, in press

Hofmann JF, Laroche T, Brand GH, Gasser SM (1989) RAP1 factor is necessary for DNA loop formation *in vitro* at the silent mating type locus *HML*. Cell 57,725-737

Jakobsen BK, Pelham HRB (1991) A conserved heptapeptide restrains the activity of the yeast heat shock transcription factor. EMBO J. 10, 369-375

Kimmerly WJ, Buchman AR, Kornberg RD, Rine J (1988) Roles of two DNA-binding factors in replication, segregation and transcriptional repression mediated by a yeast silencer. EMBO J. 7, 2241-2253

Kraakman LS, Mager WH, Grootjans JJ, Planta RJ (1991) Functional analysis of the promoter of the gene encoding the acidic ribosomal protein L45 in yeast. Biochem. Biophys. Acta, 1090, 204-210

Larkin JC, Thompson JR, Woolford Jr. JL (1987) Structure and expression of the *Saccharomyces cerevisiae CRY1* gene: a highly conserved rp-gene. Mol. Cell. Biol. 7, 1764-1775

Larson GP, Rossi J (1991) Altered response to growth rate changes in *Kluyveromyces lactis* versus *Saccharomyces cerevisiae* as demonstrated by heterologous expression of ribosomal protein 59 (*CRY1*) Nucl. Acids Res. 19, 4701-4707

Leer RJ, v. Raamsdonk-Duin MMC, Kraakman P, Mager WH, Planta, RJ (1985) The genes for yeast rp-genes S24 and L46 are adjacent and divergently transcribed. Nucl. Acids Res., 13, 701-709

Leonardo JM, Bhairi SM, Dickson RC (1987) Identification of upstream activator sequences that regulate induction of the ß-galactosidase gene in *Kluyveromyces lactis*. Mol. Cell. Biol. 7, 4369-4376

Mager WH, Planta RJ (1991) Coordinate expression of ribosomal protein genes in yeast as a function of cellular growth rate. Mol. Cell. Biochem. 104, 181-187

Planta RJ, Raué HA (1988) Control of ribosome biogenesis in yeast. Trends in Genet. 4, 64-68

Ruzzi M, Breunig KD, Ficca AG, Hollenberg CP (1987) Positive regulation of the ß-galactosidase gene from *Kluyveromyces lactis* is mediated by an upstream activation site that shows homology to the GAL upstream activation site of *Saccharomyces cerevisiae*. Mol. Cell. Biol. 7, 991-997

Shore D, Nasmyth K (1987) Purification and cloning of a DNA-binding protein from yeast that binds to both silencer and activator elements. Cell 51, 721-732

Vignais ML, Woudt LP, Wassenaar GM, Mager WH, Sentenac A, Planta RJ (1987) Specific binding of TUF factor to upstream activation sites of yeast ribosomal protein genes. EMBO J. 6, 1451-1457

# THE EVOLUTION OF RIBOSOMAL PROTEINS AND YEAST

E. Otaka, T. Hashimoto*, K. Mizuta, and K. Suzuki
Department of Biochemistry and Biophysics,
Research Institute for Nuclear Medicine and Biology, Hiroshima University,
Kasumi 1-2-3, Minami-ku, Hiroshima, 734 Japan.
*The Institute of Statistical Mathematics,
4-6-7 Minami-Azabu, Minato-ku, Tokyo, 106 Japan.

**Introduction**: We have been interested in analyzing sequence similarities of ribosomal proteins from various organisms from the view point of biological evolution, focusing on yeast. Ribosomes contain many protein species. In the well-studied *E. coli*, 54 ribosomal proteins (21 in the small subunit and 33 in the large subunit) are known with few others believed to exist (Wada and Sako, 1987). According to the analysis of sequence similarities, some of them have their equivalents in organisms from all three primary kingdoms: eukaryotes, metabacteria and prokaryotes, including the organelles like chloroplasts and mitochondria. It is estimated that some have their equivalents only in prokaryotes, and the others in prokaryotes and metabacteria and/or in prokaryotes and eukaryotes. Similar situations are also seen for ribosomal proteins in eukaryotic and metabacterial species (Fig. 1).

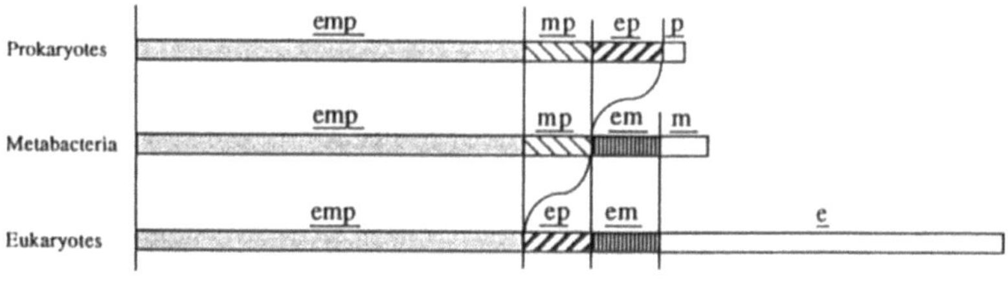


**Figure 1:** Schematic grouping. e: eukaryotics; m: metabacterials; p: prokaryotics.

Thus the ribosome has changed depending on the evolutionary step of the respective organism but has kept playing, albeit conservatively, an important role in the mechanism of protein synthesis. The ribosome is a highly complicated supramolecule requiring cooperation of the

NATO ASI Series, Vol. H 71
Protein Synthesis and Targeting in Yeast
Edited by A. J. P. Brown, M. F. Tuite, and J. E. G. McCarthy

constituents. When it is regarded as a gigantic single molecule, that is, a supramolecule, the individual proteins must correspond to partial fragments of the gigantic molecule. It should be understood that to study ribosomal protein species is to analyze the evolutionary changes that have occurred locally in a single molecule.

To thoroughly understand the evolution of such a gigantic molecule, we undertook a compilation of equivalents among ribosomal protein species from various organisms. Ultimately, such a task should yield a universal nomenclature for ribosomal protein species. Considering the accumulation of ribosomal protein sequence data and the confusion of nomenclatures variously devised by individual laboratories, such a compilation is a pressing need. In parallel with introducing a preliminary compilation, we will mention the evolutionary place of yeast.

**Compilation System (Universal Code System, UCS)**

In brief (see Otaka *et al.*, 1992), the numbering system for *E. coli's* ribosomal protein species was mainly adopted with the prefix UL or US. As shown in Fig. 1, species members of groups emp, ep, mp and p are able to accept *E. coli's* numbering system. Species members of other groups em, e, m, and others (such as mitochondrial proteins), are coded by proper numberings. To search for the equivalent of a given protein, it is most important that sequences should show significant correlation. To search for significant correlations, we adapted a computer program first devised by Lipman and Pearson, to pick up sequences which show considerable correlation scores, and then used RELATE and ALIGN programs to further check the previous selection. Finally a computer program that detected weak similarities between distantly related sequences was employed. Recording of the analytical data was done with given names of equivalents compiled by US or UL codes as well as constructed alignments. This compilation record, which will be made available as a database, is currently in preparation.

**Compilation within US and UL codes**

Currently the US or UL codes to which the *E. coli's* numbering system is universally applicable are assigned to some of the groups mentioned above, as shown in Table 1. In Table 2, members of the group emp are shown by names of organisms. As a matter of course, except for group emp, any grouping can be assigned within the present known data.

**Table 1** Current grouping of species members of US1 to US21, and UL1 to UL36, except for numbers 8 and 26. e: eukaryotics; m: metabacterials; p: prokaryotics.

| group | emp | ep | mp | p |
|---|---|---|---|---|
| | US 3 | US 6 | | US 1 |
| | US 4 | US13 | | US 2 |
| | US 5 | | | US16 |
| | US 7 | | | US18 |
| | US 8 | | | US20 |
| | US 9 | | | US21 |
| | US10 | | | |
| | US11 | | | |
| | US12 | | | |
| | US14 | | | |
| | US15 | | | |
| | US17 | | | |
| | US19 | | | |
| | /13 | /2 | /0 | /6 |

| group | emp | ep | mp | p |
|---|---|---|---|---|
| | UL 2 | UL13 | UL 1 | UL 9 |
| | UL 3 | | UL 4 | UL16 |
| | UL 5 | | | UL17 |
| | UL 6 | | | UL19 |
| | UL7/L12 | | | UL20 |
| | UL10 | | | UL21 |
| | UL11 | | | UL25 |
| | UL14 | | | UL27 |
| | UL15 | | | UL28 |
| | UL18 | | | UL31 |
| | UL22 | | | UL32 |
| | UL23 | | | UL33 |
| | UL24 | | | UL34 |
| | UL29 | | | UL35 |
| | UL30 | | | UL36 |
| | /15 | /1 | /2 | /15 |

These groupings may change as a consequence of new data though the code numbers will never change. Concerning other groups; several species have been tentatively assigned to groups em or m with many species assigned to group e including to our data, namely, YL16 (Hashimoto *et al.*, 1992), YL35 (Otaka *et al.*, 1985) and YL41 (Suzuki *et al.*, 1990) from yeast. There are not many families of sequences from such species in any group. Taking account of the fact that the sequence data from eukaryotes and metabacteria are not sufficient on their own, the species members of group emp must be greater in number, meaning that a major part of the prokaryotic "gigantic molecule" is evolutionarily maintained in all the organisms on the earth (see Fig. 1).

After detailed analyses of sequence similarities, alignments for US1 to US21 and UL1 to UL36 in which prokaryotic members are contained (see Table 1) have been entirely or preliminarily constructed, including those for US4 (Mizuta *et al.*, 1991), US5 (Otaka *et al.*, 1992), US6 (Otaka *et al.*, 1986), UL7/L12 (Otaka *et al.*, 1989), UL10 (Otaka *et al.*, 1990), UL23 (Raué *et al.*, 1989) and UL30 (Mizuta *et al.*, 1992) according to our method. Remarkably, all the patterns of newly treated alignments represent 'one to one correlations' This is because we identified three irregular patterns; 'double-transposition' of US6 (Otaka *et al.*, 1986), 'single-transposition' of UL7/L12 (Otaka *et al.*, 1989), and 'fusing of two species' of UL10 (Otaka *et al.*, 1990) which confirmed the impression that such irregular cases may not be so rare. Actually such drastic events are likely to have been very unusual.

Table 2: Sequence data materials and residue numbers. p: prokaryotics: m: metabacterials; e: eukaryotics; chl: chloroplast; mit: mitochondrion.

US 3 p :*E coli* 232, Spinach chl 218, Maize chl 224
m:*H hal* 302, *M vann* 211
e :Rat 243, Yeast fragment, *Xen leavis* subunit

US 4 p :*E coli* 205, Tobacco chl 201 *etc*
m:*H cut* fragment
e :Yeast 197, Slime mold 185

US 5 p :*E coli* 116, *M capr* 250, *B stear* 166
m:*M vann* 225
e :Yeast 254, Mouse 221

US 7 p :*E coli* 177, Rice chl 156 *etc*
m:*M vann* 146, *H hal* 210, *Sulf acid* 195
e :Rat 204,

US 8 p :*E coli* 129, Tobacco chl 134 *etc*
m:*M vann* 130, *H mar* 126 *etc*
e :Yeast 129,

US 9 p :*E coli* 128, *B stear* 129
m:*H mar* 130
e :Rat 145, Mouse 145

US10p :*E coli* 103, *M capr* 102, *C parad* 105
m:*M vann* 91, *Sulf acid* 102
e :Rat 119

US11p :*E coli* 128, *B subt* 131, Pea chl 138 *etc*
m:*H mar* 129
e :Yeast 137, *Dr melan* 151, Human 151 *etc*

US12p :*E coli* 123, *S plat* 124, Soybean chl 123 *etc*
m:*M vann* 147, *H morrh* 142, *Sulf acid* 118 *etc*
e :Yeast pers comm, *Tetrahymena* 142

US14p :*E coli* 98, Broad bean mit 100 *etc*
m:*M vann* 53
e :Rat 56

US15p :*E coli* 87, *B stear* 88, Rice chl 90 *etc*
m:*H mar* 155
e :Yeast fragment

US17p :*E coli* 83, *M capr* 85 *etc*
m:*M vann* 109, *H mar* 109
e :Soybean 141, Rat 158, Human 158

US19p :*E coli* 91, *B stear* 91, Spinach chl 92 *etc*
m:*H mar* 115
e :Rat 145, (Soybean , Human)

UL 2p:*E coli* 272, *C parad* 275, Maize chl 273 *etc*
m:*H cut* fragment, *H mar* fragment
e :Slime mold 237

UL 3 p :*E coli* 209, *M capr* 223, *C parad* 209 *etc*
m:*H mar* 337, (*M vann*)
e :Human 348, Yeast 387

UL 5 p :*E coli* 178, *B stear* 179, *M capr* 180 *etc*
m:*M vann* 181
e :Yeast 174, Slime mold fragment

UL 6 p :*E coli* 176, *B stear* 177, *M capr* 180 *etc*
m:*H mar* fragment, *M vann* 182
e :Yeast 191, Rat 192

UL 7 p :*E coli* 120, *B stear* 122, Spinach chl 130 *etc*
m:*M vann* 99, *H cut* 114, *Sulf acid* 105 *etc*
e :Human P1 114, Human P2 115, Yeast 110 *etc*

UL10p :*E coli* 165, *S typh* 164
m:*H cut* 352, *M vann* 336, *Sulf acid* 335
e :Yeast 312, Mouse 317, Human 317

UL11p :*E coli* 141, *S marc* 141, Spinach chl 223
m:*H cut* 162, *H mar* 161, *Sulf acid* 170
e :Yeast 166, Rat 165

UL14p :*E coli* 123, *B subt* 122, Maize chl 123, *etc*
m:*M vann* 132
e :Yeast 137

UL15p :*E coli* 144, *M capr* 145, *B stear* 146
m:*M vann* 143, *H mar* 154
e :Yeast 148, Rat 135

UL18p :*E coli* 117, *M capr* 116, *B stear* 120
m:*M vann* 195
e :Rat 296, *S pombe* 239, *X laev* 295 *etc*

UL22p :*E coli* 110, Spinach chl 199, Maize chl 126, *etc*
m:*H mar* 154, *H hal* 155
e :Yeast fragment

UL23p :*E coli* 100, *M capr* 94, Maize chl 93 *etc*
m:*M vann* 85, *H mar* 84
e :Yeasts 142

UL24p :*E coli* 103, *M capr* 108, Garden pea chl 194 *etc*
m:*H mar* 119, *M vann* 119
e :Rat 145, Yeast fragment

UL29p :*E coli* 63, *B stear* 66, *B subt* 66, *M capr* 138
m:*H hal* 71, *H mar* 69
e :Rat 122

UL30p:*E coli* 58, *B stear* 62
m:*M vann* 154, *H mar* 153
e :Yeast 244, Rat 258, Slime mold 246 *etc*

Generally members of the families of sequences of each of the five groups: eukaryotes, metabacteria, eubacteria, chloroplasts and mitochondria, are similar in sequence length (see Table 2) and are closely-related in sequence similarity making it relatively easy to align them. Some irregularity is occasionally encountered in chloroplasts' and in mitochondrials as shown by US18 (Fig. 2-A). However, if members of group emp are examined, the eukaryotics are not always longer than the metabacterials and/or prokaryotics in sequence length, but sometimes vary, though the ribosomes of eukaryotes are bigger than those of the other two groups. Those of UL10, UL15, US19, UL23, UL24 etc. correspond to extended N-termini, a typical alignment of which is shown for UL23 (Fig. 2-B). According to Raué *et al.* (1989), the elongated sequence of UL23 should result in it becoming eukaryotic since it includes nuclear import signals. Metabacterial UL23s do not have such an extension, but do have the same extensive deletion as the eukaryotics, an indicator of the evolutionary place of metabacteria. Fig. 2-C shows the elongated C-termini of over 120 residues of the eukaryotic UL18s. The metabacterial UL18 follows them with wide deletion sites. There are three well-conserved regions in all groups, where a rRNA-binding function might be inferred to exist. As shown in Fig. 2-D, the prokaryotic UL30s correspond to the metabacterials only in their N-terminal region and lack an equivalent of the metabacterial C-terminal region (Mizuta *et al.*, 1992). The metabacterials lack an equivalent of the eukaryotic N-terminal region of over 80 residues. Such elongated patterns at both ends of the molecule are shown by US5 and US17. There are several other cases in which alignments end with similar sized N- and C-termini as illustrated by US8 (Fig. 2-E). Showing such a similar alignment pattern, eukaryotic and metabacterial US14s are shorter than others because of internal deletions of 45 residues. The situation with US4, is intricate: the prokaryotics are longer than others by about 30 residues at the N-terminus while the eukaryotics contain an insertion of over 40 residues in the middle portion preceding the region by over 40 residues which are well-conserved through all sequences (Mizuta *et al.*, 1991).

Looking at the alignments in group emp, members from every kingdom, including the organelles, show grouping changes in deletion/insertion size as well as in sequence length (Fig. 2). The changes in internal transposition and in gene fusion are also seen (Fig. 3). Such changes must be regarded as extremely significant events in protein evolution, compared with the conversion of amino acids in sequences. The grouping changes come

## (A) US18

```
                     ECOLI   1        2        3        4        5        6        7
                     ARYFRRRKPCRFTAEGVQEIDYKDIATLKNYITESGKIVPSRITGTRAKYQRQLARAIKRARYLSLLPYTDRHQ
                 BSTEA   1        2        3        4        5        6        7
                 AGRKGGRGKRRK-VCYFTANNITHIDYKDVDLLKKFISERGKILPRRVTGTSAKYQRKLTVAIKRARQMALLPYVADE
                  MAPOC   1        2        3        4        5        6        7
                  MNKSKRSSRRR-MPPIRS--GEIIDYKNISLLRRFVSEQGKILSRRMNRLTSKQQRLLTIAIKRARVLALLPFLNNEN
             TOBAC   1        2        3        4        5        6        7        8                             9        0
             MDKSKRPFLKFKRSFRRR-LPPIQS--GDRIDYRNMSLISRFISEQGKILSRRVNRLTLKQQRLITLAIKQARILSLLPFLNN--EKQFER------------25-----------TESTARTTGFKARNK
ORSAC   1        2        3        4        5        6        7        8        9        0        1        2        3        4        5        6
MYTSKQPFHKSKQTFHKSKQTFRKSKQTFRKFKQPFRKPKQPFRRR--PRIGP--GDRIDYRNMSLINRFISEQGKILSRRINRLTLKQQRLITLAIKQARILSFLPFRNYENEKQFQAQSISIITGPRPRKNRHIPPLTQKFNSNRNLRNSNQTLRNTNRNLSSDC
```

## (B) UL23

```
CANUT   1        2        3        4        5        6        7        8        9        0        1                       2        3        4
MAPSTKAASAKKAVVKGSNGSKALKVRTSTTFRLPKTLKLTRAPKYARKAVPHYQRLDNYKVIVAPIASETAMKKVEDGN-TLVFQVDIKANKHQIKQAVKDLYEVDVLAVNTLIRP-----13------NGTKKAYVRLTADHDALDIANKIGYI
SACAR   1        2        3        4        5        6        7        8        9        0        1                       2        3        4
MAPSAKATAAKKAVVKGTNGKKALKVRTSATFRLPKTLKLARAPKYASKAVPHYNRLDSYKVIEQPITSETAMKKVEDGN-ILVFQVSMKANKYQIKKAVKELYEVDVLKVNTLVRP-----13------NGTKKAYVRLTADYDALDIANRIGYI
                                                         METVA   1        2        3        4        5        6                       7        8
                                                         MDAFDVIKTPIVSEKTMKLIEEEN-RLVFYVERKATKEDIKEAIKQLFNAEVAEVNTNITP-----13------KGQKKAYIKLKDEYNAEVAASLPIY
                                                           HAMAR   1        2        3        4        5                       6        7        8
                                                           SWDVIKHPHVTEKAMNDMDFQN-KLQFAVDDRASKGEVADAVEEQYDVTVEQVNTQNTM-----13------DGEKKAVVRLSEDDDAQEVASRIGVF
                                                      ECOLI   1        2        3        4        5        6        7        8        9        0
                                                      MIREERLLKVLRAPHVSEKASTAMEKSN-TIVLKVAKDATKAEIKAAVQKLFEVEVEVVNTLVVKGKVKRHGQRIGRRSDWKKAYVTLKEGQNLDFVGGAE
                                                        BSTEA   1        2        3        4        5        6        7        8        9
                                                        MKDPRDIIKRPIITENTMNLIGQKK--YTFEVDVKANKTEVKDAVEKIFGVKVEKVNIMNYKGKFKRVGRYSGYTNRRKKAIVTLTPDSKEIELFEV
                                                           WHEAC   1        2        3        4        5        6        7        8        9
                                                           MDGIKYAVFTEKSLRLLGKNQ--YTFNVESGFTKTEIKHWVELFFGVKVVAVNSHRLPGKGRRIGPILGHTMHYRRMIITLQPG-YSIPLLDREKN
                                                           MAIZC   1    .   2        3        4        5        6        7        8        9
                                                           MDGIKYAVFTEKSLRLLGKNQ--YTFNVESGFTKTEIKHWVELFFGVKVVAVNSHRLPGKGRRMGPILGHTMHYRRMIITLQPG-YSIPLLDRETN
```

## (C) UL18

```
RATL5   1        2        3        4        5        6        7        8        9        0        1        2        3        4        5        6        7
GFVKVVKNKAYFKRYQVRFRRRREGKTDYYARKRLVIQDKNKYNTPKYRMIVRVTNRDIICQIAYARIEGDMIVCAAYAHELPKYGVKVGLTNYAAAYCTGLLLARRLLNRFGMDKIYEGQVEVNGDEYNVESIDGQPGAFTCYLDAGLARTTTGNKVFGALKGAVDGGLSIPHSTKR
XLAEV   1        2        3        4        5        6        7        8        9        0        1        2        3        4        5        6        7
GFVKVVKNKAYFKRYQVKFRRRREGKTDYYARKRLVIQDKNKYNTPKYRMIVRVTNRDIICQIAYARIEGDMIVCAAYAHELPKYGVKVGLTNYAAAYCTGLLLARRLLNKFGLDKVYEGQVEVTGDEYNVESVDGEPGAFTCYLDAGLTRTTTGNKVFGALKGAVDGGLSIPHSTKR
  METVA        1       2        3                4        5        6        7        8        9        0                                           1        2        3
  MIMAQN----AKHRVPFRRRREGKTDFRQRLGLLLS-------GKPRLVARKSLNNIIAQLMAYDEKGDIVLVSAHSRELVKMGYKGHCGNLPAAYLTGLLLGKKAVEEGLE------------27-----------EAILDKGLHRATKGAAIFAVLKGALDAGMDIPCGEEI
               ECOLI   1        2                 3        4        5        6        7        8                                           9        0        1
               MDKKSARIRRATRARRKLQELG--------ATRLVVHRTPRHIYAQVI--APNGSEVLVAASTVEKAIAEQLKYTGNKDAAAAVGKAVAERALEKGIK------------27-----------DVSFDRSGFQ--YHGRVQALADAAREAGLQF
           BSTEA   1        2                 3        4        5        6        7        8        9                                           0        1        2
           MITKVDRNAVRKKRHARIRKKIFGTT-------ERPRLSVFRSNKHIYAQII--DDTKSATIVSASTLDKEF--GLDSTNNIEAAKKVGELVAKRALEKGIK------------27-----------QVVFDRGGYL--YHGRVKALADAAREAGLEF


  8 RAT    9        0        1        2        3        4        5        6        7        8        9
FLGYDSESKEFNAEVHRKHIMGQNVADYMRYLMEEDEDAYKKQFSQYIKNNVTPDMMEEMYKKAHAAIRENPVYEKKPKREVKKKRWNRPKMSLAQKKDRVAQKKASFLRAQERAAES
  8 XLAEV  9        0        1        2        3        4        5        6        7        8        9
FPGYDSESKEFNPEVHRKHIFAQNIAEYMRLLMEEDEDAYKKQFSQYIKNGVAADQLEDIYKKAHAGIRENPVHEKKPKKEVKKKRWNRAKLSLEQKKDRVAQKKASFLRAQEKA-DS
METVA            4        5        6        7                                        8        9
------13-----IADEERLNGTHIKQYAELLK-EDEEAYKKQFSKYLEKGLNPE-------------------40-----------------DLPEHFEELKGKILNL
```

## (D) UL30

```
RAT      1         2         3         4         5         6         7         8
MEAVPEKKKKVAAALGTLKKKKVPAVPETLKKKRRNFAELKVKRLRKKFALKTLRKARRKLIYEKAKHYHKEYRQMYRTEIRMA
YEAST                         1         2         3         4         5         6
MAAE--------18--------KILTPESQLKKSAKQQKTAEQVAAERAARKAANKEKRAIILERNAAYQKEYETAERNIIQAK

RAT  9         0         1         2         3         4         5         6         7         8         9         0         1         2         3         4         5
RMARKAGNFYVPAEPKLAFVIRIRGINGVSPKVR-KVLQLLRLRQIFNGTFVKLNKASVNMLRIVEPYIAWGYPNLKSVNELIYKRGYGKINKKRIALTDNSLVARSLGKFGIICMEDLIHEIYTVGKRFKEANNFLWPFKLSSPRGGMK--KKTTHFVEGGDAGNREDQIEHPIKR
  7 YEAST  8         9         0         1         2         3         4         5         6         7         8         9         0         1         2         3         4
RDAKAAGSYYVEAQHKLVFVVRIKGINKIPPKPR-KVLQLLRLTRINSGTFVKVTKATLELLKLIEPYVAYGYPSYSTIRQLVYKRGFGKINKQRVPLSDNAIIEANLGKYGILSIDDLIHEIITVGPHEKQANNFLWPFKLSNPSGGWGVPRKFKHFIQGGSFGNREEFINKLVKSMN
               METVA  1         2         3         4         5         6         7                 8         9         0         1         2         3         4         5
               MAYAVVRVRGSVGVRGDIA-DTMKMLRLHRVNHCVVIPENDHYTGMIKKVKDYVTYGEIDKETLVSLILKRGRL--------AGNKRLSEELLKELVELPVDALAEKV-LAGEIKLKDTPIKPVFRLHPPRRGYDRGGIKKGFSIGGALGYRAGKINDLLNKMM
                HAMAR 1         2         3         4         5         6         7                8         9         0         1         2         3         4         5
                MHALVQLRGEVNMHTDIQ-DTLEMLNIHHVNHCTLVPETDAYRGMVAKVNDFVAFGEPSQETLETVLATRAEP-------LEGDADVDDELVAEHTDYDDISGLAFALLSEETTLREQGLSPTLRLHPPRGGHD--GVKHPVKEGGQLGKHDTEGIDLLEAMR
               ECOLI  1         2         3         4         5
               AKTIKITQTRSAIGRLPKHKATLLGLGLRRIGHTVEREDTPAIRGMINAVSFMVKVEE
               BSTEA  1         2         3         4         5         6
               AKKLAITLTRSVIGRPEDQRITVRTLGLRKMHQTVVHNDNPAIRGMINKVAHLVKVKEIEEE
```

## (E) US8

```
 YEAST   1         2         3         4         5         6         7           8.        9         0         1         2
 TRSSVLADALNAINNAEKTGKRQVLIRPSSKVIIKFLQVMQKHGYIGEFEYIDDHRSGKIVVQLNGRLN--KCGVI----SPRFNVKIGDIEKWTANLLPAR--QFGYVILTTSAGIMDHEEARRKHVSGKILGFVY
METVA    1         2         3         4         5         6         7           8         9         0         1         2         3
MSLMDPLANALNHVSNCEGVGKNVAYLKPASKLIGRVLKVMQDQGYIGNFEYIEDGKAGVYKVDLIGQIN--KCGAV----KPRYAVKYQEFEKFEKRYLPAK--GFGLLIVSTPKGLMTHDEARTAGVGGRLISYVY
 HAMAR   1         2         3         4         5         6         7           8         9         0         1         2
 TGNDPFANALSALNNAESVGHLEQTVSPASNEIGSVLEVFYDRGYIDGFSFVDDGKAGEFEVELKGAIN--ECGPV----KPRYSAGAD---EFEKRFLPAR--DYGTLVVTTSVGIMSHYEAREQGVGGQVIAYVY
 ECOLI   1         2         3         4         5         6         7           8         9         0         1         2
 SMQDPIADMLTRIRNGQAANK-AAVTMPSSKLKVAIANVLKEEGFIEDFKVEG-DTKPELELTLKYFQG--KAVVE----SIQRVSRPGLRIYKRKDQLPKVMAGLGIAVVSTSKGVMTDRAARQAGLGGEIICYVA
 MCAPR   1         2         3         4         5         6         7           8         9         0         1         2
 MTTDVIADMLTRIRNANQRYL-KTVSVPSSKVKLEIARILKEEGFISDFTVEG-DVKKTINIELKYQGK--TRVIQ----GLKKISKPGLRVYAQANEIPQVLNGLGISIVSTSQGIMTGKKARLANAGGEVLAFIW
 CYPAC   1         2         3         4         5         6         7           8         9         0         1         2         3
 MVNDTIADMLTRIRNANLAKH-KVARVKATKITRCLANVLKEEGLIQNFEEIENNLQNELLISLKYKGKKRQPIIT----ALKRISKPGLRVYANHKELPRVLGGLGIAILSTSSGIMTDQTARHKGCGGEVLCYIW
 MAIZC   1         2         3         4         5         6         7         8         9         0         1         2         3
 MGKDTIADLLTSIRNADMNKK-GTVRVVSTNITENIVKILLREGFIESVRKHQESNRYFLVSTLRHQRRKTRKGIYRTRTFLKRISRPGLRIYANYQGIPKVLGGMGIAILSTSRGIMTDREARLNRIGGEVLCYIW
```

**Figure 2**: Alignments. Some delegations out of the family sequences were selected. To save space, references were by an irregular format. (A) ECOLI: *E coli* S18 (Schnier J *et al* 1986 Mol Gen Genet 204:126-132); BSTEA: *B stear* S18 (McDougall J *et al* 1989 FEBS LEtt 245:253-260); MAPOC: liverwort chl S18 (Ohyama K 1986 submitted to EMBL/GenBank); TOBAC: *Nicotiana tabacum* chl S18 (Sugiura M 1989 submitted to EMBL/GenBank); ORSAC: rice chl S18 (Sugiura M 1986 submitted to EMBL/GenBank). (B) CANUT: *Candida utilis* L25 (Woudt LP *et al* 1987 Curr Genet 12:193-1980; SACAR: *S carlsberg* L25 (Leer RJ *et al* 1984 Nucl Acids Res 12:6685-6700, see Woudt LP *et al* 1987); METVA: *Methan vann* L23 (Koepke AKE *et al* 1988 FEBS Lett 239:313-318); HAMAR: *Hal marrism* L25 (Hatakeyama *et al* 1988 Eur J Biochem 172:703-711); ECOLI: *E coli* L23 (Wittmann-Liebold B *et al* 1979 FEBS Lett 108:69-74); BSTEA: *B stear* L23 (Kimura *et al* 1984 Eur J Biochem 150:491-497); WHEAC: wheat chl L23 (Bowman CM *et al* 1988 Curr Genet 14:127-136); MAIZC: maize chl L23 (McLaughlin WE *et al* 1988 Nucl Acids Res 16:8183-8183). (C) RAT: rat L5 (Tamura S *et al* 1987 Eur J Biochem 168:83-87); XLAEV: *Xen laevis* L5B (Wormington WM 1989 Mol Cell Biol 9:5281-5288); METVA: *Methan vann* L18 (Auer J *et al* 1989 J Mol Biol 209.21-36); ECOLI: *E coli* L18 (Brosius J *et al* 1975 FEBS Lett 56:359-361); BSTEA: *B stear* L18 (Kimura J *et al* 1987 FEBS Lett 210:85-90). (D) RAT: rat L7 (Lin *et al* 1987 J Biol Chem 262:12665-12671); YEAST: *S cerevisiae* YL8 (Mizuta K *et al* 1992 Nucl Acids Res 20:1011-1016); METVA: *Methan vann* L30 (Auer J *et al* 1989 J Mol Biol 209:21-36); HAMAR: *Hal marism* L20 (Hatakeyama *et al* 1989 Eur J Biochem 185:685-693); ECOLI: *E coli* L30 (Ritter E *et al* 1975 B FEBS Lett 60: 153-155); BSTEA: *B stear* L30 (Kimura M *et al* 1984 259:1051-1055). (E) YEAST: *S cerevisiae* S24 (Leer RJ *et al* 1985 Nucl Acids Res 13:701-709); METVA: *Methan vann* S8 (Auer J *et al* 1989 J Mol Biol 209:21-36); HAMAR: *Hal marism* S16 (Kimura J *et al* 1987 J Biol Chem 262:12150-12157); ECOLI: *E coli* S8 (Allen G *et al* 1978 Hoppe-Seyler's Z 359:1509-1525); MCAPR: Ohkubo S *et al* 1987 Mol Gen Genet 185:239-244); CYPAC: *Cyan paradoxa* cyanelle S8 (Bryant DA *et al* 1990 FEBS Lett 259:273-280); MAIZC: maize chl S8 (Markmann-Mulisch U *et al* 1988 Eur J Biochem 170:507-514).

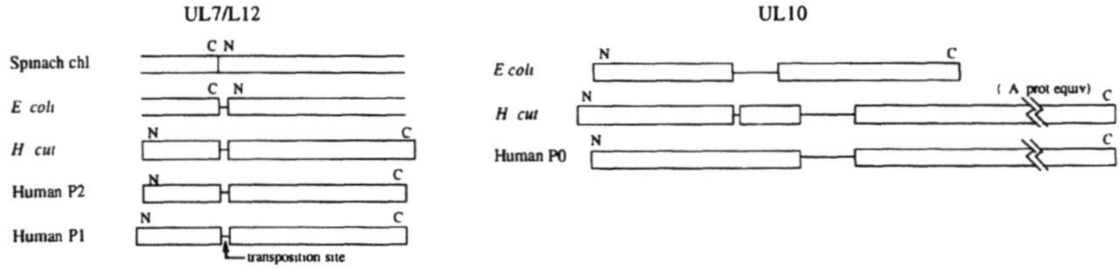


**Figure 3:** Schematic alignments. Open boxes correlate with the lower three sequences by transposition. N and C identify the amino- and carboxyl-termini, respectively.

out so tightly that even higher eukaryotes, like mammals, maintain in common such important changes as have occurred in lower eukaryotes, like the unicellular yeasts. One of the rare expections to this phenomenon is seen between *S. cerevisiae* YL35 (88 residues) and rat L37 (111 residues) which are tentatively assigned to group e (Otaka *et al.*, 1985). On the other hand, we know the great difference found between mammals and yeasts in analytical patterns of ribosomal proteins by 2D-PAGE (Otaka & Osawa, 1981). The amino acid differences might be the main cause of such differences. This will be discussed in the next section.

**'A' proteins and the related species, coded as UL7/L12 and UL10**

According to Doolittle (1981) the ultimate goal of the study of protein evolution is the reconstruction of the past events that gave rise to the vast inventory of proteins in existence today. Similarly, an evolutionary examination of ribosomal proteins, as revealed here, demonstrates the various past events on these proteins. Our studies have shown us that the transposition and gene fusion events that occurred in *E. coli* L7/L12 equivalent species are significant for their infrequency as well as for their importance (Fig. 3).

In *E. coli,* protein L7 is shown as acetylated protein L12. This species forms a dimer and the two dimers form a complex with protein L10 to build the stalk of the large ribosomal subunit (Subramanian 1975; Petterson *et al.*, 1976; Möller & Massen, 1986). Being acidic and containing many alanines, L7/L12 equivalent proteins seen in other organisms are usually called 'A' proteins. The 'A' proteins from eukaryotes and metabacteria are classified as "transposition type" in contrast to "prototype" for prokaryotes, as identified by examination of the N-terminal sequences which correspond to the C-terminal 36 residues of *E. coli* L7/L12 protein, resulting from the internal transposition of the prototype sequence (Otaka *et al.*, 1989). The transposition type in eukaryotes comprises P1 and P2 line proteins, which are thought to form a tetrameric complex $(P1)_2(P2)_2$. For the pentameric complex, as mentioned in *E. coli*, a third phosphoprotein PO is present. In metabacteria, the transposed 'A' protein is a single protein class termed meta-line. It is reasonable to consider that the meta-line and the PO line proteins in metabacteria must work similarly to others.

Examination of the 'A' proteins through the prototype and the transposition type (Otaka *et al.*, 1989; see Shimmin *et al.*, 1989), can identify an alignment of all the sequences although, 11 gaps as deletion/insertion sites of amino acid residues are necessary. Collating them with the crystallographic data for the C-terminal fragment (CTF) from *E. coli* L7/L12 (Leijonmarck & Liljas, 1987), all the gaps identified in the CTF are located between segments that correspond to structural and functional elements such as α-helix, ß-strand, turning loop or hinge regions (Fig. 4). This suggests that there are specific "preservation units" in protein molecules which are conserved in evolution. In contrast, the transposition event is very progressive. The site is exactly located at the centre of an α-helial element that is part of a folding domain. On the other hand, sequence similarities between these lines were examined by means of correlation coefficient values. Sequences of the P1 line are closer to those of metabacteria than to those of the P2 line. The P1 line might be close to an original type which may have existed before the transposed 'A' protein had been duplicated in the eukaryotics. Yeasts contain multiple species of P1 and P2 line genes and their P1 sequences show low values compared to other P1 line sequences, indicating a great evolutionary distance between lower and higher eukaryotes within the eukaryotes.

Together with the P1 and P2 line proteins, the PO line proteins react with antibodies from some patients with systemic lupus erythematosus (SLE) (Elkon *et al.*, 1985; Francoeur *et al.*, 1985). The common epitope exists in the C-terminal region of these proteins, 22 amino acids in length according to Elkon *et al.* (1986). In fact, about 20 residues at the C-termini of PO proteins from metabacteria, yeast and human are homologous with those of their own transposition-type 'A' proteins. As shown in the right panel of Fig. 3, the N-terminal three-quarters of the sequences align naturally and the first two-thirds of the alignment could

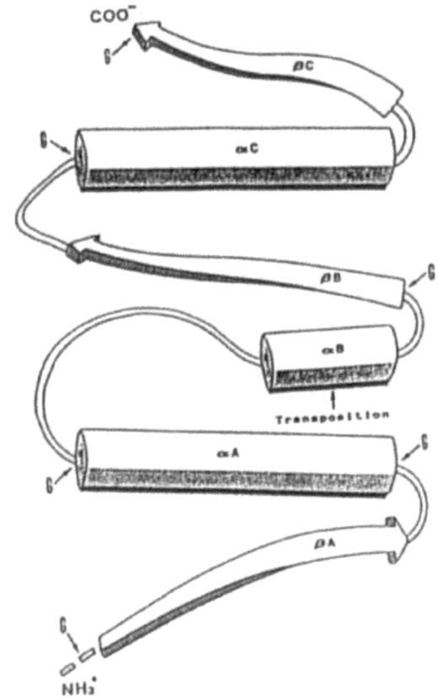


Figure 4: Schematic organization of various elements of the C-terminal fragment. G:gap

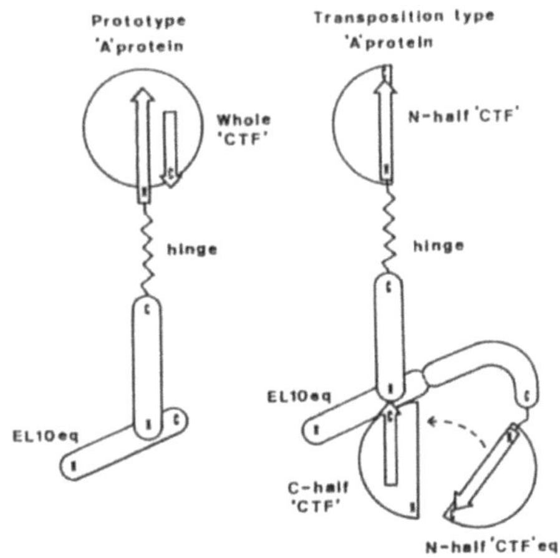


**Figure 5:** Possible arrangements of *E. coli* L7/L12 and L10 equivalents in two types

involve the *E. coli* L10 sequence. An alignment of the remaining sequences of the C-termini was established, relying on the well-matching sequence similarities between the metabacteria 'A' protein and their PO protein sequences. Finally, the C-terminal halves of PO protein sequences correspond with most of the overall length of the transposition-type 'A' proteins. Thus, the gene fusion of a protein equivalent to EL10 and a transposition-type 'A' protein might have resulted in the formation of the PO proteins. A coupling of this gene fusion and the transposition of prototype 'A' proteins may have given rise to the complementary molecular transformations required for the development toward the mode of higher organisms. As shown in Fig. 5, the C-termini of PO proteins, homologous to those of transposition-type 'A' proteins, are considered to form the whole 'CTF' conformation together with the N-termini of transposition-type 'A' proteins within a pentameric complex $(L7/L12)_2(L7/L12)_2$L10-like $(P1)_2(P2)_2$PO which might be involved in given functions as the so-called 'CTF'.

**Evolutionary position of yeast ribosomal proteins**

Based on the evidence described in the above section, we have inferred evolutionary courses of protein species UL7/L12 and UL10 as illustrated in Fig. 6. On the *E. coli* genome, EL10 and EL7 genes are adjacent (Post *et al.*, 1979; Brückner & Matzura, 1981). The first major event in their development, the transposition, occurred within the prototype 'A' protein gene, after which the adjoining two genes branched off toward metabacteria and eukaryotes. In the events leading to the metabacteria, the 'A' protein gene was duplicated only once; one of the duplicated genes being termed 'metabacteria line' of 'A' proteins and the other one forming the metabacteria PO line by a gene fusion with the adjoining gene encoding the EL10 equivalent. Such a gene fusion event could have taken place easily within the prokaryotic gene organization of polycistronically transcribed components.

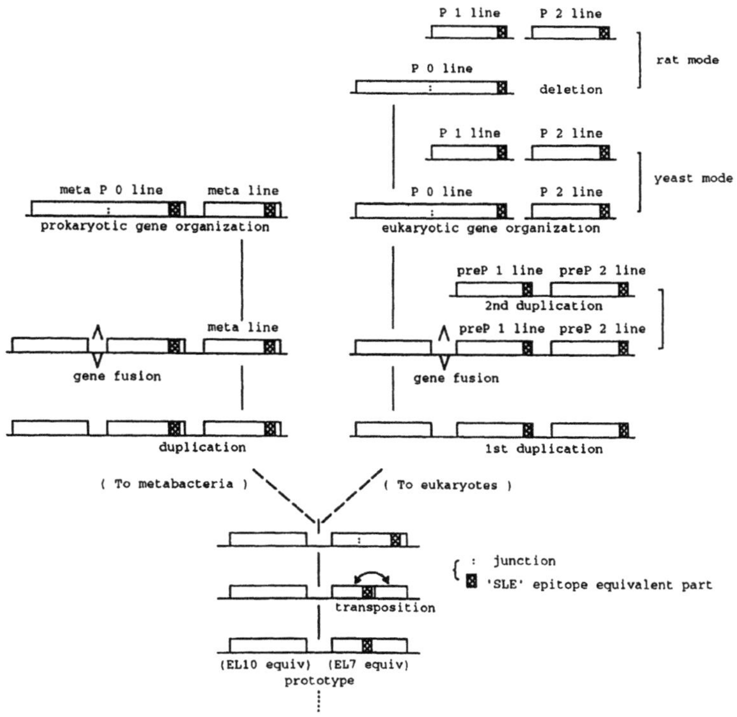


**Figure 6:** Evolutionary course of ribosomal protein genes equivalent to *E. coli* L7/L12 and L10 genes.

On the other hand, the 'A' protein gene was first duplicated, developing into two lines, pre-P1 and pre-P2, and once again or even several times duplicated. This repeated duplication has generated a great disparity between metabacteria and eukaryotes in their present evolutionary position. One of pre-P1 line genes fused with the adjacent gene encoding the EL10 equivalent to produce a eukaryotic PO line gene while the prokaryotic gene organization was still intact. As a result of eukaryotes creating a gene organization of monocistronically transcribed components, the lower eukaryotes still retained multiple species of P2 (or P1 and P2) line proteins, but finally, in the higher eukaryotes they readjusted to become single species. Thus we can see the great evolutionary distance between the lower and the higher eukaryotes besides the changes in molecules. As shown in Fig. 2, the drastic changes in the eukaryotic proteins were mentioned as grouping changes. With the eukaryotes, the development of gene organization was revealed here. It is still a major problem that the various events during protein evolution as described here are yet without numerical evaluation.

Summary

We have attempted to systematically compile equivalents of ribosomal protein species from various organisms from the view point of molecular evolution. Even though we are undertaking the task of compilation, the evolutionary profile of ribosomal proteins has begun to emerge: (1) there are many protein species which exist throughout the three primary kingdoms, (2) almost all the alignments must show 'one to one correlation' to align regularly, that is, irregular correlation via internally-transposition appears to be rarely found, and (3) considering the organelles, grouping changes of sequences from every kingdom appear as wide insertion/deletions, elongation in length and so on. Through these preliminary profiles, the evolutionary place of yeast has emerged.

References:

Brückner, R., Matzura, H. (1981) In vivo synthesis of a polycistronic messenger RNA for the ribosomal proteins L11, L1, L10 and L7/L12 in *Escherichia coli*. *Mol. Ge n. Genet.* 183: 277-282.

Doolittle, R.F. (1981) Similar amino acid sequences: chance or common ancestry? *Science* 214: 149-159.

Elkon, K., Parnassa, A.P., Foster, C.L. (1985) Lupus autoantibodies target ribosomal P proteins. *J. Exp. Med.* 162, 459-471.

Elkon, K., Skelly, S., Parnassa, A., Moller, W., Danho, W., Weissbach, H., Brot, N. (1986) Identification and chemical synthesis of a ribosomal protein antigenic determinant in systemic lupus erythematosus. *Proc. Natl. Acad. Sci.* USA 84: 5580-5584.

Francoeur, A.M., Peebles, C.L., Heckman, K.J., Lee, J.C., Tan, E.M. (1985) Identification of ribosomal protein autoantigens. *J. Immunol.* 135 , 2378-2384.

Hashimoto, T., Suzuki, K., Mizuta, K., Otaka, E. (1992) Yeast ribosomal proteins: XIV. Complete nucleotide sequences of the two genes encoding *Saccharomyces cerevisiae* YL16. Biochem. Biophys. Acta, 1132, 195-198.

Leijonmarck, M., Liljas, A. (1987) Structure of the C-terminal domain of the ribosomal protein L7/L12 from *Escherichia coli* at 1.7Å. *J. Mol. Biol.* 195: 555-580.

Mizuta, K., Hashimoto, T., Suzuki, K., Otaka, E. (1991) Yeast ribosomal proteins: XII. YS11 of *Saccharomyces cerevisiae* is a homologue to *E. coli* S4 according to the gene analysis. *Nucl. Acids Res.* 19: 2603-2608.

Mizuta, K., Hashimoto, T., Otaka, E. (1992) Yeast ribosomal proteins: XIII. *Saccharomyces cerevisiae* YL8A gene, interrupted with two introns, encodes a homolog of mammalian L7. *Nucl. Acids Res.* 20: 1011-1016.

Möller, W., Massen, J.A. (1986) On the structure, function, and dynamics of L7/L12 from *Escherichia coli* ribosomes In: Structure, Function, and Genetics of Ribosomes. (Hardesty, B., Kramer, G., eds.) Springer-Verlag, New York, 309-325.

Otaka, E., Osawa, S. (1981) Yeast ribosomal proteins: V. Correlation of several nomenclatures and proposal of a standard nomenclature. *Mol. Gen. Genet.* 181: 176-182.

Otaka, E., Ooi, T., Kumazaki, T., Itoh, T. (1985) Examination of protein sequence homologies: II. Six *Escherichia coli* L7/L12-type ribosomal "A" protein sequences from eukaryotes and metabacteria, contrasted with those from prokaryotes. *J. Mol. Evol.* 22: 342-350.

Otaka, E., Ooi, T., Itho, T., Kumazaki, T. (1986) Examination of protein sequence homologies: III. Ribosomal protein YS25 from *Escherichia coli* and its counterparts from *Schizosaccharomyces pombe*, liver, rat and *Escherichia coli*. *J. Mol. Evol.* 23: 337-342.

Otaka, E., Ooi, T., Suzuki, K. (1989) Examination of protein sequence homologies: VI. The evolution of *Escherichia coli* L7/L12 equivalent ribosomal proteins ('A' proteins), and the tertiary structure. *Prot. Seq. Data Anal.* 2: 395-402.

Otaka, E., Suzuki, K., Hashimoto, T. (1990) Examination of protein sequence homologies: VII The complementary molecular coevolution of ribosomal proteins equivalent to *Escherichia coli* L7/L12 and L10. *Prot. Seq. Data Anal.* 3: 11-19.

Otaka, E., Hashimoto, T., Mizuta, K. (1992) The ribosomal proteins: I. An introduction to a compilation of the protein species equivalents from various organisms by a universal code system. *Prot. Seq. Data Anal.* in press.

Pettersson, I., Hardy, S.J.S., Liljas, A. (1976) The ribosomal protein L8 is a complex of L7/L12 and L10. *FEBS Lett* 64: 135-138.

Post, L.E., Strycharz, G.D., Nomura, M., Lewis, H., Dennis, P.P. (1979) Nucleotide sequence of the ribosomal protein gene cluster adjacent to the gene for RNA polymerase subunit ß in *Escherichia coli*. *Proc. Natl. Acad. Sci.* USA, 76, 1697-1701.

Raué, H.A., Otaka, E., Suzuki, K. (1989) Structural comparison of 26S rRNA-binding ribosomal protein L25 from two different yeast strains and the equivalent proteins from three bacteria and two chloroplasts. *J. Mol. Evol.* 28: 418-426.

Shimmin, L.C., Ramirez, C., Matheson, A.T., Dennis, P.P. (1989) Sequence alignment and evolutionary comparison of the L10 equivalent and L12 equivalent ribosomal proteins from Archaebacteria, Eubacteria, and Eucryotes. *J. Mol. Evol.* 29: 448-462.

Subramanian, A.R. (1975) Copies of proteins L7 and L12 and heterogeneity of the large subunit of Escherichia coli ribosome. *J. Mol. Biol.* 95: 1-8.

Suzuki, K., Hashimoto, T., Otaka, E. (1990) Yeast ribosomal proteins: XI. Molecular analysis of two genes encoding YL41, an extremely small and basic ribosomal proteins, from Saccharomyces cerevisiae. *Curr. Genet.* 17: 185-190.

Wada, A., Sako, T. (1987) Primary structures of and genes for new ribosomal proteins A and B in *Escherichia coli*. *J. Biochem.* 101: 817-820.

# THE ACIDIC RIBOSOMAL PROTEINS AND THE CONTROL OF PROTEIN SYNTHESIS IN YEAST

J.P.G.Ballesta, M. Remacha, T. Naranda, C. Santos, B. Bermejo, A. Jimenez-Diaz, and B. Ortiz-Reyes.
Centro de Biología Molecular
Consejo Superior de Investigaciones Científicas y Universidad Autónoma.
Canto Blanco, 28049 Madrid, Spain

The protein L7/12 of the large bacterial ribosomal subunit, is unique among the components of the particle due to its strongly acidic pI and to the fact that it is the only multicopy components of the bacterial ribosome. Two dimers of this protein form the L7/12 stalk one of the most typical structural features of the subparticle, that shows a noticeable flexibility and seems to be implicated in the interaction of the supernatant factors with the ribosome during protein synthesis (see (Möller and Maassen, 1986) for review).

The eukaryotic ribosomes contains a set of proteins with similar physicochemical properties and molecular weight as their bacterial counterparts,which also seem to play a similar functional role in the protein synthesis mechanism (van Agthoven et al., 1978; Sanchez-Madrid et al., 1979; Lavergne et al., 1987). In fact, proteins from both biological systems have been shown to be interchangeable when tested in some *in vitro* tests (Wool and Stöffler, 1974; Möller et al., 1975; Sanchez-Madrid et al., 1981b). However, eukaryotic cells ribosomes have developed a more complicated system of acidic proteins which show several clearly distinctive characteristics:

a) While in bacteria protein L7 and L12 correspond simply to the blocked and unblocked amino terminal forms of the same polypeptide, which are functionaly identical, the eukaryotic ribosomes having a family of acidic polypeptides encoded by independent genes. In addition, a protein called P0, with an overall basic pI and a higher molecular weight but with an amino acid sequence at the carboxyl terminus highly homologous to the typical acidic proteins, has been found in eukaryotic cells and seems to be functionally related to them.

NATO ASI Series, Vol. H 71
Protein Synthesis and Targeting in Yeast
Edited by A. J. P. Brown, M. F. Tuite, and J. E. G. McCarthy

b) The eukaryotic acidic proteins are found phosphorylated in the ribosome.

c) An unexpectedly large cytoplasmic pool of acidic ribosomal proteins has been found in the different eukaryotic species tested.

d) The amount of acidic proteins found in the ribosomes changes with the ribosome preparation and it seems to be related to the metabolic conditions of the cell.

e) An exchange between the acidic proteins in the cytoplasmic pool and in the ribosomes has been found to take place.

We are going to review recent results concerning each one of these specific features of the acidic ribosomal proteins from *Saccharomyces cerevisiae,* which seem to support the idea that these ribosomal components appear to have gained a new role during evolution. Protein P0, although in some aspects seems to be functionally related to the acidic proteins, represents a clearly different ribosomal component and will not be specifically considered here.

Ribosomal acidic protein diversity

Acidic proteins from a number of eukaryotic organimsms have been sequenced (Amons et al., 1979; Lin et al., 1982; Qian et al., 1987; Rich and Steitz, 1987; Wigboldus, 1987; Beltrame and Bianchi, 1990; Schijman et al., 1990; Prieto et al., 1991; Vazquez et al., 1992). The number of components in the eukaryotic acidic protein family varies from two in mammals, up to eight in *Trypanosoma cruzi* (Schijman et al., 1990; Vazquez et al., 1992). The acidic proteins can be classified in two groups according to their sequence homology to mammal proteins P1 and P2 (Wool et al., 1991). All the organisms studied have at least one protein from each group.

In *S. cerevisiae* four acidic protein genes have been reported (Mitsui and Tsurugi, 1988; Remacha et al., 1988; Newton et al., 1990), encoding four polypeptides, called YP1α,YP1ß,YP2α and YP2ß, two corresponding to each group (Wool et al., 1991). However, only three polypeptides are found in detectable amounts when proteins from yeast ribosomes are resolved by isoelectrophocusing. These proteins have been biochemically characterized as proteins L44′, L44 and L45 (Juan-Vidales et al., 1984) and correspond to YP1ß, YP2α and YP2ß respectively. Protein YP1α has been identified by an *in vitro* transcription-translation coupled system (Mitsui and Tsurugi, 1989) as the most acidic band in the isoelectrofocusing systems. This protein which is stained very poorly by the usual staining techniques, is found in very low amounts when analyzed by isoelectrofocusing but can be purified by HPLC (C. Santos,B. Ortiz-Reyes, T. Naranda, M. Remacha y J.P.G.Ballesta, manuscript in preparation).

Function of the different yeast acidic proteins.

The substantial differences among the acidic proteins suggest that they may play different functions and, in the case of *S.cerevisiae,* this possibility has been studied using gene disruption techniques. Yeast strains were obtained in which each one of the four genes encoding these proteins was disrupted, and by appropriate crosses among them, strains containing multiple disrupted genes were prepared (Remacha et al., 1990; Remacha et al., 1992). The disruption of only one out of the four genes affects differently, but not very dramatically, the growth of the cells, depending on the protein eliminated (Fig 1). The absence of YP1α is the most damaging, increasing the doubling time of the culture up to 70%; the disruption of the YP2α(L44) gene seems to be the most innocuous for cell growth, causing only a 10% increment in the generation time of the cells. In general, inactivation of one of the members of the two groups, YP1α and YP2ß(L45), is more damaging than inactivation of the other, YP1ß(l44′) and YP2α(L44), but none of them is esential for cell growth although the presence of the four proteins is required for optimal growth. This is in contrast with the situation in *S. pombe*. In this yeast species one of the components of the two groups is essential for cell viability and the other is dispensable (Beltrame and Bianchi, 1990).

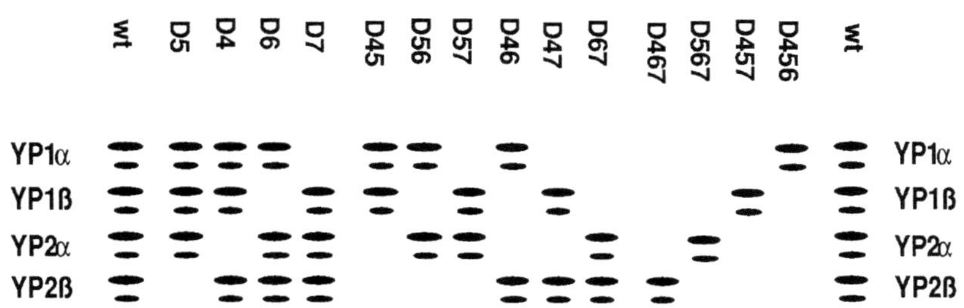


| | wt | D5 | D4 | D6 | D7 | D45 | D56 | D57 | D46 | D47 | D67 | D467 | D567 | D457 | D456 | wt |
|---|---|---|---|---|---|---|---|---|---|---|---|---|---|---|---|---|
| YEPD | 85 | 125 | 95 | 110 | 145 | 215 | 155 | 155 | 115 | 155 | 235 | 300 | 310 | 285 | 275 | 85 |
| YNBD | 135 | 185 | 160 | 170 | 220 | 290 | 215 | 230 | 175 | 225 | 310 | 325 | 375 | 300 | 300 | 135 |

Fig. 1.- Doubling times of *S.cerevisiae* strains carrying disrupted acidic ribosomal proteins genes. The acidic proteins expressed in each strain are schematically represented as double bands, corresponding to the phosphorylated and dephosphorylated forms, as they appear in isoelectrofocusing gels. Doubling times in rich YEPD and minimal YNBD are expressed in minutes.

The effect of a double disruption is specially important when the two constituents of the same protein group, either P1 or P2, are missing. In this cases (disruptant strains D45 and D67), the cells are still viable but the generation time is increased more than 2.5 times and the average polysome size found in cellular extracts is considerably shorter than in control strains. Moreover, strain D45 is unable to grow in galactose and to sporulate. These effects are reverted by the introduction of an active YP2ß(L45) gene, suggesting that translation of some components of the galactose metabolism and of the sporulation pathways are specifically affected by the defective protein synthesizing mechanism of this strain (Remacha, Bermejo and Ballesta, unpublished results)

Surprisingly, the ribosomes obtained from these two double disruptants are totally deprived of acidic proteins, even of those whose genes are unaffected and consecuently are present in the cells (Remacha et al., 1992). Neither in the singly nor in the doubly disrupted strains is the absence of one proteins compensated by a parallel increase in the others, which are present in roughly the original proportion in the defective particles, in spite of being in excess in the cytoplasmic pool (Remacha et al., 1990; Remacha et al., 1992).

The removal of one more protein from the double disruptants, yielding strains with a unique acidic protein, causes an additional negative effect on the growth rate of the cells, but they are still viable (fig 1).

These results indicate that the structure of the acidic proteins complex requires the presence of at least one protein of the P1-type and another of the P2-type in order to bind to the large subunit. Both members of the same group seem to be equally effective in stimulating the binding of the proteins of the other group, from this point of view being functionally equivalent (Remacha et al., 1992).

It seems also clear that ribosomes are able to participate in protein synthesis in the absence of firmely bound acidic proteins, as indicated by the viability of D45 and D67, and of the triply disrupted strains. However, in these conditions the synthetic process is much less efficient and it seems to carry out unequally the translation of different mRNAs, some of them being more affected than others.

These data, therefore, indicate that the acidic proteins may not be an essential component of the basic protein synthesis machinery and they could, instead, somehow modulate the function of the ribosome. This conclusion is supported by results obtained in bacterial systems showing that in certain conditions ribosomes can perform elongation factor-dependent reactions in the absence of the acidic bacterial protein L7/12 (Glick, 1977; Kotelansky et al., 1977). Interestingly, the elongation factors have also been shown not to be absolutely required for polypeptide synthesis and they have been considered as a late evolutionary acquisition of the protein synthesis machinery that made this process more efficient (Gavrilova et al., 1976). The acidic proteins might be part of this relatively recent evolutionary event, and their peculiar physicochemical charactersitics, quite different from those of the typical ribosomal proteins, are compatible with this notion. They could be considered as a component of the supernatant factor mechanism that, in a later step, acquired the ribosome binding capacity to improve the efficiency of the system.

Amount of acidic proteins present in the ribosomes.

The presence of four copies of the acidic proteins L7/12 has been well documented in bacterial ribosomes (Subramanian, 1975). Contrary to this, the estimation of the equivalent proteins in eukaryotic ribosomes has resulted in values far from four molecules per ribosome. In *Artemia salina* and yeast, 2.7 and 3 copies per ribosome of acidic proteins have been reported found (Kruiswijk et al., 1978; van Agthoven et al., 1978), and in our studies, values ranging from 2.5 to 3.7 were obtained depending on the ribosome preparation (Juan-Vidales et al., 1984).

This variability has been attributed to a tendency of the acidic proteins to be released from the ribosomes, especially upon dissociation of the subunits (Kruiswijk et al., 1978; van Agthoven et al., 1978). However, when the amount of acidic proteins was estimated by ELISA, using acidic protein-specific rabbit antibodies, in ribosomes from cultures at different growth stages, it was found that particles from exponentially growing cells contained almost twice as much acidic proteins than those from stationary phase (Saenz-Robles et al., 1990). Moreover, similar estimations in polysomes and 80S ribosomes, yielded nearly 4 copies per ribosome in the first case and only 3 in the free particles (Saenz-Robles et al., 1990).These results indicate that ribosomes directly implicated in translation, from either metabolically active cells or polysomes, contain notably more acidic proteins than particles not involved in protein synthesis, and suggest that the amount firmly bound acidic proteins is related to the active state of the particles.

Functional role of the eukaryotic acidic protein phosphorylation

Phosphorylation of acidic proteins was initially reported in yeast (Zinker and Warner, 1976) and afterwards confirmed in other eukaryotic species (see Bielka, 1982 for a review).

The acidic protein phosphorylation does not seem to be notably affected by the state of the cells, at least in the case of lower eukaryotes, likeS. *cerevisiae*, in which these proteins are found monophosphorylated in all conditions (Sanchez-Madrid et al., 1981a). However, it has been found that phosphorylation is functionally important since, *in vitro*, dephosphorylated proteins were unable either to activate ethanol-$NH_4Cl$ washed particles (McConnell and Kaplan, 1982) or to participate in reconstitution of dimethylmaleic acid treated 60S subunits (Lavergne et al., 1987). This effect has been shown to be due to the inability of non-phosphorylated polypeptides to bind to the ribosomal core particles (Juan-Vidales et al., 1984). In fact, phosphorylation causes an important change in the protein conformation as revealed by the different susceptibility of the phopshorylated and non-phosphorylated proteins to iodination (Juan-Vidales et al., 1984).

We have recently been able to show that modification of the phosphorylatable serine by site-directed mutagenesis makes the acidic proteins unable to bind to the ribosome particles and forces them to accumulate in the cytoplasm (Naranda and Ballesta, 1991), confirming *in vivo* the functional relevance of this modification for optimal ribosome activity and protein synthesis.

The location of the phosphorylation site is not the same in the four yeast acidic proteins. Thus, while protein YP2ß(L45) is modified at serine in position 19, close to the amino terminus, the other three polypeptides are modified around position 72, closer to the carboxyl end (T. Naranda, M. Remacha and J.P.G. Ballesta, submitted manuscript). These results are surprising in the case of YP2-type proteins which, as indicated previously, seem to play a similar role and, therefore, were expected to have analogous structural features. It is possible, therefore, that the functions of the two members of the YP2 acidic protein family in yeast are not completely identical, wht is also supported by the different size of YP2ß and YP2α.,110 and 106 amino acids respectively.

The differences in the acidic protein phosphorylation site also extend to their amino acid enviroment. The YP2ß site has the typical consensus structure for a casein kinase II site, while the phosphorylatable serines are followed by a number of alanine and glycine residues in the other three polypeptides (Table 1).

This second sequence does not correspond to any identified protein kinase modification site.

It is worth noting that a serine present in the highly conserved carboxyl terminus which forms part of a typical casein kinase II site, is not modified in the yeast proteins either *in vivo* or *in vitro* (T. Naranda, M. Remacha and J.P.G. Ballesta, submitted manuscript). This serine is present in the acidic proteins from all species except protoza and is found phosphorylated in insects (van Agthoven et al., 1978) and in mammals (Hasler et al., 1991). In this last case, it has been shown to be susceptible to casein kinase II (Hasler et al., 1991). On the other hand, acidic proteins of different origin have potential phosphorylation sites of similar structure and location than the yeast polypeptides. Indeed, it is known that, at least in mammals, the acidic proteins are found multiphosphorylated when isolated from ribosomes although the position and functional relevance of the different sites has not been studied.

TABLE 1
**AMINO ACID SEQUENCE AROUND POTENTIAL PHOSPHORYLATION SITES IN YEAST RIBOSOMAL ACIDIC PROTEINS**

| PROTEIN | SEQUENCE | |
|---|---|---|
| | SITE 1 | SITE 2 |
| **YP2ß** | **14-GNAAPSAADI-24** | **68-TGGASSAAAGA-78** |
| **YP2α** | **13-GNTPDATKIK-23** | **67-AGPASAGGAAAASGDAAA-84** |
| **YP1ß** | **17-GLEITSDNLL-26** | **68-GAGAASGAAAA-78** |
| **YP1α** | **18-EIEISSEKLL-27** | **57-LLVNFSAGAAA-67** |

Amino acid sequence around position 19 (site 1) and position 72 (site2) in the four acidic proteins from *S. cerevisiae*. Phosphorylated serines are underlined and marked in larger type. Phosphorylation takes place at only one of the two marked serines in YP2α site 2, but precise identification of the modified residue has not been performed.

Little is known about the enzymes that modify the acidic proteins. An acidic protein-specific protein kinase purified from ribomose wash has been reported (Kudlicki et al., 1980; Pilecki et al., 1992). Peliminary experiments indicate that this enzyme modifies preferentially protein YP2α*in vitro,,* affecting protein YP2ß and YP1α to a much lesser extent (R. Szyszka, T. Naranda, E. Gasior and J.P.G. Ballesta). These results, together with the differences in the phosphorylation site structure, suggest the existence of more than one kinase able to modifiy the acidic proteins.

The cytoplasmic pool of acidic proteins.

The ribosomal proteins are immediatly transported after synthesis to the nucleolus to be assembled into the ribosome structure. Therefore, they are not detected free in the cytoplasm, where they are, in fact, are degraded when artificially overproduced (Warner, 1989). The acidic proteins are, however, an exception, and an important cytoplasmic pool of them has been detected in different eukaryotic species (van Agthoven et al., 1977; Zinker, 1980; Sanchez-Madrid et al., 1981a; Elkon et al., 1986; Mitsui et al., 1988). In yeast, using specific antibodies that recognize all the acidic proteins, two different values for the pool size have been reported. The free acidic proteins have been reported to represent 0.3 % of the acidic proteins in the ribosomes by immunobloting estimations (Mitsui et al., 1988). This value constrasts with ELISA results showing as much, or even more, free acidic proteins as those bound to the particles (Saenz-Robles et al., 1990). The results obtained in *Artemia salina* by radioimmunoassay are in good agreement with the yeast ELISA estimations, with 72% of the total immunoreactive material showing up in the cytosolic pool (van Agthoven et al., 1977).

It seems that, regardless of the exact dimension of the pool, the acidic proteins are unique among the ribosomal components in being free in large amounts in the cytoplasm. This accumulation of free acidic proteins indicates that some structural features preserve them from degradation. In this sense it is interesting that mutation of serine at position 2 in protein YP1ß(L44′) results in the disappearance of the protein from the cytoplasm but not from the ribosomes (Naranda et al., 1992), indicating that the conformation of the amino end of the protein is important in determining its sensitivity to degradation.

An important characteristic of the free acidic proteins is their dephosphorylated state. This has been shown by immunoprecipitation (Zinker, 1980) as well as by double labeling experiments (Sanchez-Madrid et al., 1981a). The non-phosphorylated form of the supernatant proteins is not due to dephosphorylation during the preparation since the addition of purified $^{32}P$-labeled acidic proteins to cellular extracts does not affect the state of the exogenously added polypeptides (Naranda, 1991).

Exchange of acidic proteins between ribosomes and cytoplasmic pool

Zinker and Warner (1976) found that acidic proteins become incorporated into the ribosomes in conditions of ribosome synthesis blockade, indicating that these ribosomal components must come from exchange with the cytoplasmic pool. Exchange between free and ribosome bound acidic proteins was also confirmed in regenerating rat livers, and it was observed that in the process not only newly synthesized acidic proteins but also polypeptides from the pool participate (Tsurugi and Ogata, 1985). A similar exchange has been proposed to takes place in plant cell cultures under heat-shock conditions (Scharf and Nover, 1987).

An exchange between free acidic proteins and ribosomes takes also place *in vitro*. When exogenous radioactively labeled proteins are added to a cell-free protein synthesizing system, incorporation of radioactive proteins into the particles takes place. This incorporation process is dependent on the implication of the ribosome in protein synthesis since it is strongly inhibited by the addition of cycloheximide (A. Jimenez-Diaz and J.P.G. Ballesta, umpublished results).

Precise data on the rate of either *in vivo* or *in vitro* acidic protein exchanges are, unfortunately, scarce. However, the process must be rather fast since in rat liver, specific activity of pulse-labeled ribosome-bound acidic proteins is reduced 50% in 30 min. after a chase, in conditions that increase the specific activity of basic ribosomal proteins (Tsurugi and Ogata, 1985).

Eukaryotic acidic proteins as controllers of ribosome activity

Taken all together, the data available on the acidic proteins from eukaryotic cells are compatible with the existence of a control mechanism for the regulation of the number of active large ribosomal subunits able to participate in polysomes, as presented in fig 2. The ratio between the phosphatase and protein kinase enzymes acting on the acidic proteins would affect the equilibrium between particles containing firmly bound phosphorylated acidic proteins, and protein-deficient subunits. In conditions of low protein synthesis requirements like in stationary phase, either the phosphatase would be stimulated or the protein kinase inhibited, resulting in an accumulation of acidic protein deficient ribosomes in the cytoplasm.

The acidic protein defective particles are apparently able to support protein synthesis, as shown in the case of the acidic protein disrupted strains, but the efficiency of the synthesizing machinery is notably affected and, more importantly, the mRNA selection and the pattern of protein produced is altered. In this way the proposed mechanism can regulate not only the overall protein synthesizing capacity of the cell but also the expression of some proteins affecting specific metabolic pathways.

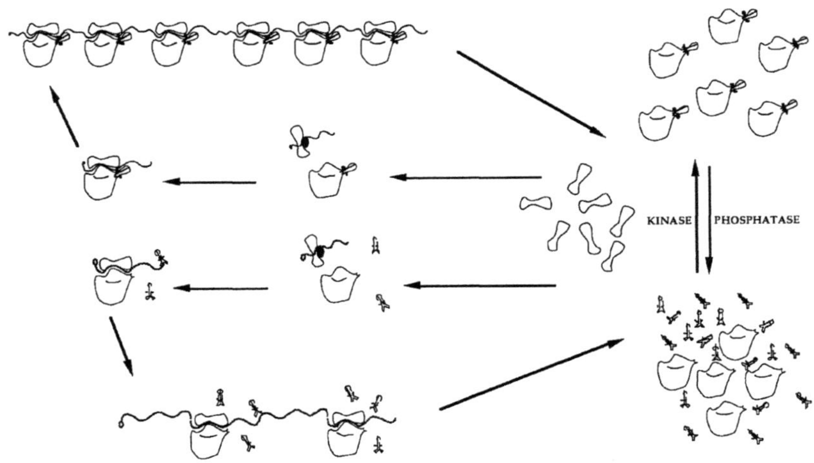


Fig 2.-Schematic model of a possible regulatory mechanism of protein synthesis through acidic ribosomal protein phosphorylation.
60S subunits carrying phosphorylated acidic proteins forming the lateral stalk are in equilibrium with acidic protein-deficient particles and free dephosphorylated acidic proteins. The equilibrium state is determined by the ratio of kinase to phosphatase activities. Complete 60S subunits are involved in regular protein synthesis forming large polysomes; defective subunits carry out protein synthesis less efficintly preferentially translating a different set of mRNAs than standard ribosomes

The existence of the proposed control mechanism is compatible with the data available but a number of aspects require further experimental confirmation. The acidic protein dephosphorylation is the less known part of the process. Preliminary evidence indicates that the acidic proteins are not immediatly released from the ribosome upon dephosphorylation and some additional step, probably subunit dissociation, is required for them to be liberated to the cytoplasm. Moreover, ribosomes totally deprived of acidic components have not been found indicating that, perhaps, not all acidic proteins are involved in the exchange process. The release of only one acidic protein type is an attractive possibility, and the finding of only one of them in purified rat liver subunits (Arpin et al., 1978) is compatible with that hypothesis. In this sense, an exploration of the relative sensitivity of the different acidic proteins to dephosphorylation is important.

The phosphorylation-dephosphorylation enzymatic system is an aspect that requires further exploration. Very little is known about the protein kinases acting on the acidic proteins and practically nothing is known about the protein phosphatases. The location of these enzymes is obviously very important in order to understand the mechanistic aspects of the process. Protein kinases acting on the acidic proteins have been found associated to the ribosomes (Kabat, 1970; Kudlicki et al., 1980; Hasler et al., 1991; Pilecki et al., 1992). This association suggest that the acidic proteins are modified on the particle and, therefore, that an interaction between the ribosome and the proteins must take place in the absence of phosphorylation and this interaction is stabilized upon modification. Similarly, the phosphatases must be bound to the ribosome but nothing is known about them apart from the fact that acidic proteins in ribosomes become slowly dephosphorylated when ribosome preparations are incubated.

On the other hand, if this proposal is correct, one or both modifying enzymes must also be somehow regulated in response to the metabolic cell requirements, which constitues an additional reason to pursue their identification and characterization.

## CONCLUSIONS

The experimental evidence clearly indicates that in the eukaryotic ribosomes the acidic proteins have evolved to carry out more sophisticated functions than those played by their counterparts in the bacterial ribosome. They are, indeed, implicated in the activity of the supernatant factors, although more experimental data are required to understand their role, which became complicated by the presence of different types of polypeptides. However, it is, perhaps, more relevant at this moment to focus our attention on the possible regulatory function of the ribosome activity that, as indicated above, can affect quantitatively as well as qualitatively, the translation capacity of the protein synthesis machinery. However, it is probable that, after all, both aspects of the acidic protein functions are closely related and, therefore, both must be explored to obtain a clear picture of the way these peculiar ribosomal components work.

## ACKNOWLEDGEMENTS

We thank M.C. Fernandez Moyano for expert technical assistance.This work was partially supported by an institutional grant to the Centro de Biología Molecular from the Fundación Ramon Areces and by grant PB04550 from the Dirección General de Investigación Científica y Técnica

REFERENCES

Amons, R., Pluijms, W. and Möller, W. (1979). The primary structure of ribosomal protein eL12/eL12P from *Artemia salina* 80S ribosomes. FEBS Lett. 104 : 85-89.

Arpin, M., Madjar, J.J. and Reboud, J.P. (1978). Occurrence of phosphorylated forms of an acidic protein in large ribosomal subunit of rat liver. Biochim. Biophys. Acta. 519 : 537-541.

Beltrame, M. and Bianchi, M.E. (1990). A gene family for acidic ribosomal proteins in *Schizosaccharomyces pombe*: Two essential and two nonessential genes. Mol. Cell. Biol. 10 : 2341-2348.

Bielka, H. (1982). The eukaryoitc ribosome. Springer-Verlag (Berlin, Heidelberg, New York).

Elkon, K.B., Skelly, S., Parnassa, A., Möller, W., Danho, W., Weissbach, H. and Brot, N. (1986). Identification and chemical synthesis of a ribosomal protein antigenic determinant in systemic lupus erythematosus. Proc. Natl. Acad. Sci. USA. 83 : 7419-7423.

Gavrilova, L.P., Kostiashkina, O.E., Koteliansky, V.E., Rutkevitch, N.M. and Spirin, A.S. (1976). Factor-free ("non-enzymic") and factor-dependent systems of translation of polyuridykic acid by *Escherichia coli* ribosomes. J. Mol. Biol. 101 : 537-552.

Glick, B.R. (1977). The role of Escherichia coli ribosomal proteins L7 and L12 in peptide chain propagation. FEBS Lett. 73 : 1-5.

Hasler, P., Brot, N., Weisbach, H., Parnassa, A.P. and Elkon, K.B. (1991). Ribosomal proteins P0, P1, and P2 are phosphorylated by casein kinase II at their conserved carboxyl termini. J. Biol. Chem. 266 : 13815-13820.

Juan-Vidales, F., F., S.-M. and Ballesta, J.P.G. (1981). The acidic proteins of eukaryotic ribosomes. A comparative study: Biochim. Biophys. Acta. 656 : 28-35.

Juan-Vidales, F., Saenz-Robles, M.T. and Ballesta, J.P.G. (1984). Acidic proteins of the large ribosomal subunit in *Saccharomyces cerevisiae*. Effect of phosphorylation. Biochemistry. 23 : 390-396.

Kabat, D. (1970). Phosphorylation of ribosomal proteins in rabbit reticulocytes. Characterization and regulatory aspects. Biochemistry. 9 : 4160-4174.

Kotelansky, V.E., Domagatsky, S.P., Gudkov, A.T. and Spirin, A.S. (1977). Elongation factors-dependent reactions on ribosomes deprived of proteins L7 and L12. FEBS Lett. 73 : 6-11.

Kruiswijk, T., Planta, R.J. and Mager, W.H. (1978). Quantitative analysis of protein composition of yeast ribosomes. Eur. J. Biochem. 83 : 245-252.

Kudlicki, W., Szyszka, R., Palén, E. and Gasior, E. (1980). Evidence for a highly specific protein kinase phosphorylating two strongly acidic proteins of yeast 60S ribosomal subunit. Biochim. Biophys. Acta. 633 : 376-385.

Lavergne, J.-P., Conquet, F., Reboud, J.P. and Reboud, A.-M. (1987). Role of acidic phosphoproteins in the partial reconstitution of the active 60S ribosomal subunit. FEBS Lett. 216 : 83-88.

Lin, A., Wittman-Leibold, B., McNelly, T. and Wool, I.G. (1982). The primary structure of the acidic phopshoprotein from rat liver 60S ribosomal subunit. J. Biol. Chem. 257 : 9189-9197.

McConnell, W.P. and Kaplan, N.O. (1982). The activity of the acidic phophoproteins from the 80S rat liver ribosome. J. Biol. Chem. 257 : 5359-5366.

Mitsui, K., Nakagawa, T. and Tsurugi, K. (1988). On the size and the role of a free cytosolic pool of acidic ribosomal proteins in yeast *S. cerevisaie*. J. Biochem. 104 : 908-911.

Mitsui, K. and Tsurugi, K. (1988). cDNA and deduced amino acid sequence of acidic ribosomal lprotein A2 from *Saccharomyces cerevisiae*. Nucleic Acids Res. 16 : 3575.

Möller, W. and Maassen, J.A. (1986). On the structure, function and dynamics of L7/12 from *Escherichia coli* ribosomes. In Structure, function and genetics of ribosomes. B. Hardesty and G. Kramer, ed. (New York: Springer-Verlag)

Möller, W., Slobin, L.I., Amons, R. and Richter, D. (1975). Isolation and characterization of acidic proteins of 60S ribosomes from *Artemia salina* cysts. Proc. Natl. Acad. Sci. USA. 72 : 4744-4748.

Naranda, T. and Ballesta, J.P.G. (1991). Phosphorylation of acidic proteins controls the activity of the ribosome. Proc. Natl. Acad. Sci. USA. 88 : 10563-10567.

Newton, C.H., Shimmin, L.C., Yee, J. and Dennis, P.P. (1990). A family of genes encode the multiple forms of the *Saccharomyces cerevisiae* ribosomal proteins equivalent to the *Escherichia coli* L12 protein and a single form of the L10-equivalent ribosomal protein. J. Bacteriol. 172 : 579-588.

Pilecki, M., Grankowski, N., Jacobs, J. and Gasior, E. (1992). Specific protein kinase from *Saccharomyces cerevisiae* cells phosphorylating 60S ribosomal subunit. Eur. J. Biochem. 206: 259-267.

Prieto, J., Candel, E. and Coloma, A. (1991). Nucleotide sequence of a cDNA encoding riboosmal acidic protein P1 from Dictyostelium discoideum: Identification of a novel carboxyl-terminal sequence in "A"proteins. Nucleic Acids res. in press :

Qian, S., Zhang, J.-Y., Kay, M.A. and Jacobs-Lorena, M. (1987). Structural analysis of the *Drosophila* rPA1 gene, a member of the eukaryotic "A"type ribosomal protein family. Nucleic Acids Res. 15 : 987-1003.

Remacha, M., Saenz-Robles, M.T., Vilella, M.D. and Ballesta, J.P.G. (1988). Independent genes coding for three acidic proteins of the large ribosomal subunit from *Saccharomyces cerevisae*. J. Biol. Chem. 263 : 9094-9101.

Remacha, M., Santos, C. and Ballesta, J.P.G. (1990). Disruption of single-copy genes encoding acidic ribosomal proteins in *Saccharomyces cerevisiae*. Mol. Cell. Biol. 10 : 2182-2190.

Remacha, M., Santos, C., Bermejo, B., Naranda, T. and Ballesta, J.P.G. (1992). Stable binding of the eukaryotic acidic phosphoproteins to the ribosome is not an absolute requirement for *in vivo* protein synthesis. J. Biol. Chem. 267: 12061-12067.

Rich, B.E. and Steitz, J.A. (1987). Human acidic ribosomal phospoproteins P0,P1 and P2. Analysis of cDNA clones,in vitro synthesis and assembly. Mol. Cell. Biol. 7 : 4065-4074.

Saenz-Robles, M.T., Remacha, M., Vilella, M.D., Zinker, S. and Ballesta, J.P.G. (1990). The acidic ribosomal proteins as regulators of the eukaryotic ribosomal activity. Biochim. Biophys. Acta. 1050 : 51-55.

Sanchez-Madrid, F., Juan-Vidales, F. and Ballesta, J.P.G. (1981a). Effect of phosphorylation on affinity of acidic proteins from *Saccharomyces cerevisiae* for the ribosome. Eur. J. Biochem. 114 : 609-613.

Sanchez-Madrid, F., Juan-Vidales, F. and Ballesta, J.P.G. (1981b). Functional role of acidic ribosomal proteins. Interchangeability of proteins from bacterial and eukaryotic cells. Biochemsitry. 20 : 3263-3266.

Sanchez-Madrid, F., Reyes, R., Conde, P. and Ballesta, J.P.G. (1979). Acidic ribosomal proteins from eukaryotic cells. Effect on ribosomal functions. Eur. J. Biochem. 98 : 409-416.

Scharf, K.-D. and Nover, L. (1987). Control of ribosome biosynthesis in plant cell cultures under heat shock conditions. II. Ribosomal proteins. Biochi. Biophys. Acta. 909 : 44-57.

Schijman, A., Dussetti, N.J., Vazquez, M.P., Lafon, S., Levy-Yeyati, P. and Levin, M.J. (1990). Nucleotide cDNA and complete deduced amino acid sequrence of *Trypanosoma cruzi* ribosomal P protein (P-JL5). Nucleic Acids Res. 18 : 3399.

Subramanian, A.R. (1975). Copies of proteins L7 and L12 and heterogeneity of the large subunit of *Escherichia coli* ribosome. J. Mol. Biol. 95 : 1-8.

Tsurugi, K. and Ogata, K. (1985). Evidence for the exchangeability of acidic ribosomal proteins on cytoplasmic ribosomes in regenerating rat liver. J. Biochem. 98 : 1427-1431.

van Agthoven, A., Kriek, J., Amons, R. and Möller, W. (1978). Isolation and characterization of the acidic phosphoproteins of 60S ribosomes from *Artemia salina* and rat liver. Eur. J. Biochem. 91 : 553-556.

van Agthoven, A.J., Maassen, J.A. and Möller, W. (1977). Structure and phosphorylation of an acidic protein from 60S ribosomes and its involvement in elongation factor-2 dependent GTP hydrolysis. Biochem. Biophys. Res. Commun. 77 : 989-998.

Vazquez, M.P., Schijman, A.G. and Levin, M.J. (1992). Nucleotide sequence of a cDNA encoding a *Trypanosoma cruzi* acidic ribosomalP1 type protein. Nucleic Acids Res. en prensa :

Warner, J.R. (1989). Synthesis of ribosomes in *Saccharomyces cerevisiae.* Microbiol. Rev. 53 : 256-271.

Wigboldus, J.D. (1987). DNA and deduced amino acid sequence of Drosophila rp21C, another "A"type ribosomal protein. Nucleic Acids Res. 15 : 10064.

Wool, G. and Stöffler, G. (1974). Structure and function of eukaryotic ribosomes. In Ribosomes. M. Nomura et al., ed. (Cold Spring Harbor: Cold Spring Harbor Laboratory)

Wool, I.G., Chan, Y.L., Glück, A. and Suzuki, K. (1991). The primary structure of rat ribosomal proteins P0, P1, and P2 and a proposal for a uniform nomenclature for mammalian and yeast ribosomal proteins. Biochimie. 73 : 861-870.

Zinker, S. (1980). P5/P5′the acidic ribosomal phosphoprotein from *S. cerevisiae.* Biochim. Biophys. Acta. 606 : 76-82.

Zinker, S. and Warner, J.R. (1976). The ribosomal proteins of Saccharomyces cerevisiae. Phosphorylated and exchangeable proteins. J. Biol. Chem. 251 : 1799-1807.

# *SUP35* AND *SUP45* GENES CODE FOR RIBOSOME-BOUND PROTEINS INVOLVED IN THE CONTROL OF TRANSLATIONAL FIDELITY IN YEAST.

Michael D. Ter-Avanesyan, Svetlana A. Didichenko,
Vitaly V. Kushnirov and Adilya R. Dagkesamanskaya.

Institute of Experimental Cardiology
Cardiology Research Center
3rd Cherepkovskaya Str., 15A,
121552 Moscow, Russia

## INTRODUCTION

The study of suppressor mutations that affect translational fidelity provides an efficient approach for identifying new components of the translational machinery in eukaryotes. Some of the identified genes controlling fidelity in lower eukaryotes presumably code for ribosomal proteins (Dequard-Chablat et al, 1986; All-Robyn et al, 1990; Hinnebusch and Liebman, 1991) and translation factors (Sandbaken and Culbertson, 1988; Kushnirov et al, 1988; Breining and Piepersberg, 1986). However, functions of the majority of these genes have not been established.

We present here a brief review of our recent studies of the omnipotent suppressors *SUP35* and *SUP45* representing the evolutionarily conserved genes that code for previously unidentified proteins involved in translation. Mutations in these genes were identified in several laboratories as omnipotent suppressors *sup1* and *sup2* (Inge-Vechtomov and Andrianova, 1970), *sup45* and *sup35* (Hawthorne and Leupold, 1974), frameshift suppressor *suf12* (Culbertson et al, 1982), allosuppressors *sal4* and *sal3* (Cox, 1977), cell cycle arresting mutation *gst1* (Kikuchi et al., 1988) and mutation of resistance to novobiocin *nov1* (Pocklington et al, 1990). In addition, not only mutations in but also overexpression of the *SUP35* gene can generate an omnipotent suppressor effect (Chernoff et al, 1988). It was shown that *SUP1*, *SUP45*, *SAL4* and *NOV1* are identical genes as *SUP2*, *SUP35*, *SUF12*, *SAL3* and *GST1* (Kushnirov et al, 1988; Wilson and Culbertson, 1988; Tuite et al, 1988; Pocklington et al, 1990; Kikuchi et al., 1988). It was suggested that these genes code for protein factors participating in translation (Himmelfarb et al, 1985; Breining and Piepersberg, 1986; Kushnirov et al, 1988). Disruption of *SUP45* and *SUP35* genes reveals that they are essential yeast genes (Himmelfarb et al, 1985; Wilson and Culbertson, 1988).

NATO ASI Series, Vol. H 71
Protein Synthesis and Targeting in Yeast
Edited by A. J. P. Brown, M. F. Tuite, and J. E. G. McCarthy

## Sup35 AND Sup45 PROTEINS ARE TRANSLATION FACTORS ASSOCIATED WITH THE RIBOSOME.

Two lowly abundant 49 kDa and 79 kDa proteins corresponding to the Sup45 and Sup35 proteins, respectively, were identified in yeast cell extracts using the polyclonal antibody raised against Sup45-chloramphenicol acetyltransferase and Sup35-β-galactosidase fusion proteins. Sucrose sedimentation analysis revealed that the Sup45 and Sup35 proteins are associated with ribosomes in a substoichiometric ratio respectively to ribosomal proteins (Didichenko et al, 1991; and unpublished). We have examined the effect of the inhibition of different steps of protein synthesis on the distribution of these proteins among ribosomal population.

In cells treated with cycloheximide, an inhibitor of elongation of translation, these proteins were nonuniformly distributed among ribosomes and did not appear in a free form at the top of sucrose gradient. Both proteins had the same distribution, being detected in polysomes and monosomes, but the predominant portion of them was found in the region of sedimentation of native 40S ribosomal subunits. It also can be noted that distribution of Sup proteins through the gradient correlates with the distribution of the poly (A) mRNA, suggesting that they are included into the free cytoplasmic mRNA-containing particles (Fig. 1).

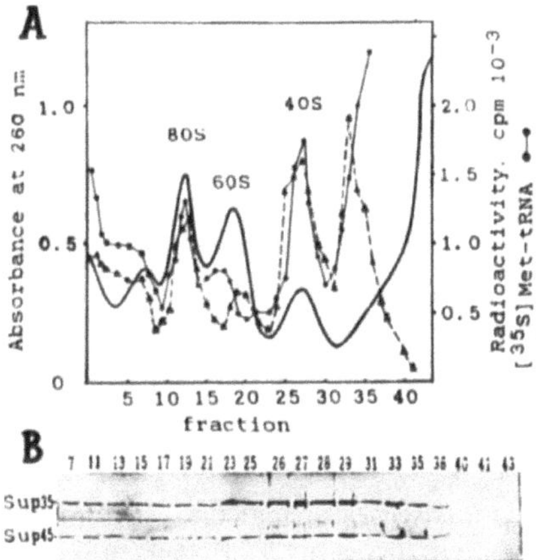


**Figure 1. Distribution of Sup35 and Sup45 proteins in cycloheximide-treated lysate.** Yeast cells of the wild-type strain were incubated with cycloheximide (100 μg/ml) before harvesting. (A) Sucrose sedimentation analysis. Distribution of poly (A) mRNA and [$^{35}$S] Met-tRNA was analyzed as described by Safer et al. (1979).
(B) Immunoblot analysis of the Sup35 and Sup45 proteins in sucrose gradient fractions. Fraction numbers are indicated above the lines.

Inhibition of initiation of translation by treatment of cells with $NaN_3$ leads to polysome degradation, accumulation of inactive monosomes and increase in the amounts of native ribosomal subunits. The native 40S subunits are heterogeneous with respect to their banding in equilibrium density gradient and in their sedimentation characteristics producing an asymmetrical double peak of absorbance in sucrose gradient (Fig. 2) and constitute two major groups with buoyant densities 1.42 and 1.49 g/cm$^3$. The lower density group designated $40S_L$ and supposed to bind to mRNA and directly participate in initiation is associated with additional proteins (Henshaw et al, 1973). Among these accessory proteins we have detected the Sup35 protein. However, we failed to detect this protein in the high density group designated $40S_H$ which is suggested to be released from the polyribosomes after termination of translation (Henshaw et al, 1973; Hirsch et al, 1973). Trace amounts of the Sup35 protein also sediment in overlapping peaks of 80S ribosomes and 60S ribosomal subunits.

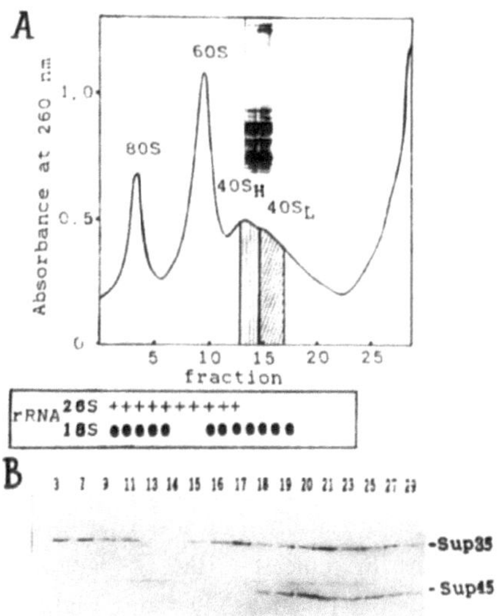


**Figure 2. Distribution of the Sup35 and Sup45 proteins in $NaN_3$-treated lysate.** (A) Sucrose sedimentation analysis of ribosomes from wild-type cells incubated with 1mM $NaN_3$ before harvesting. Positions of free 40S (designation of $40S_L$ and $40S_H$ is described in text), 60S ribosomal subunits and 80S monoribosomes are indicated. SDS PAGE analysis of proteins of the two forms of 40S ribosomal subunit is shown on insert. Analysis of rRNA is shown below the absorbance profile. (B) Immunoblot analysis of the Sup45 and Sup35 proteins in sucrose gradient fractions. Fraction numbers are indicated above the lines.

Analysis of the distribution of the Sup45 protein leads us to a conclusion that under the conditions of inhibition of translation initiation this protein loses the ability to associate with the ribosome since it was detected in the region of gradient where mRNP is expected to migrate as well as at the top of gradient. In the same region we have also detected a significant amount of the Sup35 protein (Fig. 2). The data obtained may indicate a possible interaction of Sup proteins not only with ribosomes but also with poly (A) mRNA.

## FUNCTIONAL DOMAINS OF THE *SUP35* ENCODED PROTEIN.

The structure of the *SUP35* gene is of particular interest because of similarity of its gene product to the elongation factor EF-1 . The nucleotide sequences of the *SUP35* genes of the distantly related yeasts *S. cerevisiae*, *P. pinus* and man (*GST1*) are now available (Kushnirov et al, 1990; Hoshino et al, 1989). The comparison of predicted protein sequences of the Sup35 protein from these organisms reveals a highly conserved C-terminal domain of 432 amino acids similar to the EF-1α family of elongation factors (domain **C**) and the N-terminal region that varies in both length and sequence. The sequence of the N-terminal region of Sup35 protein of *S. cerevisiae* may be divided into N-terminal (**N**) and middle (**M**) domains of 123 and 130 amino acids, respectively, separated by the second methionine residue. We have elucidated the functional significance of the N- and C-terminal regions of the Sup35 protein through the analysis of *SUP35* deletion mutants generated *in vitro* (Fig. 3).

Deletion analysis of the *SUP35* gene showed that the essential function of the protein is contained within the evolutionary conserved **C** region. The first 114 amino acids of the **N** domain are able to cause an efficient "multicopy" suppressor effect (e.g. the "ΔEco" construct). The remaining part of the protein, beginning from the Met-124, does not demonstrate such "multicopy" suppressor activity (e.g. "2ATG", "2ATG-ΔBcl" and "2ATG-ΔSal"). The deletion of the middle 44 amino acids of the **N** region (i.e. the "ΔBst-ΔHind" construct) reduces, but does not eliminate the ability to cause a "multicopy" suppressor effect. The ability of the fragments of the **N** region to demonstrate functional activity indicates that this region is not organized in an integral structural unit, but represents an unstructured array of small elements similar in structure and amino acid composition. For this reason the function of this region can tolerate small deletions. In contrast to the **N** region, all deletions affecting the **C** region eliminated its essential function, suggesting that the C-terminal 432 amino acids comprise a single structural unit whose integrity is necessary for function. The **M** region is not responsible for any of the two tested functions indicating that it either performs some unknown function or serves as a spacer between the **N** and **C** domains.

Immunoblot analysis of ribosomal protein samples has shown that both the **N** and **C** domains of the Sup35 protein possess sites for ribosome binding since its truncated versions encoded by the constructs "2ATG", "3ATG", "ΔHpa", "ΔSal", "ΔBcl" and "ΔEco" were able to interact with the ribosome. In contrast, the **M** region, whose functional significance we have failed to reveal, could not associate with the ribosome ("2ATG-ΔBcl" construct).

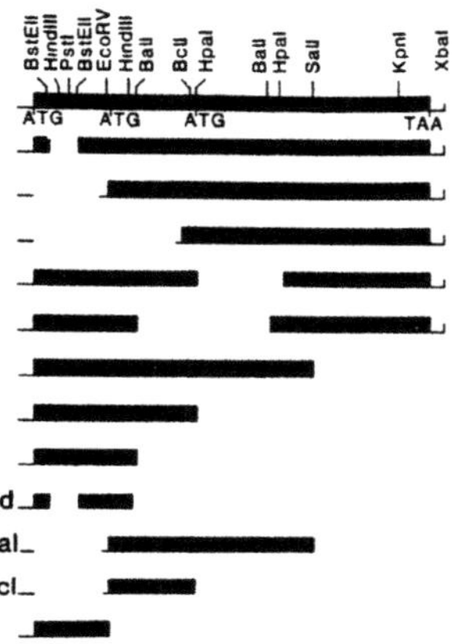


| | VIABILITY | | SUPPRESSION | ANTISUPP-RESSION | | RIBOSOME BINDING |
|---|---|---|---|---|---|---|
| | M | L | | M | L | |
| SUP35* | + | + | + | − | − | + |
| ΔBst | + | + | − | + | + | NT |
| 2ATG | + | + | − | + | + | + |
| 3ATG | + | + | − | + | + | + |
| ΔHpa | − | NT | + | − | NT | + |
| ΔBal | − | NT | + | − | NT | + |
| ΔSal | − | NT | ++ | − | NT | + |
| ΔBcl | − | NT | + | − | NT | + |
| ΔBal2 | − | NT | + | − | NT | + |
| ΔBst-Hind | − | NT | +/− | − | NT | NT |
| 2ATG-ΔSal | − | NT | − | − | NT | + |
| 2ATG-ΔBcl | − | NT | − | − | NT | − |
| ΔEco | − | NT | + | − | NT | + |

**Figure 3.** ***SUP35*** **deletion constructs and their properties.** Twelve truncated versions of th *SUP35* gene were generated by deletion of fragments internal to the coding sequence. All thes constructs retaining the original reading frame were cloned into both multicopy (M) and low-cop number (L) plasmids. Coding regions of the constructs are represented by solid bars. Restrictio endonucleases sites are shown for the wild-type *SUP35* gene. ATG codons corresponding t methionines that divide the Sup35 protein into three regions are shown under the restriction ma Each construct was tested for the ability (i) to compensate for the lethal effect of a *sup35::TRP1* nu allele. For this purpose the meiotic segregation of diploid heterozygous for the null allele transforme with plasmids that carry deletion versions of the *SUP35* were studied. "+", plasmid can rescue *SUP35*-disrupted haploid segregants, "-", plasmid cannot rescue these segregants; (ii) to induce th multicopy suppressor effect. The multicopy suppressor effects was studied in transformants of test strain as the suppression of the ochre mutation *his7-1* and mutation *lys2-187*, which type is n identified. "+", growth on selective media; "-", no sign of growth by 9 days); (iii) to inhibit th suppression (antisuppressor effect) of the *ade2-1* ochre mutation caused by the combined action of th weak ochre-specific *SUQ5* suppressor and the non-Mendelian [*psi*$^{+}$] determinant (Cox, 1965). "+ antisuppressor phenotype, "-", suppressor phenotype; (iv) of the encoded truncated protein to eith bind ("+") or not ("-") to the ribosome.

All 5'-deletions affecting the N-terminal part (but leaving intact the **C** region) c the Sup35 protein eliminate the "multicopy" suppressor effect and lead to the ant suppressor phenotype, i.e. to increased fidelity of translation. This indicates th even though the N-terminal part of the Sup35 protein is not essential, it is involve in the codon recognition process. The antisuppressor effect caused by deletio constructs mentioned above is dominant and thus it is not a consequence of mut tions that simply delete or reduce the *SUP35* function.

## INTERACTION OF THE *SUP45* AND *SUP35* OMNIPOTENT SUPPRESSORS WITH THE NON-MENDELIAN DETERMINANT [*PSI*$^+$].

It was shown that the heterozygous *sup45::TRP1* null mutation exhibits the dominant effects in the [*psi*$^+$] diploid strains: slow growth and inability to sporulate (Dagkesamanskaya and Ter-Avanesyan, 1991). These effects were not due to the toxicity of the anomalous protein encoded by the disrupted *sup45* allele since transformation of the [*psi*$^+$] diploid strain with the autonomously replicating plasmid carrying the *sup45::TRP1* allele did not noticeably alter the growth rate and sporulation efficiency. Therefore we conclude that the *SUP45* gene dosage is critical in the [*psi*$^+$] background.

Another type of interaction with the [*psi*$^+$] determinant was observed for the *SUP35* gene. It was demonstrated that transformants of [*psi*$^+$] strains which carry multicopy plasmids with the *SUP35* gene grow slowly on selective medium and are extremely unstable in non-selective conditions. Since both, the extra copies of the *SUP35* gene and the [*psi*$^+$] determinant, increase translational misreading, one can suggest that combination of these factors leads to a high level of inaccuracy incompatible with cell viability.

Replacement of the wild-type *SUP35* allele for its "antisuppressor" versions "2ATG", "3ATG" or "ΔBst" (denoted as *sup35-ΔN*) in the *ade2-1 SUP16* [*psi*$^+$] strain leads to a non-suppressed phenotype (pink colour and adenine auxotrophy). Obtained *sup35-ΔN* mutants were crossed with the [*psi*$^+$] and [*psi*$^-$] variants of the *ade2-1 SUQ5* tester strain. Hybrids formed by these crosses were pink and Ade$^-$ due to the dominant antisuppressor effect of the *sup35-ΔN* deletion mutations. Diploids from crosses with the [*psi*$^+$] tester yielded as a rule tetrads with segregation 2 Ade$^+$ : 2 Ade$^-$. Surprisingly, the diploids from crosses with the [*psi*$^-$] tester did not segregate suppressed spore cultures. This observation suggests that the *sup35-ΔN* deletion alleles cause elimination of the [*psi*$^+$] determinant in a haploid strain. The elimination of the [*psi*$^+$] is a recessive trait, since hybrids from crosses with the [*psi*$^+$] tester can produce the [*psi*$^+$] meiotic segregants. Thus, the N-terminal region of the Sup35 protein is essential for the [*psi*$^+$] determinant. Dominant and recessive mutations causing the [*psi*$^+$] loss were described in *S. cerevisiae* (McCready and Cox, 1973; McCready, Cox and McLaughlin, 1977; Cox, Tuite and Mundy, 1980). The *sup35-ΔN* alleles differ from these mutations by their ability to cause both phenotypes simultaneously. Interestingly, the *PNM2* gene, mutation in which causes elimination of the [*psi*$^+$] appeared to be identical with the *SUP35* gene (B.S. Cox, personal communication). Despite the fact that the *SUP35* gene is essential for the maintenance of the [*psi*$^+$], its expression is necessary but not sufficient for the [$psi^+$] phenotype since strains carrying the wild-type *SUP35* gene may either possess or not the [$psi^+$] determinant. It is noteworthy that in contrast to a dominant antisuppressor phenotype, the recessive *pnm*-like effect of the *sup35-ΔN* alleles suggests the loss of some function that is necessary for the [$psi^+$] maintenance. This allows us to suggest that the Sup35 protein serves as a trans-acting factor essential for the [$psi^+$] maintaining machinery.

## *SUP35* AND *SUP45* PROTEINS INTERACT WITH EACH OTHER.

The sucrose sedimentation analysis of the distribution of the Sup35 and Sup45 proteins presented above indicates that they may be involved in the same protein complexes interacting with ribosomes and poly (A) mRNA. This means that the Sup35 and Sup45 proteins may directly or functionally interact with each other. Actually, synthetic lethal phenotype produced by some but not all combinations of suppressor mutations in the *SUP35* and *SUP45* genes, suggests that the product of these two genes could interact with each other (Inge-Vechtomov and Andrianova, 1970; Hawthorne and Leupold, 1974). Moreover, we have shown that the presence of extra plasmid-borne copies of the *SUP45* gene relieves the lethality of temperature-sensitive *sup35* mutation and *vise versa* multiple copies of the *SUP35* protein can suppress the temperature-sensitivity of *sup45* mutation. This suppression does not occur via a bypass of *SUP35* and *SUP45* genes function since null alleles of these genes cannot be cross-suppressed by the wild-type genes cloned on multicopy plasmids. Interestingly that multicopy plasmid with the "3ATG" deletion construct of the *SUP35* gene in contrast to those with the "2ATG" and "ΔBst" deletion alleles (Fig. 3), cannot suppress temperature-sensitive defect of growth mediated by the *sup45* mutation. This suggests a specific role of the region **M** of the Sup35 protein in interaction with the Sup45 protein. Finally, mutations in the *SUP35* and *SUP45* genes influence the distribution of corresponding proteins through the sucrose gradient. It was observed that in either *sup35* or *sup45* mutants, both proteins appear in a free form at the top of sucrose gradient (data not shown). Therefore, mutations in the *SUP35* and *SUP45* genes can cross-influence the association of corresponding proteins with ribosomes.

## CONCLUSIONS.

Although the data available at present allow no suggestion on the function of the Sup35 and Sup45 proteins, some general features of their suppressor action may be considered. The suppressor or antisuppressor effect of *sup35* mutations may be due to decrease or increase in either fidelity of elongation or efficiency of termination. The latter possibility, however, does not explain the frameshift suppression in *suf12* mutants and the decreased fidelity of poly (U)-directed translation in the cell-free extracts of *sup35* mutants (Culbertson et al, 1982; Surguchov et al, 1984; Eustice et al, 1986). Therefore it seems more likely that the suppressor effect of *sup35* mutations is caused by the decreased fidelity of elongation. The same consideration is true for the *SUP45* gene, since mutations in this gene also cause a high level of misreading in poly (U)-directed *in vitro* translation system (Surguchov et al, 1984). The important question then is whether the decrease in the fidelity of the elongation step of translation is caused by direct or indirect action of the Sup35 and Sup45 proteins. In the first case the translational misreading requires direct participation of the Sup35 and Sup45 proteins in particular elongation step. The second suggestion presumes that mutations in the *SUP35* and *SUP45* gene alter translational fidelity, causing some changes of other components of translational apparatus in yeast. It was shown that there is less than one Sup35 molecule per twenty ribosomes in actively growing yeast cells (Didichenko et al, 1991). The Sup45 protein is also substantially less abundant then ribosomal proteins. Moreover, it was observed that the major portion of both Sup35 and Sup45 proteins is associated not with elongating ribosomes, but with free native 40S ribosomal subunits. These results indicate that the Sup35 and Sup45 proteins cannot participate in any significant proportion of elongation cycles. Given this, the suggestion about the direct involvement in codon recognition process is especially unprobable for the Sup35 protein since it cannot explain either the antisuppressor effect of certain *SUP35* deletion alleles (e.g." 3ATG", " 2ATG" and "ΔBst" constructs) or their dominant expression. Actually, if the accuracy of codon recognition will be increased in a small proportion of elongating cycles, the overall fidelity of translation will still be determined by the majority of the less "accurate" cycles. The "direct participation" hypothesis means dominance of the "less accurate" alleles over the "more accurate" alleles of suppressor genes. This is confirmed in the cases of suppressor mutations in the *TEF2* gene coding for elongation factor EF-1α and ribosomal protein genes *SUP44* and *SUP46* (Sandbaken and Culbertson, 1988; Hinnebusch and Liebman, 1991; All-Robyn et al, 1990). In contrast to these cases, the *SUP35* gene is characterized by the dominance of "more accurate" alleles. Taken together, these facts allow us to suggest that the Sup35 (and probably, Sup45) protein influences the fidelity of translation indirectly by changing the properties of ribosomes or other components of translational machinery in yeast. Such changes may probably be accomplished either through modification of some components of translational apparatus or by influencing their relative expression levels. It is noteworthy that the *SUP35* encoded protein probably may act as a trans-acting regulator factor required for the maintenance of the cytoplasmically inherited determinant [*psi*$^+$].

## REFERENCES

All-Robyn, J.A., Brown, N., Otaka, E. and Liebman, S. (1990) Sequence and functional similarity between a yeast ribosomal protein and the *E. coli* S5 ram protein. Mol. Cell. Biol. 10: 6544-6553.

Breining, P. and Piepersberg, W. (1986) Yeast omnipotent suppressor *SUP1(SUP45)*: nucleotide sequence of the wild type and a mutant gene. Nucleic Acids Res. 4: 5187-5197.

Chernoff, Y.O., Derkach, I.L., Dagkesamanskaya, A.R., Tichomirova, V.L., Ter-Avanesyan, M.D. and Inge-Vechtomov, S.G. (1988) Nonsense-suppression caused by the amplification of translation protein factor gene. Dokl. Acad. Nauk SSSR (Proc. USSR Acad. Sci., in Russian) 301: 1227-1229.

Cox, B.S. (1965) *psi*, a cytoplasmic suppressor of super-suppressor in yeast. Heredity 20: 505-521.

Cox, B.S. (1977) Allosuppressors in yeast. Genet. Res. 30: 159-178.

Cox, B.S., Tuite, M.F. and McLaughlin, C.S. (1988) The [*psi*] factor of yeast: a problem in inheritance. Yeast 4: 159-178.

Cox, B.S., Tuite, M.F. and Mundy, C.J. (1980) Reversion from suppression to non-suppression in *SUQ5* [*psi*$^+$] strains of yeast: classification of mutations. Genetics 95: 589-609.

Culbertson, M.R., Gaber, R.F. and Cummins, C.M. (1982) Frameshift suppression in *Saccharomyces cerevisiae*. III. Isolation and genetic properties of nongroup-specific suppressors. Genetics 102: 361-378.

Dagkesamanskaya, A.R. and Ter-Avanesyan, M.D. (1991) Interaction of the yeast omnipotent suppressors *SUP1*(*SUP45*) and *SUP2*(*SUP35*) with non-Mendelian factors. Genetics 128: 513-520.

Dequard-Chablat, M., Coppin-Raynal, E., Picard-Bennoun, M. and Madjar, J.J. (1986) At least seven rubosomal proteins are involved in the control of translational accuracy in an eucaryotic organisms. J. Mol. Biol. 190: 167- 175.

Didichenko, S.A., Ter-Avanesyan, M.D. and Smirnov, V.N. (1991) Ribosome-bound EF-1 -like protein of yeast *Saccharomyces cerevisiae*. Eur. J. Biochem. 198: 705-711.

Eustice, D.C., Walkem, L.P., Wilhelm, J.M. and Sherman, F. (1986) Altered 40S ribosomal subunits in omnipotent suppressors of yeast. J. Mol. Biol. 188: 207-214.

Hawthorne, D.C. and Leupold, L. (1974) Suppressor mutations in yeast. Curr. Top. Microbiol. Immunol. 64: 1-47.

Henshaw, E.C., Guiney, D.G. and Hirsch, C.A. (1973) Ribosome cycle in mammalian protein synthesis. Place of monomeric ribosomes and ribosomal subunits in the cycle. J. Biol. Chem. 25: 4367-4376.

Himmelfarb, H.J., Maicas, E. and Friesen, J.D. (1985) Isolation of the *SUP45* omnipotent suppressor gene of *Saccharomyces cerevisiae* and characterization of its gene product. Mol. Cell. Biol. 5: 816-822.

Hinnebusch, A.G. and Liebman, S.W. (1991) Protein synthesis and translational control in *Saccharomyces cerevisiae*. **In** The Molecular and Cellular Biology of the Yeast *Saccharomyces*., Volume 1 (Broach, J.R., Pringle, J.R., and Jones, E.W., eds), pp. 627-735. Cold Spring Habor, NY: Cold Spring Harbor Laboratory Press.

Hirsch, C.A., Cox, M.A., van Venrooij, W.J.W. and Henshaw, E.C. (1973) The ribosome cycle in mammalian protein synthesis. Association of the native smaller ribosomal subinits with protein factors. J. Biol. Chem. 25: 4377- 4385.

Hoshino, S.-i., Miyazawa, H., Enomoto, T., Hanaoka, F., Kikuchi, Y., Kikuchi, A. and Ui, M. (1989) A human homologue of the yeast *GST1* gene codes for a

GTP-binding protein and is expressed in a proliferation-dependent manner in mammalian cells. EMBO J. 8: 3807-3814.
Inge-Vechtomov, S.G. and Andrianova, V.M. (1970) Recessive super-suppressors in yeast. Genetika (in Russian) 6: 103-115.
Kikuchi, Y., Shimatake, H. and Kikuchi, A. (1988) A yeast gene required for the G1 to S transition encodes a protein containing an A-kinase targed site and GTPase domain. EMBO J. 7: 1175-1182.
Kushnirov, V.V., Ter-Avanesyan, M.D., Didichenko, S.A., Smirnov, V.N., Chernoff, Y.O., Derkatch, I.L., Novikova, O.N., Inge-Vechtomov, S.G., Neistat, M.A. and Tolstorukov, I.I. (1990) Divergence and concervation of *SUP2*(*SUP35*) gene of yeasts *Pichia pinus* and *Saccharomyces cerevisiae*. Yeast 6, 461-472.
Kushnirov, V.V., Ter-Avanesyan, M.D., Telckov, M.V., Surguchov, A.P. Smirnov, V.N. and Inge-Vechtomov, S.G. (1988) Nucleotide sequence of the *SUP2*(*SUP35*) gene of yeast *Saccharomyces cerevisiae*. Gene 66: 45-54.
McCready, S.J. and Cox, B.S. (1973) Antisuppressors in yeast. Mol. Gen. Genet. 124: 305-320.
McCready, S.J., Cox, B.S. and McLaughlin, C.S. (1977) The extra-chromosomal control of nonsense suppression in yeast: an analysis of the elimination of [$psi^+$] in the presence of a nuclear gene $PNM^-$. Mol. Gen. Genet. 150: 265- 270.
Pocklington, M.J., Johnston, L., Jenkins, J.R. and Orr, E. (1990) The omnipotent suppressor *SUP45* affects nucleic acid metabolism and mitochondrial structure. Yeast 6: 441-450.
Safer, B., Jagus, R. and Kemper, W.M. (1979) Analysis of initiation factor function in highly fractionated and unfractionated reticulocyte lysate systems. Methods in Enzymology, 60: 61-87.
Sandbaken, M.G. and Culbertson, M.R. (1988) Mutations in elongation factor EF-1 affect the frequency of frameshifting and amino acid misincorporation in *Saccharomyces cerevisiae*. Genetics 120: 923-934.
Surguchov, A.P., Smirnov, V.N., Ter-Avanesyan, M.D. and Inge-Vechtomov, S.G. (1984) Ribosomal suppression in eukaryotes. Physicochem. Biol. Rev. 4: 147- 205.
Tuite, M.F., Izgu, F., Grant, C.M. and Crouzet, M. (1988) Genetic control of tRNA suppression in *Saccharomyces cerevisiae*: allosuppressors. NATO ASI Series, Volume H14: 393-402.
Wilson, P.G. and Culbertson, M.R. (1988) *SUF12* suppressor protein of yeast. A fusion protein related to the EF1 family of elongation factor. J. Mol. Biol. 199: 559-573.

# MUTATIONS AFFECTING TRANSLATIONAL ACCURACY IN YEAST

Susan W. Liebman, Annette Vincent, Jae Mahn Song, and Jamie All-Robyn
Department of Biological Sciences
Laboratory for Molecular Biology
University of Illinois at Chicago
Chicago, Illinois 60680
U.S.A.

## INTRODUCTION

To investigate the mechanisms used in eukaryotes to maintain translational accuracy, we and others have isolated and characterized mutations that affect the fidelity of protein synthesis in yeast. These mutations include nonsense suppressors that cause misreading of specific stop codons; omnipotent suppressors, which cause general translational inaccuracy and suppression of UAG, UAA, and UGA codons; and allosuppressors and antisuppressors, which respectively enhance or lower the efficiency of certain suppressors. Using this layered approach we should identify translational components that are likely to interact. Recently the genetic work has been extended by cloning and characterizing the mutant genes. We have also identified components of the translational apparatus important for fidelity by selecting for wild-type genes that alter accuracy when present at higher copy number. Below we summarize our work in this area.

## NON-MENDELIAN MODIFIERS

Wild-type yeast differ in translational accuracy due to the presence of non-Mendelian factors. One such factor, [psi+] (see Cox et al. 1988 for review), enhances misreading induced by aminoglycoside antibiotics and increases the efficiency of various UAA, UGA and frameshift suppressors. We have described another non-Mendelian factor, [eta+], which has many similarities to [psi+] (Liebman and All-Robyn 1984). Both factors are inherited in a non-Mendelian fashion and can be cured by guanidine hydrochloride. However, the two factors differ in their pattern of inheritance. While [psi+] is inherited by all meiotic products, [eta+] is passed on to only about 70% of the meiotic offspring. Another difference is that [eta+], but not [psi+], causes the omnipotent suppressors sup35-2 and sup45-2 to be lethal, while [psi+] but not [eta+] enhances the efficiency of certain tRNA-suppressors. The molecular mode(s) of action of the [psi] and [eta] factors remain unknown.

NATO ASI Series, Vol. H 71
Protein Synthesis and Targeting in Yeast
Edited by A. J. P. Brown, M. F. Tuite, and J. E. G. McCarthy

## OMNIPOTENT SUPPRESSORS

Omnipotent suppressors were originally obtained by selecting for suppressors that act simultaneously on more than one type of nonsense codon (see Hinnebusch and Liebman 1991 for review). In a variety of such studies, only two loci were revealed, the recessive suppressors sup35 and sup45. Since it seemed possible that the [eta+] factor might affect the type of omnipotent suppressor that could be obtained, we undertook a search for new omnipotent suppressors in an [eta+] strain (All-Robyn et al. 1990b). At the same time, using a similar rationale, other investigators searched for additional omnipotent suppressors in [psi+] strains. From our search three dominant omnipotent suppressors were obtained. One is an allele of an omnipotent suppressor, SUP46, which had previously been obtained in a [psi+] strain (Ono et al. 1981, 1984); two others are new dominant suppressors that we named SUP38 and SUP39. These suppressors would not have been detected in an [eta-] strain, since their efficiencies are severely reduced in strains cured of [eta+]. We later found SUP38 to be an allele of SUP44, concurrently isolated in a [psi+] strain (Wakem and Sherman 1990). We now use the SUP44 designation. Other suppressors obtained in our search are alleles of sup35 and sup45 unaffected by [eta+].

We cloned the sup44 and sup46 genes by selecting for complementation of the recessive paromomycin-sensitivity associated with the suppressors. In order to rule out the possibility that the clones contained complementing genes unlinked to the suppressors, we performed integrative transformation of both clones. The sites of integration were shown to be linked to the sup44 and sup46 genes respectively. We have assigned the SUP44 gene to chromosome VII on the basis of OFAGE mapping and found it to be 7.5 cM from cyh2 by tetrad analysis (All-Robyn et al. 1990b). SUP46 was previously assigned to chromosome II distal to tyr1 by tetrad analysis (Ono et al 1984). A comparison of the restriction map of our sup46 clone with the unpublished restriction map of chromosome II (H. Feldmann, personal com.) places sup46 about 35 kb distal to tyr1. It should be noted that the physical map cited in Mortimer et al. (1989) which shows sup46 proximal to tyr1 is in error.

In contrast to the recessive omnipotent suppressors sup35 and sup45 which were found not to encode ribosomal proteins, our recent analyses of clones of the dominant omnipotent suppressors SUP44 and SUP46 show that they do encode ribosomal proteins. Indeed, sup46 encodes S13 (Vincent and Liebman in press), consistent with the biochemical data showing that the SUP46 suppressor mutation causes an electrophoretic mobility shift of the 40S ribosomal protein S13 (formerly called YS11) (Ishiguro et al. 1981); sup44 encodes S4 (All-Robyn et al. 1990a). Thus both of the dominant omnipotent suppressors thus far analyzed encode ribosomal proteins. These are the first sequenced ribosomal protein genes shown to control translational ambiguity in a eukaryote. Sequence similarity indicates that these yeast ribosomal proteins are clearly related to the E. coli ribosomal proteins S4 and S5. Ram (ribosomal ambiguity) mutations in the S4 and S5 genes increase translational errors (for review see Piepersberg et al. 1980). The conservation of both structure and function between these proteins is interesting particularly since it is frequently impossible to correlate ribosomal proteins between prokaryotes (or chloroplasts) and eukaryotes (Wool et al. 1990).

**SUP44 :** The DNA sequence of the sup44 gene (All-Robyn et al. 1990a) contains features characteristic of yeast ribosomal protein genes. The predicted protein is small (27.5 kD) and basic (pI 11.23) and the gene is highly expressed. It has one GF1 upstream activating sequence, previously associated with a few other ribosomal protein genes (Woolford and Warner 1991). The gene is unique, essential and does not contain an intron. While the SUP44 sequence does not match any of the 50 yeast ribosomal protein sequences that are completely or partially known (Warner 1989), its predicted molecular weight and amino acid composition are very similar to those of a small subunit ribosomal protein identified as S4 (Higo and Otaka 1979). We therefore had the Harvard Microchemistry facility isolate and sequence a proteolytic fragment of purified S4 protein (kindly provided by E. Otaka). The sequence exactly matched a region of the predicted SUP44 protein establishing that the sup44 gene encodes the yeast S4 ribosomal protein.

The predicted amino acid sequence of the sup44 product has significant similarity to ribosomal protein S5 from both E. coli and Bacillus stearothermophilus. Extensive similarity was also found to a mouse protein translated from a functional cDNA clone, LLrep3. The identity between the SUP44 protein and the S5 proteins of E. coli and B. stearothermophilus is 26% (52% similarity considering conservative changes) and 31% (53% similarity), respectively, across the overlapping sequence. The LLrep3 predicted protein is 55% identical (73% similar) to SUP44 over the entire overlapping sequence.

The LLrep3 mouse gene is a member of a 200-copy repeated sequence family consisting almost entirely of pseudogenes. Recently, the human homolog of LLrep3 was cloned and sequenced. Although the researchers who sequenced LLrep3 did not report any homology with proteins in the data base, we find significant similarity between LLrep3 and the bacterial S5. Mammalian ribosomal protein genes are generally encoded by one functional gene and are associated with many pseudogene copies. We propose that sup44 in yeast, and LLrep3 in mouse and human, encode ribosomal proteins equivalent to the bacterial S5.

A comparison of the SUP44 and LLrep3 open-reading-frames suggested that the LLrep3 clones isolated from mouse and human were truncated and rearranged in the 5' region. Specifically, we noted that the open reading frame 5' to the proposed LLrep3 initiator AUG was similar to the yeast sequence and contained an interesting Arg/Gly-rich sequence. The recently cloned rat homolog of SUP44 (Suzuki et al. 1991) contains the entire coding region including the Arg/Gly-rich sequence in the N-terminal region. Such sequences have been found repeated in many nucleolar genes including nucleolin, fibrillarin, and yeast single-stranded nucleic acid-binding protein. To our knowledge, only two other sequenced ribosomal proteins contains this motif, L7 from Dictyostelium discoideum and S10 from rat (Robyn et al. 1990a, Suzuki et al. 1991).

We also sequenced a SUP44 suppressor allele that contains an amino acid substitution of Tyr for Ser at position 200. The position of this mutation corresponds to position 126 in the E. coli S5 protein, which is not far from the known alterations at positions 103 and 111 in S5 that affect fidelity. The crystal structure of the B. stearothermophilus S5 protein has recently been solved (S. White and V. Ramakrishnan, pers. com.) and shown to be composed of two clearly separate domains. Both the E. coli and yeast ram mutations fall within the same domain. In contrast, E. coli mutations in S5 that cause resistance to the antibiotic spectinomycin, but no suppressor phenotype, fall in the other domain of the protein.

**SUP46:** The DNA sequence of sup46 (Vincent and Liebman in press), like sup44, has many features characteristic of yeast ribosomal protein genes. The gene has a predicted open reading frame of 585 bp interrupted by a 413-bp intron near its 5' end. The open reading frame encodes a protein of 22.3 kD with a predicted pI of 10.7. sup46 has a codon bias index of 0.90 indicative of a highly expressed gene. The promoter contains two GRF1 upstream activating sequences approximately 200 bp upstream of the initiator AUG codon. Such sequences are frequently found in front of ribosomal protein genes in one or two copies, and were previously called ribosmal protein gene (RPG) boxes (Woolford and Warner 1991). The amino terminal peptide sequence matches residues 2-18 of the S13 N-terminus previously sequenced by Edman degradation (Otaka et al. 1982). This identifies the sup46 gene as the S13 ribosomal protein and explains why mutations in sup46 cause alteration of the S13 ribosomal protein.

Using an internal 450-bp EcoRI-HindIII fragment from sup46, we showed that there are two copies of the gene. Mizuta et al. (1991) have cloned the other copy of the S13 ribosomal protein gene (YS11). However, only the sup46 gene copy can fully complement the recessive paromomycin-sensitivity associated with the SUP46 mutation. A disruption of the sup46 gene causes very slow growth, while the YS11A gene disruption is not deleterious. The double disruption is lethal. These results are consistent with RNA blot analyses indicating that sup46 is the predominantly transcribed gene (Mizuta et al. 1991).

A search of the NBRF data bank revealed that SUP46 is similar to the 70S S4 family of ribosomal proteins. SUP46 and the S4 maize chloroplast protein are 29.3% identical over a 133-amino acid overlap in the central region of the protein. Tobacco chloroplast, liverwort chloroplast, and E. coli S4 also exhibit similarity to SUP46. SUP46 is also similar to the Dictyostelium ribosomal protein 1024.

The E. coli S4 ribosomal protein is a primary 16S rRNA binding protein. The portion of S4 known to be required for binding rRNA is largely contained within the central highly conserved region of the protein (Conrad and Craven 1987). We have sequenced two SUP46 mutations and found leucine to trptophan and aspartate to asparagine changes, respectively, in the conserved region. Similarly, several prokaryotic ram mutations fall within this conserved region (Henkin et al. 1990).

Both copies of the yeast S13 genes contain introns with noncanonical TACTAAC boxes. Divergent splicing signals have been found to be associated with genes that are regulated at the level of RNA splicing (Eng and Warner 1991; Fragapane et al. 1992), suggesting that S13 may likewise be regulated at this level. Furthermore, while the introns of other duplicated ribosomal protein genes are highly diverged, the S13 introns contain two conserved sequences of 25 and 31 bp in length. Since the E. coli S4 ribosomal protein is a primary rRNA binding protein it is likely that the analogous S13 yeast protein binds to 18S rRNA. Possibly the conserved S13 intron sequences have a three-dimensional structure similar to that at the rRNA binding site. Indeed, E. coli S4 binds to both 16S rRNA and a similar structure in its own mRNA, thereby controlling its own translation (Nomura et al. 1984). In the S13 case, binding at the intron could promote autoregulation at the level of splicing. Such an autoregulatory mechanism has been demonstrated for one yeast ribosomal protein (Eng and Warner 1991).

## ANTISUPPRESSORS

To learn more about the functional roles of the recessive omnipotent suppressors, sup35 and sup45, and to define other components affecting fidelity, we isolated mutations in unlinked "antisuppressor" genes, which reduced the efficiency of these suppressors. At the time of this search these were the only known omnipotent suppressors. We identified mutations in two loci, asu9 and ASU10, that had allele-specific effects on sup35 and sup45 (Liebman and Cavenagh 1980; Liebman et al. 1980). We further showed that one of these mutations, asu9-1, caused an alteration in the 2D-gel pattern of 40S ribosomal proteins (Liebman and Cavenagh 1981). This indicates that the layered approach will work to identify new components of the translational apparatus. The ASU9 gene has been cloned but no sequence data is yet available (Vincent and Liebman unpublished results).

In addition to the antisuppressors asu9 and ASU10 we found other unlinked antisuppressors that reduced the efficiency of sup35. These antisuppressors cause a concomitant increase in UAG-specific nonsense suppression. We later identified these antisuppressors as dominant tRNA suppressors that interact in an allele-specific manner with sup35 (Song and Liebman 1985). These allele-specific effects predict a direct interaction between the SUP35 protein and the tRNA. Recent support for this prediction has been obtained from the sequence of SUP35, which contains a putative tRNA binding domain within the EF-1α-like region (Kushnirov et al. 1988; Wilson and Culbertson 1988).

## ALLOSUPPRESSORS

In a parallel search to the antisuppressor study described above, we isolated allosuppressors that enhance the efficiency of sup35 and sup45 (Song and Liebman 1987). Again, the goal was to learn more about the functional roles of the recessive omnipotent suppressors, sup35 and sup45, and to define other components affecting fidelity. We isolated mutations in two genes, sal6 and sal2. They enhance the efficiency of the omnipotent suppressors sup45-2, sup35-4, SUP44-1 and SUP39-1, and the serine-inserting tRNA-suppressor, SUQ5-o. The sal6 mutant in conjunction with sup45-2 is cold-sensitive suggesting that it is part of a complex.

We cloned SAL6 by transforming a sal6-1 sup45-2 ura3-52 strain that also contained suppressible markers, with a yeast library made in YCp50. Transformants were screened for complementation of the increased suppression efficiency associated with sal6-1 and for complementation of the sal6-1 cold-sensitive phenotype. Integrative mapping established that the complementing clones contained SAL6. We have assigned SAL6 to the left arm of chromosome XVI between TPK2 and pep4 using OFAGE mapping and the yeast prime clone grid filters (Link and Olson 1991; L. Riles pers. com.).

Preliminary results indicate that the C-terminus of the protein predicted from the SAL6 sequence shows 60% identity to several PP-1 serine/threonine phosphatases. Unlike other known PP-1 phosphatases, the predicted protein also has a large N-terminal region. This region is unusually serine-rich and has no significant homology with other protein sequences in the data base (Vincent and Liebman unpublished results).

## GENE DOSAGE EFFECTS ON TRANSLATIONAL FIDELITY

Since the accurate synthesis of proteins requires the coordinate action of many gene products, the relative levels of these products may be important for the control of translational accuracy. If this were true, changing the dosage of such genes may lead to alterations in translational accuracy. We tested this hypothesis by searching for clones from wild-type YEp24 and YCp50 yeast libraries that alter translational fidelity. Four non-overlapping plasmids were isolated that increase the efficiency of the sup45-2 omnipotent suppressor and the level of paromomycin-induced misreading. Two other non-overlapping plasmids were obtained that had the opposite effect, causing a decrease in misreading. We localized the portion of each insert responsible for the effect on translational accuracy by deletion mapping (Song 1987).

One of the inserts that increases translational misreading was studied in detail (Song et al. 1989). The active portion of this insert contains TEF1, one of two redundant genes coding for the translation elongation factor, EF-1$\alpha$. The fact that extra copies of the EF-1$\alpha$ gene decrease translational accuracy supports our hypothesis that the relative levels of components of the translational apparatus are important for accuracy. The finding further suggests that this approach will identify other translational components whose concentrations are important for accuracy. The types of genes that are predicted to enhance misreading of stop codons when present in increased concentration include new specialized EF-1$\alpha$ "like" genes and genes that interfere with termination. Genes expected to reduce stop codon misreading should encode products that enhance termination, such as a termination factor.

We directly demonstrated that changes in the number of genes coding for EF-1α in yeast cause parallel changes in the misreading of nonsense codons by suppressors and paromomycin. Misreading was measured by the level of growth permitted on appropriate omission media, and with a gene fusion assay that requires misreading of an in-frame UAG codon for β-galactosidase activity. We found that a deletion of one of the TEF genes, tef2-Δ, is associated with antisuppressor activity. We showed that the number of genes coding for EF-1α is reflected in the level of EF-1α protein actually present in the cell. This establishes that the level of EF-1α is critically involved in translational fidelity.

## LITERATURE CITED


All-Robyn JA, Brown N, Otaka E and Liebman SW (1990a) Sequence and functional similarity between a yeast ribosomal protein and the Escherichia coli S5 ram protein. Mol. Cell. Biol. 10:6544-6553

All-Robyn JA, Kelley-Geraghty DK, Griffin E, Brown N and Liebman SW (1990b) Isolation of omnipotent suppressors in an [eta+] yeast strain. Genetics 124:505-514

Conrad RC and Craven GR (1987) A cyanogen bromide fragment of S4 that specifically rebinds 16S RNA. Nucleic Acids Res. 15:10331-10343

Cox BS, Tuite MF and McLaughlin CS (1988) The [psi] factor of yeast: A problem in inheritance. Yeast 4:159-178

Eng FJ, and Warner JR (1991) Structural basis for the regulation of splicing of a yeast messenger RNA. Cell 65:797-804

Fragapane P, Caffarelli E, Lener M, Prislei S, Santoro B and Bozzoni I (1992) Identification of the sequences responsible for the splicing phenotype of the regulatory intron of the L1 ribosomal protein gene of Xenopus laevis. Mol. Cell. Biol. 12:1117-1125

Henkin TM, Chambliss GH and Grundy FJ (1990) Bacillus subtilis mutants with alterations in ribosomal protein S4. J. Bact. 172:6380-6385

Higo K and Otaka E (1979) Isolation and characterization of fourteen ribosomal proteins from the small subunits of yeast. Biochem. 18:4191-4196

Hinnebusch AG and Liebman SW (1991) Protein synthesis and translational control in Saccharomyces cerevisiae, pp. 627-735 in The molecular and cellular biology of the yeast Saccharomyces: genome dynamics, protein synthesis, and energetics, vol. I, (eds) Broach JR, Pringle JR and Jones EW. Cold Spring Harbor Laroratory, Cold Spring Harbor, N.Y.

Ishiguro J, Ono B, Masurekar M, Mclaughlin CS and Sherman F (1981) Altered ribosomal protein S11 from the SUP46 suppressor in yeast. J. Mol. Biol. 147:391-397

Kushnirov VV, Ter-avanesyan M, Telckov MV, Surguchov AP, Smirnov VN and Inge-vechtomov SG (1988) Nucleotide sequence of the sup2 (sup35) gene of Saccharomyces cerevisiae. Gene 66: 45-54

Liebman SW and All-Robyn JA (1984) A new non-Mendelian element, [eta+], causes lethality of yeast omnipotent-suppressor strains. Current Genetics 8: 567-573

Liebman SW and Cavenagh M (1980) An antisuppressor which acts on omnipotent suppressors in yeast. Genetics 95:49-61

Liebman SW and Cavenagh MM (1981) 40S ribosomal protein from a S. cerevisiae antisuppressor mutant exhibiting a unique 2D gel pattern. Current Genetics 3: 27-29

Liebman SW, Cavenagh MM and Bennett LN (1980) Isolation and properties of an antisuppressor in S. cerevisiae specific for an omnipotent suppressor. J.Bact. 143:1527-1529

Link AJ and Olson MV (1991) Physical map of the Saccharomyces cerevisiae genome at 110-kb resolution. Genetics 127:681-698.

Mizuta K, Hashimoto T, Suzuki K and Otaka E (1991) Yeast ribosomal proteins: XII. YS11 of Saccharomyces cerevisiae is a homologue of E. coli S4 according to the gene analysis. Nucleic Acids Res. 19:2603-2608

Mortimer RK, Schild D, Contopoulou R and Kans JA (1989) Genetic map of Saccharomyces cerevisiae, edition 10. Yeast 5:321-403

Nomura M, Gourse R and Baughman G (1984) Regulation of the synthesis of ribosomes and ribosomal components. Annu. Rev. Biochem. 53:75-117

Ono B, Stewart JW and Sherman F (1981) Serine insertion caused by the ribosomal suppressor SUP46 in yeast. J. Mol. Biol. 147:373-379

Ono B, Moriga N, Ishihara K, Ishiguro J, Ishino Y and Shinoda S (1984) Omnipotent suppressors effective in [psi+] strains of Saccharomyces cerevisiae recessiveness and dominance. Genetics 107:219-230

Otaka E, Higo K and Osawa S (1982) Isolation of seventeen proteins and amino-terminal amino acid sequences of eight proteins from cytoplasmic ribosomes of yeast. Biochemistry 21: 4545-4550

Piepersberg W, Geyl D, Hummel H and Bock A (1980) Physiology and biochemistry of bacterial ribosomal mutants, pp. 359-377 in Genetics and evolution of RNA polymerases, tRNA and ribosomes, (eds) Osawa S, Ozeki H, Uchida H, and Yura T. Tokyo University Press, Tokyo.

Song JM (1987) Genetic and molecular studies of genes that affect translational fidelity in Saccharomyces cerevisiae. PhD Thesis, Univ. of Illinois.

Song JM and Liebman SW (1985) Interaction of UAG-suppressors and omnipotent suppressors in Saccharomyces cerevisiae. J. Bact. 161:778-780

Song JM and Liebman SW (1987) Allosuppressors that enhance the efficiency of omnipotent suppressors in Saccharomyces cerevisiae. Genetics 115: 451-460

Song JM, Picologlou S, Grant CM, Firoozan M, Tuite MF and Liebman SW (1989) Elongation factor EF-1α gene dosage alters translational fidelity in Saccharomyces cerevisiae. Mol. Cell. Biol. 9:4571-4575

Suzuki K, Oliver J, and Wool IG (1991) The primary structure of rat ribosomal protein S2. J. Biol. Chem. 266: 2007-20010

Vincent A and Liebman SW The yeast omnipotent suppressor SUP46 encodes a ribosomal protein which is a functional and structural homolog of the E. coli S4 ram protein. Genetics in press

Wakem LP and Sherman F (1990) Isolation of omnipotent suppressors in the yeast Saccharomyces cerevisiae. Genetics 124:515-522

Warner JR (1989) Synthesis of ribosomes in Saccharomyces cerevisiae. Microbiol. Rev. 53:256-271

Wilson PG and Culbertson MR (1988) SUF12 suppressor protein of yeast: a fusion protein related to the EF-1 family of elongation factors. J. Mol. Biol. 199:559-573

Wool IG, Endo Y, Chan YL and Gluck A (1990) Structure, function, and evolution of mammalian ribosomes, pp. 203-214 in The ribosome: structure, function, and evolution, (eds) Hill WE, Dahlberg A, Garrett RA, Moore PB, Schlessinger D, and Warner JR. American Society for Microbiology, Washington, D.C.

Woolford JL and Warner JR (1991) The ribosome and its synthesis, pp. 587-626 in The molecular and cellular biology of the yeast Saccharomyces: genome dynamics, protein synthesis, and energetics, vol. I, (eds) Broach JR, Pringle JR, and Jones EW. Cold Spring Harbor Laboratory, Cold Spring Harbor, N.Y.

## ACKNOWLEDGMENTS

We would like to thank our colleagues who contributed to the results summarized here, S. White, V. Ramakrishnan, M. Tuite and E. Otaka. We also thank other members of our laboratory who participated in this research M. Cavenagh, N. Brown, G. Newnam, E. Griffin and D. Kelley-Geraghty. This work was supported by American Cancer Society Grant MV-553 to SWL.

# DOSAGE-DEPENDENT MODIFIERS OF PSI-DEPENDNET OMNIPOTENT SUPPRESSION IN YEAST

Yury O. Chernoff[1,2*] and Bun-ichiro Ono[1]
[1]Laboratory of Environmental Hygiene Chemistry, Okayama University, Okayama 700, Japan and [2]Department of Genetics, St.-Petersburg University, St.-Petersburg 199034, Russia

## Omnipotent Suppressors - an Introduction

*S. cerevisiae* codon-nonspecific (omnipotent) suppressors, like *E. coli ram* and *tuf* mutations (Gorini 1974; Vijgenboom et al. 1985), affect translational fidelity, therefore they are useful for understanding the mechanism of eucaryotic translation. The efficiency of omnipotent suppressors vary between loci and to some extent within a locus; some act as suppressors, and others as allosuppressors. In order to elucidate omnipotent suppression, some omnipotent suppressors or their wild type alleles have been cloned (cf Hinnebusch and Liebman 1991). One of the most interesting and surprising outcomes of such works is that the cloned wild type genes manifest omnipotent suppressor activity by themselves (Liebman et al. 1988; Song et al. 1989; Kushnirov et al. 1990; Chernoff et al. 1991, 1992). It is therefore interesting to see how wild type genes are involved in translational fidelity. From this viewpoint, we began to look for wild type genes that affect the efficiency of *sup111* when their dosage is changed by cloning. In this study, we used *sup111* because it was one of the least efficient omnipotent suppressors and appeared to be useful for studying gene interactions; *sup111* acts as a weak omnipotent suppressor in the $psi^+$ cytoplasm (Ono et al. 1982, 1986) and as an allosuppressor in the $psi^-$ cytoplasm (Ono et al. 1991). Our another aim was to clone the wild type *SUP111* gene to study its role in translation. Here, we describe the results of our work and present a model for omnipotent suppression.

---------------
* Present address: Department of Biological Sciences, University of Illinois, Chicago, Illinois 60680, USA

NATO ASI Series, Vol. H 71
Protein Synthesis and Targeting in Yeast
Edited by A. J. P. Brown, M. F. Tuite, and J. E. G. McCarthy

## Cloning of Genes That Affect the Efficiency of sup111

Strain YC13-6C (psi$^+$ MATalpha ade2-1 aro7-1 can1-100 ilv1-2 leu2-1 met8-1 trp1-delta ura3-52 sup111) grew slowly in the absence of leucine (Leu$^{+/-}$) due to inefficient suppression of leu2-1 by sup111. This strain was treated with *S. cerevisiae* genomic libraries constructed with low-copy (centromeric) vectors, YCp50(URA3) and YCp19(URA3-TRP1). Ura$^+$ transformants were selected, and they were tested for growth on medium lacking leucine. Clones that grew like wild type (Leu$^+$) and those did not grow at all (Leu$^-$) were picked. Although most clones gave rise to subclones of the original phenotype (Ura$^-$ Leu$^{+/-}$) at frequencies of 3 to 30 % after the growth in nonselective conditions, a few clones gave rise to no such subclone. They were presumed to be revertants rather than transformants and were excluded from further study. By this way, we obtained 18 Leu$^+$ clones and 1 Leu$^-$ clone from 6400 Ura$^+$ transformants.

## Physical Properties of Plasmids Recovered from the Leu$^+$ Transformants

Plasmids isolated from the 18 Leu$^+$ transformants were classified into four groups by their transforming properties and restriction patterns (Table 1). Class 1 (1 clone) gave rise to Leu$^+$ transformants to strain YC13-6C as well as strain 33G-D373 (psi$^-$ MATalpha ade2-144,717 leu2-3,112 his7-1 lys9-21 pheA10 trp1-289 ura3-233). We therefore concluded that this plasmid contained the LEU2 gene, and excluded it form further analysis. Class 2 (11 clones) gave rise to Leu$^+$ transformants to strain YC13-6C but not to strain 33G-D373. Restriction patterns of them overlapped with each other and with that of the TEF2 clone (Fig. 1a). The 1.3 kbp XbaI-XbaI subfragment of R8 hybridized with chromosomes II and XVI (Fig. 2a), indicating that the R8 fragment in fact contains the TEF2 gene; it should be noted that TEF2 (on chromosome II) and TEF1 (on chromosome XVI) differ by only 2 out of 1374 nucleotides, while the flanking regions differ significantly (Nagata et al. 1984; Schirmaier

Table 1. Classification of DNA clones that obtained from the Leu$^+$ transformants.

| Class | Growth on -Leu medium[a] | | Vector | No. of clones | Representative clone |
|---|---|---|---|---|---|
| | YC13-6C | 33G-D373 | | | |
| 1 | + | + | YCp19 | 1 | R11 |
| 2 | + | - | YCp50 | 10 | R1 |
| | | | YCp19 | 1 | R8 |
| 3 | + | - | YCp50 | 5 | R14 |
| 4 | - | - | YCp50 | 1 | R16 |

[a] Strains YC13-6C (Leu$^{+/-}$) and 33G-D373 (Leu$^-$) were used as recipients.
[b] NT represents not tested.

and Philippsen 1984; Cottrelle et al. 1986; Sandbaken and Culbertson 1988).

Class 3 (5 clones), like the TEF2 clones, gave rise to Leu$^+$ transformants to strain YC13-6C but not to strain 33G-D373. However, their restriction patterns, which overlapped with each other, clearly differed from those of the TEF2 clones (Fig. 1b). The responsible gene, designated ESU1 (for Enhancement of SUppression), was confined in the 2.9 kbp BamHI-HindIII subfragment. This subfragment hybridized with chromosome VII (Fig. 2b). We are now sequencing this subfragment to elucidate the gene product of ESU1. Class 4 (1 clone), unexpectedly and surprisingly, gave rise to Leu$^-$ transformants when retransformed to strain YC13-6C. This result indicates that plasmid YCp50-R16 contains a gene that inhibits sup111, designated ISU1. An identical plasmid was isolated from the Leu$^-$ transformant obtained in the initial transformation experiment (see below).

## Physical Properties of a Plasmid Recovered from the Leu$^-$ Transformant

Plasmid YCp50-T6 recovered from the Leu$^-$ transformant gave rise to Leu$^-$ clones when retransformed to strain YC13-6C. Restriction pattern of T6 was identical with that of R16 mentioned

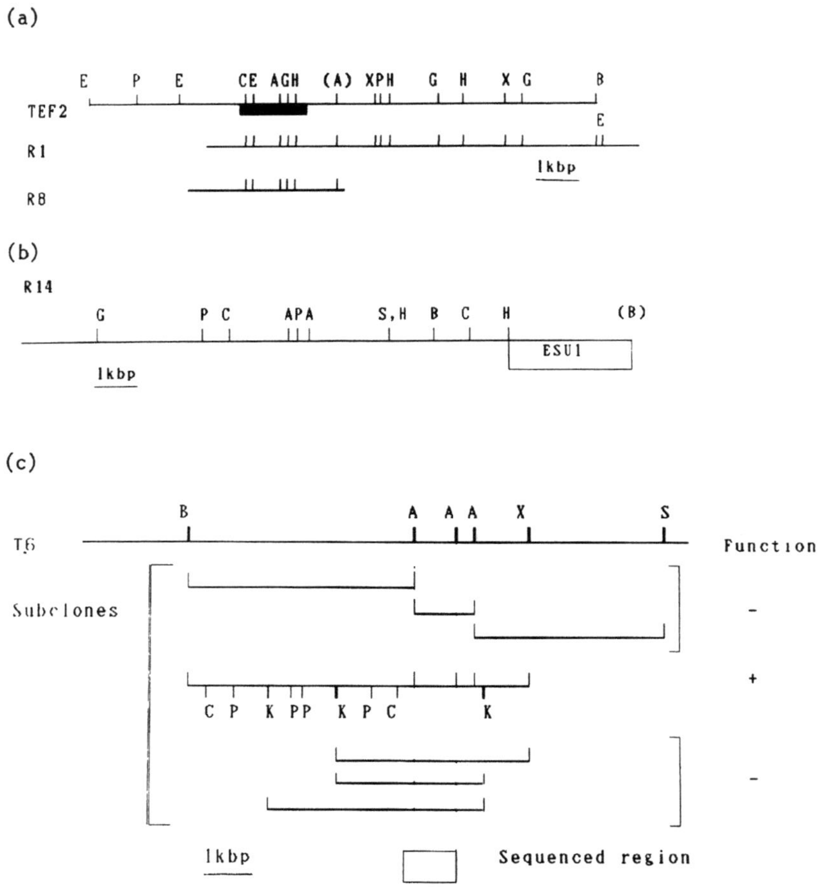


Fig. 1 Restriction patterns. (a) Comparison of a TEF2 clone (Schirmaier and Philippsen 1984) and representative class 2 clones; the solid box represent the TEF2 coding region. (b) A representative class 3 clone. (c) A class 4 clone. Restriction sites are as follows; A = XbaI, B = BamHI, C = ClaI, E = EcoRI, G = BglII, H = HindIII, K = KpnI, P = PstI, and X =XhoI.

above (Fig. 1c), leading to the conclusion that both contain ISU1. Fragment T6 was about 15 kbp in size, and we confined ISU1 in the 9 kbp BamHI-XhoI subfragment. Although we were unsuccessful in obtaining a functional subfragment smaller than

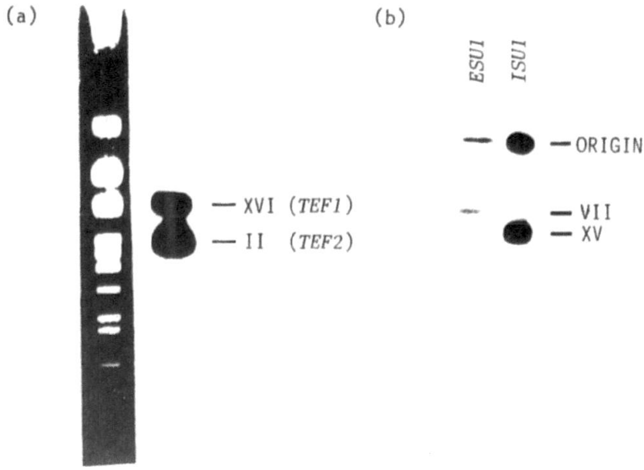


Fig. 2 Chromosome assignment. Chromosome DNAs were separated by transversionally-alterred field gel electrophoresis, blotted to nitrocellulose membrane. The membrane was probed with pBR322 containing the followings (a) the 1.3 kbp XbaI-XbaI subfragment of R8 (see Fig. 1a) and (b) the 2.9 kbp BamHI-HindIII fragment of R14 (see Fig. 1b) or the 1.0 kbp XbaI-XbaI subfragment of T6 (see Fig. 1c). Hybridization was detected by the universal probe method (Yamane et al. 1979) (a) or by the conventional autoradiography method (for procedures, see Sambrook et al. 1989) (b). While chromosomes VII and XV overlapped in the strain used for (a), they were well separated in the strain used in (b); chromosome band patern is not shown for (b).

this, we were convinced that the 1.0 kbp Xba1-Xba1 subfragment was indispensable for the ISU1 function. We sequenced this subfragment and a segment next to it (Fig. 1c) and detected promoter signals and a continuous amino acid sequence (more than 245 residues). The deduced amino acid sequence was not sufficiently homologous to any ribosomal protein nor translational protein factor sequenced by now. Thus, it appears that ISU1 is a gene whose effect on translation has not been described before. Using the same subfragment, we localized ISU1 on chromosome XV (Fig. 2b).

Table 2. Actions of dosage-dependent suppression modifiers.

| Background | Plasmid[a] | Suppression[b] of ade 2-1 | his 7-1 | ilv 1-2 | leu 2-1 | lys 9-21 | met 8-1 |
|---|---|---|---|---|---|---|---|
| psi⁻ | YCp50 (None) | - | 0 | - | - | 0 | - |
| | pFL44(None) | - | 0 | - | - | 0 | - |
| | YCp50-R1 (TEF2) | - | 0 | - | - | 0 | - |
| | YCp50-R14 (ESU1) | - | 0 | - | - | 0 | - |
| | pFL44-HB (ESU1) | - | 0 | - | - | 0 | - |
| | YCp50-T6 (ISU1) | - | 0 | - | - | 0 | - |
| psi⁺ | YCp50 (None) | - | 0 | - | - | 1 | - |
| | pFL44 (None) | - | 0 | - | - | 1 | - |
| | YCp50-R1 (TEF2) | - | 0 | - | - | 2 | - |
| | YCp50-R14 (ESU1) | - | 2 | - | - | 2 | - |
| | pFL44-HB (ESU1) | - | 2 | - | - | 2 | - |
| | YCp50-T6 (ISU1) | - | 0 | - | - | 1 | - |
| psi⁻ sup111 | pFL44 (None) | 0 | - | 0 | 0 | - | 0 |
| | pFL44-HB (ESU1) | 0 | - | 0 | 0 | - | 0 |
| psi⁺ sup111 | YCp50 (None) | 0 | - | 3 | 3 | - | 5 |
| | pFL44 (none) | 0 | - | 3 | 3 | - | 5 |
| | YCp50-R1 (TEF2) | 1 | - | 4 | 5 | - | 6 |
| | YCp50-R14 (ESU1) | 0 | - | 4 | 5 | - | 6 |
| | pFL44-HB (ESU1) | 0 | - | 4 | 5 | - | 6 |
| | YCp50-T6 (ISU1) | 0 | - | 1 | 1 | - | 4 |
| psi⁺ sup112 | YCp50 (None) | - | 2 | - | 4 | - | - |
| | YCp50-R1 (TEF2) | - | 4 | - | 5 | - | - |
| | YCp50-R14 (ESU1) | - | 4 | - | 5 | - | - |
| | YCp50-T6 (ISU1) | - | 2 | - | 4 | - | - |
| psi⁺ sup113 | YCp50 (None) | - | - | - | 3 | - | - |
| | YCp50-R1 (TEF2) | - | - | - | 4 | - | - |
| | YCp50-R14 (ESU1) | - | - | - | 4 | - | - |
| | YCp50-T6 (ISU1) | - | - | - | 1 | - | - |

a YCp50 and pFL44 were a low-copy vector and a high-copy vector, respectively.

b Suppression levels were graded as follows: 0 = absence of growth, 1 = very poor and heterogeneous growth after 7-10 days, 2 = poor and heterogeneous growth after 7-10 days, 3 = moderate growth after 4-6 days, 4 = good growth after 4-6 days, 5 = moderate growth after 1-2 days, and 6 = good growth after 1-2 days (incubation temperature was 30°C). Other nonsense-mutations were only poorly suppressed, if any, in the above shown backgrounds. Therefore, they were not systematically tested.

Functional Properties of TEF2-, ESU1- and ISU1-Clones

The results are summarized in Table 2. TEF2- and ESU1-clones enhanced the suppressor efficiency of psi⁺. While ESU1, presumably TEF2 also, on a high-copy plasmid did not detectably affect the efficiency of sup111 in the $psi^-$ cytoplasm, they clearly enhanced the efficiencies of sup111, sup112 and sup113 in the $psi^+$ cytoplasm. It should be stressed that ESU1 imposed the same effect on sup111 whether it was on a low-copy vector or on a high-copy vector. Thus, we conclude that the effect of ESU1 does not linearly correlate with dosage.

ISU1 did not affect the efficiency of psi. But, it reduced the suppression efficiencies of sup111 and sup113, but not sup112, in the $psi^+$ cytoplasm (Table 2), indicating that its action is locus-specific. The 1.0 kbp XbaI-XbaI subfragment bearing a part of putative ISU1 hybridized to an RNA species of about 3.4 kb (data not shown). The transcript is comparable with 26S rRNA in size and is unusually, if not extremely, large in yeast. It should be mentioned that this transcript was detected in both middle logarithmic phase and stationary phase. It was also found in the heat-shocked cells, where majority of yeast genes are switched off. These features strongly suggest that ISU1 plays an important role in cell viability and/or stress response. Interestingly, *Pichia pinus* (i.e. *P. methanolica*) had a homologue of ISU1. That is, ISU1 appears to be an evolutionally conserved gene despite of its unusual size.

## Omnipotent Suppression - a Model

Our results show that TEF2, ESU1 and ISU1 act as modifiers of omnipotent suppressors if their dosage is changed by means of cloning. It has been previously reported that a plasmid containing TEF2, and TEF1 also, enhances the efficiency of codon-specific and omnipotent suppressors (Liebman et al. 1988; Song et al. 1989). It is also known that the cloned SUP35 and SUP45 genes act as omnipotent suppressors (Kushnirov et al. 1990; Chernoff et al. 1991, 1992). We therefore conclude that a group of wild type yeast genes change translational fidelity by alteration of their dosage. In other words, the wild type level of translational fidelity is the result of gene dosage balance of these genes. Based on this notion, we rationalize the observed peculiar behavior of ISU1 as follows; ISU1 acts as a down-modifier in a usual situation of YCp50 (i.e. one copy per cell), but it acts as an up-modifier in an unusual situation of YCp50 (i.e. many copies per cell). It is also possible that in the primary transformation YCp50-R16 coexisted with plasmid(s) that affected the interaction of ISU1 with other genes.

It is now evident that some wild type genes, if cloned and transformed, act like cytoplasmic modifiers of omnipotent suppression. It is our speculation that the cytoplasmic determinants such as psi (Cox 1965), eta (Liebman and All-Robyn 1984) and psi-like factors (Chernoff et al. 1992) are extrachromosomally amplified forms of the wild type genes involved in the determination of translational fidelity.

## Concluding Remarks

Yeast omnipotent suppressors and their modifiers were previously identified by the conventional mutant selection procedure. However, the recent development of gene manipulation procedures makes it rather easy to change dosage of wild type genes. As we have shown in this work, a new set of wild type genes determining translational fidelity is uncovered by this approach. We have described two presumably new genes, ESU1 and ISU1, that alter the efficiencies of psi-dependent weak omnipotent suppressors. They are wild type genes which act as modifiers of omnipotent suppression by the change of their dosage. They may provide a clue to the origin of cytoplasmic omnipotent suppressors and omnipotent suppressor modifiers. At the same time, they clearly indicates that the wild type level of translational fidelity is dependent on a more complex system than we have anticipated. Further characterization of ESU1 and ISU1 will shed light on the control of translational fidelity. It should also be stressed that we need to clone SUP111, SUP112 and SUP113 to understand their roles in translational fidelity.

Acknowledgements

We thank A. Tho-e (Tokyo University) and F. Lacroute (Gif-sur Yvette) for their kind gifts of a YCp19-based genomic library and pFL44, respectively. We also thank Wakunaga Pharmaceuticals Co., Ltd. (Hiroshima) for their supply of the universal probe (i.e. biotinated single stranded form of pUCf1). YOC was a recipient of the fellowship provided by the Japanese Ministry of Science, Education and Culture in accordance with the USSR-Japan scientific exchange program. A preliminary sequence experiment was carried out in the Gene Recearch Laboratory, Okayama University.

References

Chernoff YO, Derkach IL, Dackesamanskaya AR, Tikhomirova VL, Ter-Avanesyan MD, Inge-Vechtomov SG (1991) Nonsense-suppression by amplification of translational protein factor gene. Doklady Akad. nauk SSSR (Biol Sci) 301:1227-1229 (in Russian)

Chernoff YO, Inge-Vechtomov SG , Derkach IL, Ptyushkina MV, Tarunina OV, Dackesamanskaya AR, Ter-Avanesyan MD (1992) Dosage-dependent translational suppression in yeast Saccharomyces cerevisiae. Yeast (in press)

Cottrelle P, Thiele D, Price VL, Memet S, Micouin JY, Marck C, Buhler JM, Sentenac A, Fromageot P (1985) Cloning, nucleotide sequence, and expression of one of two genes coding for yeast elongation factor 1-alpha. J Biol Chem 260:3090-3096

Cox, BS (1965) Psi, a cytoplasmic suppressor of super-suppressors in yeast. Heredity 20:505-521

Gorini, L (1974) Streptomycin and misreading of the genetic code. In Ribosomes (Nomura M et al. eds), p791-803, Cold Spring Harbor Laboratory, Cold Spring Harbor, New York

Hinnebuch AG, Liebman SW (1991) Protein synthesis and translational control in Saccharomyces cerevisiae. In The molecular and Cellular biology of the yeast Saccharomyces : genome dynamics, protein synthesis, and engineering. Vol. 1 (Strathern JN et al. eds), p627-735, Cold Spring Harbor Laboratory, Cold Spring Harbor, New York

Kushnirov, VV, Ter-Avanesyan MD, Didichenko SA, Smirnov VN, Chernoff YO, Derkach IL, Novikova ON, Inge-Vechtomov SG, Neistat MA, Tolstorukov II (1990) Divergence and conservation of SUP2(SUP35) gene of yeasts Pichia pinus and Saccharomyces cerevisiae. Yeast 6:461-472

Liebman SW, All-Robyn JA (1984) A non-Mendelian factor, [eta$^{+}$], causes lethality of yeast omnipotent strains. Curr Genet 8:567-573

Liebman SW, Song JM, All-Robyn J, Griffin E, Kelley-Geraghty D (1988) Omnipotent suppressors, allosuppressors and antisuppressors of yeast. In Genetics and translation (Tuite MF et al. eds), p403-414, Springer-Verlag, Berlin

Nagata S, Nagashima K, Tsunetsugu-Yokota Y, Fujimura K, Miyasaki M, Kaziro Y (1984) Polypeptide chain elongation factor, 1alpha (EF-1alpha), from yeast: nucleotide sequence of one of the two genes for EF-1alpha from Saccharomyces cerevisiae. EMBO J 3:1825-1830

Ono B, Tanaka M, Kominami M, Ishino Y, Shinoda S. (1982) Recessive UAA suppressors of the yeast Saccharomyces cerevisiae. Genetics 102:653-664

Ono B, Ishino-Arao Y, Tanaka M, Awano I, Shinoda S. (1986) Recessive nonsense suppressors in Saccharomyces cerevisiae: action spectra, complementation groups and map positions. Genetics 114:363-374

Ono B, Chernoff YO, Ishino-Arao Y, Yamagishi N, Shinoda S, Inge-Vechtomov SG (1991) Interactions between chromosomal omnipotent suppressors and extrachromosomal effecters in Saccharomyces cerevisiae. Curr Genet 19:243-248

Sambrook J, Fritsch EF, Maniatis T (1989) Molecular cloning. A laboratory manual. Cold Spring Harbor Laboratory, Cold Spring Harbor, New York

Sandbaken MG, Culbertson MR (1988) Mutations in elongation factor EF-1alpha affect the frequency of frameshifting and amino acid misincorporation in Saccharomyces cerevisiae. Genetics 120:923-934

Schirmaier F, Philippsen P (1984) Identification of two genes coding for the translation elongation factor EF-1alpha of Saccharomyces cerevisiae. EMBO J 3:3311-3315

Song JM, Picologlou S, Grant CC, Firoozan M, Tuite MF, Liebman S (1990) Elongation factor EF-1alpha gene dosage alters translational fidelity in Saccharomyces cerevisiae. Mol Cell Biol 9:4571-4575

Vijgenboom E, Vink T, Kraal B, Bosch L (1985) Mutants of the elongation factor EF-Tu, a new class of nonsense suppressors. EMBO J 4:1049-1052

Yamane A, Nakagami S, Miyoshi K (1988) Application of universal probe on DNA hybridization. Nucleic Acid Res Symp Ser 19:93-95

# ABERRANT mRNA DECODING IN THE DIMORPHIC YEAST CANDIDA ALBICANS

**Manuel A Santos, Gerad Keith[1] and Mick F Tuite**
Biological Laboratory, University of Kent, Canterbury, Kent CT2 7NJ, UK

[1] Institut de Biologie Moleculaire et Cellulaire, CNRS, 15 rue Rene Descartes, 67084 Strasbourg Cedex, France.

## 1 INTRODUCTION

The role tRNAs play in protein synthesis has been the subject of considerable scientific effort and debate during the last 30 years. From the early studies, the consensus view that tRNAs are primarily responsible for decoding genetic information was established. It was then assumed that tRNAs decoded only the codons allowed by correct Watson-Crick base pairing between the codon and the anticodon. Later developments allowed the formulation of the "wobble" hypothesis, to explain non-Watson-Crick base pairing at the first anticodon position (Crick, 1966). This concept of decoding perpetuated the idea that an anticodon, with a certain degree of freedom at the first base, was the only structural determinant of decoding. However, recent studies point to the fine detail of tRNA tertiary structure as also being an important determinant of accurate and efficient decoding, as is well illustrated by misreading and frameshifting tRNAs whose phenotypes can be determined by specific alterations in their primary structures outside the anticodon (Hatfield et al, 1990; Atkins et al, 1991; reviews).

These unusual tRNA phenotypes are excellent tools to aid our understanding of the relationship between tRNA structure and mRNA decoding. Perhaps one of the most interesting tRNA phenotypes so far studied is the phenotype of a tRNA species from the dimorphic fungus Candida albicans that was originally thought to readthrough both the UAG and UGA termination codons based on in vitro translation studies using defined mRNAs (Tuite et al, 1986; Santos et al, 1990). In an attempt to understand the structural determinants of this unusual tRNA-mediated phenotype and perhaps shed more light on the relationship between tRNA structure and function, this "suppressor" tRNA was purified to homogeneity, sequenced and the in vitro phenotype studied in more detail.

NATO ASI Series, Vol. H 71
Protein Synthesis and Targeting in Yeast
Edited by A. J. P. Brown, M. F. Tuite, and J. E. G. McCarthy

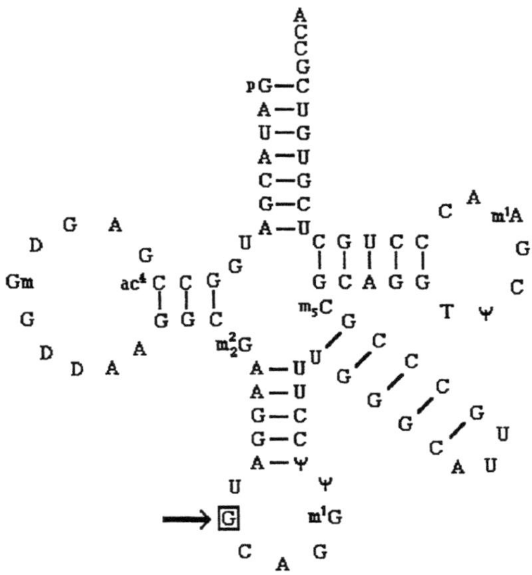

Figure 1 Primary structure of the C. albicans $tRNA_{CAG}$ arranged in the clover-leaf form

## 2. STRUCTURE OF AN UNUSUAL C. ALBICANS $tRNA_{CAG}$

The cloverleaf structure of the C. albicans tRNA responsible for the apparent "omnipotent nonsense suppressor" activity in vitro shows that this tRNA is a typical class-II tRNA with a 5'-CAG-3' ($tRNA_{CAG}$) anticodon and a 13 nucleotide variable extra-arm (Fig. 1). However, detailed analysis of the cloverleaf structure shows that the $tRNA_{CAG}$ has a G at position 33, a position that is invariably occupied by a conserved pyrimidine (being a U in all tRNAs except for the higher eukaryotic initiator methionyl-tRNAs where U is replaced by a C; Spinzl et al, 1989). Furthermore, this $tRNA_{CAG}$ shows much higher nucleotide homology with yeast seryl-tRNAs (77-79%) than with yeast leucyl-tRNAs (51-64%). Also important is the fact that the $tRNA_{CAG}$ has the seryl discriminator base $G_{73}$; Crothers et al, 1972) which suggests that the C. albicans $tRNA_{CAG}$ is a "hybrid" tRNA molecule that might have evolved from seryl-tRNAs via a leucyl-anticodon replacement. If so one has to question the identity of this tRNA, i.e. is the $tRNA_{CAG}$ a leucyl- or a seryl-tRNA? Aminoacylation studies carried out with both S. cerevisiae and C. albicans aminoacyl-tRNA synthetases showed that the $tRNA_{CAG}$ can be charged with serine but not leucine (data not shown). What remains to be shown is tRNA identity in vivo and under the conditions used in the in vitro translation systems (see below). Thus the C. albicans "suppressor" tRNA is an unusual seryl-tRNA that, based on standard concepts of mRNA decoding by tRNAs, should not be able to translate any of the three known termination codons, UAA, UAG and UGA.

## 3 IN VITRO CHARACTERISATION OF THE $tRNA_{CAG}$ PHENOTYPE

To study the in vitro translational activity of the purified C. albicans $tRNA_{CAG}$ on complex mixtures of total S. cerevisiae or C. albicans polysome-associated mRNAs, these mRNAs were translated in both S. cerevisiae and C. albicans cell-free lysates (Fig. 2). The addition of the purified $tRNA_{CAG}$ to the S. cerevisiae lysate programmed with either the C. albicans or S. cerevisiae polysome-associated mRNAs did not alter the migration pattern of the proteins synthesised in vitro (Fig. 2a, b; lane 2) which contradicts the earlier observation that the $tRNA_{CAG}$ is a highly efficient UAG and UGA suppressor tRNA (Tuite et al, 1986). If we were to have such a suppressor activity one would expect to observe the synthesis of new elongated polypeptides in its presence in the heterologous cell-free lysate. However, when the same mRNAs were translated in a C. albicans lysate (Fig. 2a, b; lanes 3,4) several proteins showed aberrant migration patterns if compared to the same mRNA sample. This result suggests that if the $tRNA_{CAG}$ was the sole translational element responsible for the aberrant migration of the proteins, its translational activity is somehow "repressed" in the S. cerevisiae lysate.

To confirm the above hypothesis the in vitro phenotype of the $tRNA_{CAG}$ was further characterized using known single mRNA templates. When the C. albicans lysate was programmed with the Brome Mosaic Virus (BMV) RNA-4, a single abnormally elongated protein was synthesised (Fig. 3a; lanes 2,3) in contrast to the normal sized protein synthesized in the S. cerevisiae lysate (Fig. 3a; lane 1). The addition of pure $tRNA_{CAG}$ to the S. cerevisiae cell-free lysate also induced the synthesis of the abnormally elongated protein (Fig. 3b) although it only represented 65-70% of the BMV RNA4-encoded products thus confirming our original hypothesis that the translational activity of the $tRNA_{CAG}$ is "repressed" in the S. cerevisiae lysate.

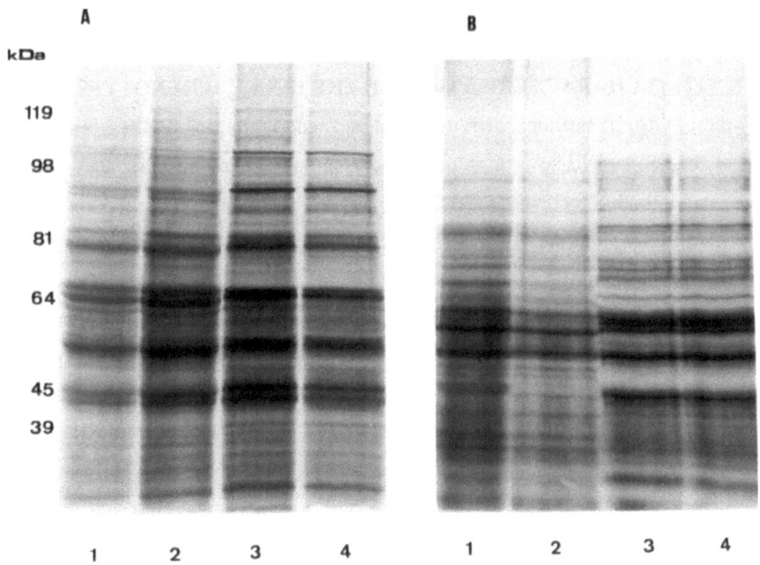


Figure 2 Translation of polysome-associated mRNAs in S. cerevisiae and C. albicans cell-free lysates

10μg of total RNA purified from polysome pellets of S. cerevisiae and C. albicans cells were used per in vitro translation reaction.

A S. cerevisiae RNAs were translated in a homologous cell-free lysate (lanes 1 and 2) or in a C. albicans lysate (lanes 3 and 4). Lanes 1 and 3, no $tRNA_{CAG}$ added; lanes 2 and 4, 100ng of pure $tRNA_{CAG}$ added.

B Same as A but with C. albicans RNAs used as template.

72h exposures of fluororagraphs of 15% SDS-PAGE gels are shown.

The $tRNA_{CAG}$ phenotype was also investigated in the rabbit reticulocyte cell-free lysate translating endogenous $\alpha$- and ß-globin mRNAs. The addition of the $tRNA_{CAG}$ to the rabbit reticulocyte lysate induced gross mobility shifts in both the $\alpha$- and ß-globin proteins synthesised (Fig. 4) and in an unknown protein with a mol. wt. of approximately 30kDa. Interestingly, one of the abnormally elongated globin proteins had the same mol. wt. as a UAA readthrough $\alpha$-globin protein that was synthesised in the presence of a S. cerevisiae UAA suppressor tRNA.

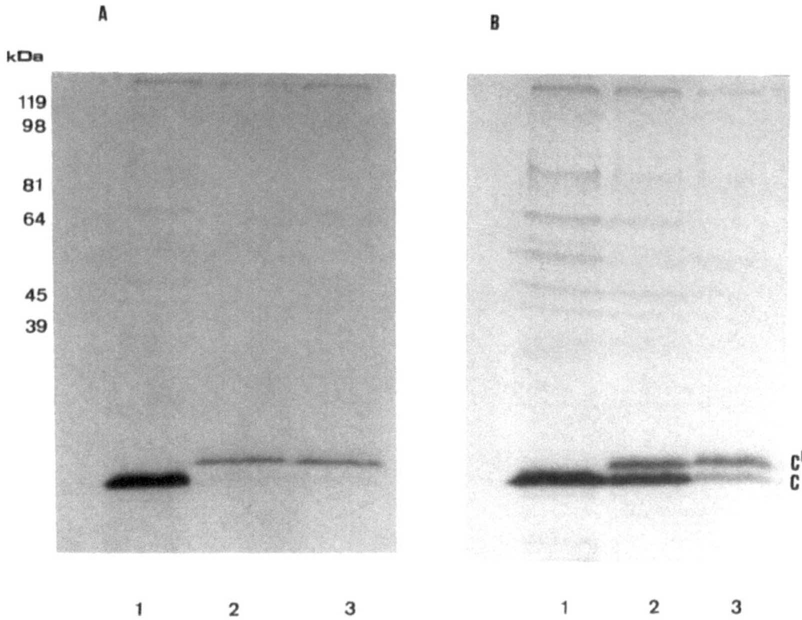


Figure 3 Translation of BMV RNA4 in S. cerevisiae and C. albicans cell-free lysates.

Total BMV RNA was translated in a S. cerevisiae lysate (panel A, lane 1 and panel B) or in a C. albicans lysate (panel A, lanes 3 and 4).

A Lane 1, 2, 3$\mu$g BMV RNA added; lane 3, 3$\mu$g of BMV RNA plus 500ng $tRNA_{CAG}$.

B Lanes 1, 2 and 3, 3$\mu$g BMV RNA added; Lanes 2 and 3, 5$\mu$g total C.albicans or 100ng of $tRNA_{CAG}$ added respectively. C and C' indicate that BMV coat protein and the slower migrating form of the coat protein respectively.

An overnight autoradiograph of a region of a 15% polyacrylamide gel is shown.

When the abnormally elongated α- and ß-globin polypeptides were further characterized by isoelectric focusing (Fig. 4b) it was found that the α- and ß-globin in vitro translation products were a very heterogenous mixture of polypeptides which suggests that the $tRNA_{CAG}$ induces high levels of misreading, as originally proposed by Tuite et al (1986). That the activity of the $tRNA_{CAG}$ is termination codon-independent can be justified by the fact that the

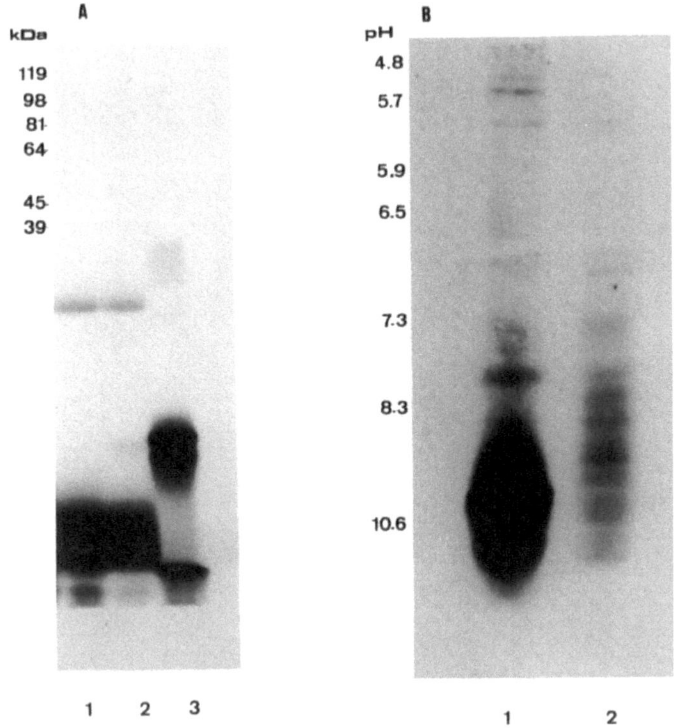


Figure 4 Translation of rabbit α- and ß-globin mRNAs in presence of $tRNA_{CAG}$ in a rabbit reticulocyte lysate.

A Lane 1 no $tRNA_{CAG}$ added; lane 2, 5μg S. cerevisiae tRNA enriched in a UAA suppressor tRNA 50ng $tRNA_{CAG}$ added; lane 3, 50ng of pure $tRNA_{CAG}$ added.

B Separation of α- and ß-globin polypeptides synthesised in vitro by isoelectric focusing. Lane 1 globin polypeptides synthesised in the absence of exogenous tRNA; lane 2, globin polypeptides synthesised in the presence of 100ng $tRNA_{CAG}$.

3 termination codons are present in the 3 mRNAs used, i.e. UAG, BMV RNA-4; UAA, α-globin; UGA, ß-globin. This was also confirmed by cyc-cleavage of the α- and ß-globin polypeptides synthesised in vitro in the presence of the $tRNA_{CAG}$; this showed that the N-terminus of the α- and ß-globin polypeptides were abnormally elongated (data not shown).

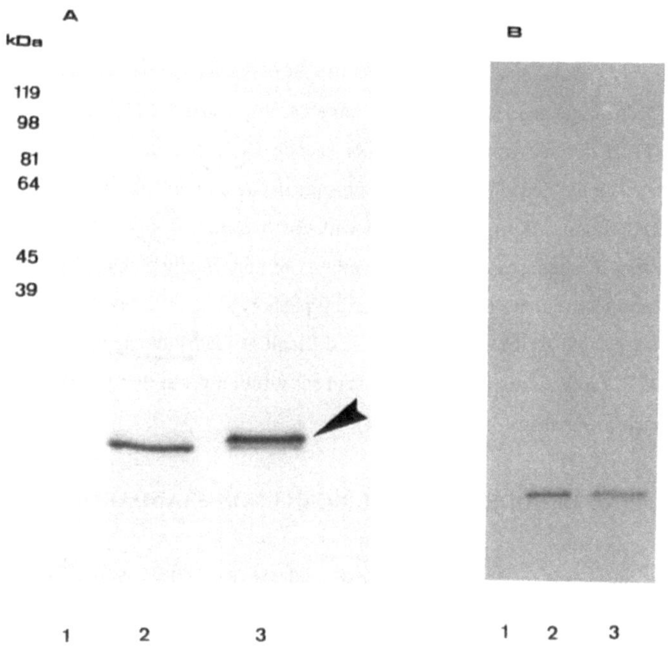


Figure 5 Translation of HSP26 and TEF1-derived mRNAs in the S. cerevisiae lysate in the presence of the $tRNA_{CAG}$.

5μg of HSP26 - (A) or TEF1-derived (B) in vitro transcripts were added to a S. cerevisiae lysate treated with micrococcal nuclease. Lane 1, no mRNA added. Lanes 2 and 3 indicate the absence or presence of the C. albicans $tRNA_{CAG}$ (100ng) respectively. Overnight exposed autoradiographs of 15% polyacrylamide gels are shown. The arrow identifies the slower migrating Hsp26 polypeptide.

The possible misreading activity of the $tRNA_{CAG}$ was further checked by translating two in vitro transcribed mRNAs of known sequence in the S. cerevisiae cell-free lysate, namely the S. cerevisiae HSP26 mRNA encoding heat shock protein Hsp26 (Bossier et al, 1989) and an artificial construct derived from the S. cerevisiae TEF1 gene encoding translation elongation factor 1$\alpha$ and kindly provided A. Hartley, University of Kent (Fig. 5a). Translation of the HSP26 mRNA in presence of the $tRNA_{CAG}$ also resulted in the synthesis of a second, slower migrating polypeptide although no such aberrant translation product was observed in the case of the truncated TEF1 protein (Fig. 5b).

A comparison of the nucleotide sequence of the 5 mRNAs used in this study and specifically the location of CUG (leucine) codons in the mRNAs indicated that the rabbit $\alpha$- and ß-globin, the BMV RNA-4 and the HSP26 mRNAs have 14, 16, 4 and 2 CUG codons respectively, while the TEF1-derived construct does not encode CUG codon. This suggests that the aberrant translational decoding by the C. albicans $tRNA_{CAG}$ is related to the decoding of its cognate CUG codon. If so, one is faced with the question of knowing what is the CUG decoding event that induces the in vitro synthesis of abnormally extended polypeptides? It would be easy to speculate that the misreading phenotype could arise from aminoacylation of the $tRNA_{CAG}$ with serine. However, it is difficult to justify the apparent mol. wt. shifts on SDS-PAGE on the basis of this amino acid substitution given that the mol. wt. of serine (105) is much lower than for leucine (131).

## 4 IS mRNA DECODING IN C. ALBICANS NON-STANDARD?

The results presented and discussed above indicate that CUG codons are abnormally translated in vitro in the presence of the C. albicans $tRNA_{CAG}$. It is not clear however what kind of translational event this tRNA mediates at the CUG (or possibly related codons) but could be partially explained by one or other of the following:

(i) The $tRNA_{CAG}$ inserts modified serine e.g. phosphoserine, in response to the leucine CUG codon. An intermediate in the biosynthesis of the selenocysteyl-$tRNA^{Sec}$ in mammalian cells appears to be a phosphoseryl-$tRNA^{Ser}$ and there is evidence from in vitro translation studies that phosphoserine can be incorporated into ß-globin in response to the natural UGA terminator (Mizutani et al, 1986). The identity of the amino acid(s) inserted in response to the CUG codon will only come from direct protein sequencing of specific polypeptides encoded by CUG-containing mRNAs; this work is currently in progress.

(ii) The $tRNA_{CAG}$ can act as an initiator tRNA in vitro causing ribsomes to initiate on CUG codons in the 5' untranslated regions (5'UTRs) of mRNAs. This would explain the synthesis of apparently elongated polypeptides observed, but three of the test mRNAs used lack CUG codons in their 5'UTRs. It is interesting to note however that the CUG codon can be used to efficiently initiate translation in mammalian cells (Dasso and Jackson, 1991).

(iii) The $tRNA_{CAG}$ causes ribosomal frameshifting at CUG codons by a four base-decoding event which encompasses the base immediately 3' to the anticodon sequence, namely the unusual $G_{33}$ residue. This would indicate context specificity of this event i.e. frameshifting only a CUGC codons. However, translational frameshifting at CUGC "codons" in our test mRNAs would generate a number of different translation products several of which would have lower mol. wt. that the authentic full length polypeptide. Such faster migrating polypeptides have not however been observed (see Figs. 3, 4, and 5).

The in vitro studies also bring to question how mRNA is decoded in vivo in C. albicans and specially the function the $tRNA_{CAG}$ plays in C. albicans protein synthesis. The low usage of the CUG codon in C. albicans mRNAs sequenced to date (Brown et al, 1991) might be related to the unusual activity of the $tRNA_{CAG}$ we have uncovered by in vitro translation studies. If so, it is possible to hypothesise that this tRNA is only used in specialized translational events that are required for specific cellular functions. The disruption of the $tRNA_{CAG}$ gene will be of crucial importance in tackling this question. That it may be a component of the pathogenicity mechanism of this important human pathogen comes from our observations that several pathogenic Candida species also encode a tRNA or tRNAs with a similar in vitro phenotype whereas the non-pathogenic Candida species we have tested do not (Santos et al, 1990).

**ACKNOWLEDGEMENTS**

We gratefully acknowledge the financial support in the UK from Glaxo Group Research and The Wellcome Trust (MFT) and in France from the CNRS (GK). MAS was supported by studentships from the EEC, JNICT (Portugal) and the University of Kent. MAS was also a recipient of an EMBO Short Term Fellowship to visit Strasbourg.

**REFERENCES**

Atkins, J.F., Weiss, R.B., Thompson, S. and Gesteland, R.F. (1991) Towards a genetic dissection of the basis of triplet decoding and it natural subversion: programmed reading frame shifts and hops. Ann. Rev. Genet. 25, 210-228.

Bossier, P., Fitch, I.T., Boucherie, H. and Tuite, M.F. (1989) Structure and expression of a yeast gene encoding the small heat shock protein Hsp26. Gene 78, 323-330.

Brown, A.J.P., Bertram, G., Feldman, P.Y.F., Peggie, M.W. and Swoboda, R.K. (1991) Codon utilisation in the pathogenic yeast Candida albicans. Nucleic Acids Res. 19, 4298.

Crick, F.H.C. (1966) Codon-anticodon pairing: the wobble hypothesis. J. Mol. Biol. 19 548-555.

Crothers, D.M., Seno, T. and Soll, D. (1972) Is there a discriminator site in tRNA? Proc. Natl. Acad. Sci. USA, 69, 3063-3067.

Dasso, M.C. and Jackson, R.J. (1989) Efficient initiation of mammalian mRNA translation at a CUG codon. Nucleic Acids Res. 17, 6485-6497.

Hatfield, D., Lee, B.J., Smith, D.W.E. and Oroszlan, S. (1990) Role of nonsense frameshift and missense suppressor tRNAs in mammalian cells. Prog. Cell. Subcell. Biol. 11, 115-145.

Kawaguchi, Y., Honda, H., Morimura, T. and Iwasaki, S. (1989) The codon CUG is read as serine in an asporogenic yeast Candida cylindracea. Nature 341, 164-166.

Mizutani, T. and Tachibana, Y. (1986) Possible incorporation of phosphoserine into globin readthrough protein via bovine opal suppressor phosphoseryl-tRNA. FEBS Letts. 207, 162-166.

Santos, M.A., Tuite, M.F., Colthurst, D.R., Wills, N. and McLaughlin, C.S. (1990) Efficient translation of the UAG termination codon in Candida species. Curr. Genet. 17, 487-491.

Sprinzl, M., Hartmann, T., Weber, Y., Blank, Y. and Zeidler, R. (1989) Compilation of tRNA sequences and sequences of tRNA genes. Nucleic Acids Res. 17, r1-r67.

Tuite, M.F., Bower, P.A. and McLaughlin, C.S. (1986) A novel suppressor tRNA from the dimorphic fungus Candida albicans. Biochem. Biophys. Acta 866, 26-31.

# EFFECT OF INITIATION FACTOR eIF-5A DEPLETION ON CELL PROLIFERATION AND PROTEIN SYNTHESIS

Hyun Ah Kang, Hubert G. Schwelberger and John W. B. Hershey
Department of Biological Chemistry
School of Medicine
University of California
Davis, CA 95616
USA

Initiation factor eIF-5A (formerly called eIF-4D) was originally isolated (Kemper et al., 1976; Benne et al., 1978) from rabbit reticulocytes based on its activity in stimulating the synthesis of methionyl-puromycin, a model of the formation of the first peptide bond in protein synthesis. This highly conserved protein is small (ca. 18 kDa) and acidic (pI = 5.4) and is one of the most abundant initiation factors in mammalian cells. eIF-5A is unique in undergoing a covalent post-translational modification by a two-step pathway that involves the attachment of an aminobutyl group from spermidine to the ε-amino group of a lysine residue, followed by hydroxylation on the number 2 carbon of the butyl group, to form a hypusine residue (Park et al., 1984). The hypusination reaction correlates with stimulation of cell growth and protein synthesis, thereby eliciting considerable interest in the protein.

In order to better determine the role of eIF-5A and its hypusine modification in cell metabolism, we cloned the human cDNA encoding eIF-5A in a collaborative effort with the group of Dr. William C. Merrick (Case Western Reserve Univ.) (Smit-McBride et al., 1989). From the structure, we determined that $Lys^{50}$is the residue that is hypusinated. We expressed the cDNA in *E. coli* and purified the unmodified precursor of eIF-5A, called ec-eIF-5A(Lys), and demonstrated that it lacked activity in the *in vitro* assay for methionyl-puromycin (Smit-McBride et al., 1989). The result suggests that hypusine is essential for the factor's activity. This view was reinforced by the fact that the *E. coli*-produced protein regains partial activity when modified *in vitro* with the enzyme that transfers the aminobutyl group to $Lys^{51}$. Transient transfection of the eIF-5A cDNA into COS cells produced no effect on translation, suggesting that the initiation factor is present in excess in the cell. Furthermore, transfection of a mutant form of the cDNA that encodes eIF-5A with $Lys^{50}$

NATO ASI Series, Vol. H 71
Protein Synthesis and Targeting in Yeast
Edited by A. J. P. Brown, M. F. Tuite, and J. E. G. McCarthy

substituted with Arg, thereby preventing the hypusine modification, caused no effect on COS cells.

Since approaches to demonstrate a hypusine requirement in mammalian cells did not provide useful information, we turned to the yeast, *Saccharomyces cerevisiae*, to better study the protein's structure and modification. By using the mammalian cDNA as a probe, we isolated a yeast gene which we called *TIF51A* (Schnier et al., 1991). Disruption of *TIF51A* led to a slow growth phenotype, suggesting the possibility of a second gene, which we then cloned and called *TIF51B*. The two yeast gene products share 90% amino acid sequence identity, and exhibit about 63% identity with the human eIF-5A protein. Disruption of both genes is lethal, indicating the eIF-5A is essential for some aspect of cell metabolism.

While this work was in progress, the sequence of a gene, called *ANB1*, was reported whose product is homologous to mammalian eIF-5A (Mehta et al., 1990). *ANB1* and *TIF51B* are essentially identical genes. A probe from *ANB1* identifies transcripts from *ANB1* (tr1) and from a second gene, called tr2 (Lowry et al., 1983). We have determined that tr2 is transcribed from *TIF51A* (Schnier et al., 1991). It was known that *ANB1* (*TIF51B*) is expressed under anaerobic conditions, whereas tr2 (*TIF51A*) is expressed under aerobic conditions (Lowry and Lieber, 1986). However, the physiological purpose for reciprocally regulating the two genes by oxygen (and heme) is not known. To better evaluate this curious regulation, we wished to map the *TIF51A* gene. We physically mapped the gene by chromosome fragmentation and pulsed-field gel electrophoresis, and found that it is located about 90 kbp from the left end of chromosome V (Kang et al., 1992). This places *TIF51A* near *CYC7* in a region called ARC (McKnight et al., 1981). From published restriction maps and sequence information provided prior to publication by Fred Sherman and Larry Melnick, we determined that *TIF51A* maps adjacent to *ANP1* and about 5.5 kb from *CYC7*. Since *TIF51B* maps close to *CYC1* in a region called COR, and since the ARC and COR regions contain a number of homologous genes, it appears that these two regions evolved by a large duplication. However, the order of homologous genes in the two regions makes a circular intermediate in the duplication event unlikely. Since *TIF51A* and *CYC1* are aerobicly expressed, whereas *TIF51B* and *CYC7* are anaerobicly expressed, both the ARC and COR regions show reciprocal regulation by oxygen. The large separation of *TIF51A* and *CYC7* indicates that a close physical linkage is not required for such control.

With the cloned yeast genes for eIF-5A in hand, we can ask if their gene products are truly the homologues of mammalian eIF-5A. Two approaches were used to demonstrate their functional relatedness. Yeast eIF-5A was purified from yeast cells expressing *TIF51A* from a multicopy plasmid and was found to be half as active as the mammalian factor in the *in vitro* assay for methionyl-puromycin synthesis constructed with mammalian components. Furthermore, the human cDNA, when expressed in yeast where both chromosomal copies of the *TIF51* genes are disrupted, supports growth. Given the strong homology and the *in vivo* and *in vitro* activities of the yeast and mammalian proteins, there is no doubt that the *TIF51* gene products encode the yeast equivalent of mammalian eIF-5A.

There was interest to determine whether or not the two eIF-5A genes produce proteins that function differently in yeast, especially under aerobic and anaerobic conditions. This question was addressed by constructing a strain where the chromosome carries the disrupted genes, *tif51A::LEU2* and *tif51B::HIS3*, and a wildtype *TIF51A* gene driven by the GAL1,10 promoter is provided on a centromeric plasmid that carries *URA3*. In a related strain, *TIF51B* under GAL1,10 control and *TRP1* in the centromeric plasmid was substituted for the *TIF51A* plasmid by the shuffle procedure. Both the *TIF51A* and *TIF51B* strains grow at wild-type rates under both aerobic and anaerobic conditions. Since only one of the two genes is expressed in each of these strains, the results demonstrate that either gene product supports cell growth under both conditions.

In mammalian cells, two isoelectric variants of eIF-5A are seen, both of which contain hypusine. In yeast, two charge variants are produced from either *TIF51A* or from *TIF51B*. Sometimes a smaller variant is seen as well, but we have determined that this is an artifact of *in vitro* proteolysis. However, the *TIF51A* gene product is slightly smaller (20 kDa) than the *TIF51B* product (22 kDa), even though the two proteins each have 157 amino acid residues and differ by only 15 residues (90% identity), 5 of which are similar. When eIF-5A was purified from yeast, the two charge variants were separated and were shown to have identical activities in the methionyl-puromycin assay. During the purification, the more acidic form decreased in abundance, suggesting that the modification to the more acidic form is reversible. We have preliminary evidence that eIF-5A is phosphorylated *in vivo* in yeast. The role of this additional modification is not known, nor has it been shown that the modification correlates with metabolic activity in the cell.

Is the hypusine modification essential for eIF-5A activity? The double-null strain which has a wild-type *TIF51A* gene on the YCp50 plasmid was transformed with a YRp7 plasmid carrying a mutant form of *TIF51A*, namely one expressing Arg in place of $Lys^{51}$ (in yeast, the Lys site of hypusination is residue 51). The eIF-5A($K^{51}R$) protein is expressed and accumulates in the transformed cells. When the cells were plated on 5-FOA and $Trp^-$ media to select for cells that have lost the YCp50-5A plasmid and retained the YRp7 derivative, no colonies were found (Schnier et al., 1991). This indicates that the mutant form of the protein cannot function in yeast, and suggests that the hypusine modification is essential for cell viability.

In the experiments discussed above, the only evidence that eIF-5A functions in initiation comes from the *in vitro* assay for methionyl-puromycin synthesis. We have speculated that eIF-5A may be necessary for efficient formation of the first peptide bond because this reaction involves non-acylated Met-tRNA in the P site. The positive charge on the methionine may not be favorable for tight binding in the donor site of the peptidyl transferase center of the ribosome, and may require stabilization. It is possible that eIF-5A provides such a stabilization. To better characterize the function of eIF-5A and to demonstrate its involvement in translation, we wished to deprive cells of the protein and characterize the effects on protein synthesis and intermediates in the pathway.

An effective way to deplete a protein in yeast is to place its gene under control of a *GAL* promoter, grow the cells in galactose, then shift to glucose which strongly represses the *GAL* promoter. The double-null strain used above with *TIF51A* driven by the *GAL1,10* promoter on a centromeric plasmid, was shifted from galactose to glucose media. Cell growth continued normally for up to 4 to 5 generations before declining. Apparently, it is necessary to dilute eIF-5A to very low levels before a growth phenotype is detectable. eIF-5A is both abundant and very stable, so levels drop only by dilution due to cell division. This long delay in cell growth effects complicates analyses of the depleted cells, since many other secondary effects may be occuring after 5 generations.

To more rapidly deplete the cells of eIF-5A, a different strategy was used. We constructed a variant form of the *TIF51A* gene that contained protein destabilizing elements. A ubiquitin gene was placed at the 5' end of the gene, followed by a codon for either Met or Arg, then an N-terminal extension and finally the coding region of *TIF51A*, by using a vector system provided by E.-C. Park and J. W. Szostak (Mass. General Hospital). When such a fusion gene is expressed, the fusion protein is cleaved at the end of the ubiquitin portion, leaving either Met or Arg as N-termini. In yeast, the N-end rule proteolytic pathway degrades proteins at variable rates according to their N-terminal amino acid. We found that R-eIF-5A (N-terminal Arg) is subjected to rapid turnover, with a half-life of about 10 minutes. Such a construct should be suitable for obtaining rapid depletion of eIF-5A.

The R-eIF-5A gene fusion was placed behind the GAL1,10 promoter in a centromeric plasmid carrying the *URA3* marker, and the vector was introduced into the double-null strain having a wild-type TIF51A gene on a cwentromeric poasmid that carries the *TRP1* marker, allowing loss of the plasmid containing the wild-type gene by segregation. The resulting strain expressing only the R-eIF-5A protein fusion grows well in galactose, thereby demonstrating that the altered protein is active in this strain. Following a shift to glucose media, Western blot analysis of eIF-5A levels showed no detectable factor after 1 hour of growth in glucose. Cells continue to grow at a rate of about 40% of wild-type for a few hours, finally ceasing growth after 15 hours. Surprisingly, the rate of protein synthesis as measured by 5 minute pulses of [$^{35}$S]methionine was about 70% of wild-type for the first 4 hours. Continuation of protein synthesis well after essentially complete depletion of eIF-5A indicates that eIF-5A is not essential for bulk protein synthesis. Polysome profiles were obtained from eIF-5A-depleted and non-depleted cells to see if the initiation phase might be affected. There were only slight changes in the polysome profiles. A slight shift of larger polysomes towards smaller polysomes was seen. Moreover, an increase in half-mers was observed for monosomes and small polysomes (half-mers are thought to be polysomes with an extra 40S ribosomal subunit attached, presumably in the scanning mode or bound to the initiator AUG awaiting junction with the 60S ribosomal subunit). These observations are consistent with a small effect of eIF-5A loss on a late step of initiation, but rule out that eIF-5A is required for the initiation process.

The failure to demonstrate a role for eIF-5A in protein synthesis is surprising. It is well established that eIF-5A stimulates methionyl-puromycin synthesis *in vitro*. Stimulations of up to 6-fold have been obtained. However, it is troubling that quite large amounts of protein are required to obtained such stimulation. We routinely use 1 to 4 micrograms of eIF-5A, corresponding to about 50 to 200 picomoles of the factor, yet only about 1 or 2 picomoles of methionyl-puromycin are synthesized. If eIF-5A stimulation is indeed an artifact of the assay system, this fact dramatically demonstrates the hazards of relying on *in vitro* assays to describe a complicated pathway such as protein synthesis. It remains possible that eIF-5A is required for the translation of a small class of mRNAs but not for the translation of the bulk of yeast mRNAs. We are in the process of evaluating the relative rates of synthesis of individual proteins in eIF-5A-depleted and non-depleted yeast cells by using high resolution 2-dimensional gel electrophoresis.

Achnowledgments. This work was supported by a U.S.P.H.S. grant from the N.I.H., GM22135. HGS was supported by an Erwin Schrödinger Fellowship from the Fonds zur Förderung der wissenschaftlichen Forschung, Austria.

**References.**

Benne, R., Brown-Luedi, M., & Hershey, J. W. B. (1978) Purification and Characterization of Protein Synthesis Initiation Factors eIF-1, eIF-4C, eIF-4D, and eIF-5 from Rabbit Reticulocytes. J. Biol. Chem. **253**, 3070-3078.

Kang, H. A., Schwelberger, H. G., & Hershey, J. W. B. (1992) The two genes encoding protein synthesis initiation factor eIF-5A in *Saccharomyces cerevisiae* are members of a duplicated gene cluster. Mol. Gen. Genet., *in press.*

Kemper, W. M., Berry, I. W., & Merrick, W. C. (1976) Purification and Properties of Rabbit Reticulocyte Protein Synthesis Initiation Factors M2Bα and M2Bβ. J. Biol. Chem. **251**, 5551-5557.

Lowry, C. V. & Lieber, R. H. (1986) Negative Regulation of the *Saccharomyces cerevisiae ANB1* Gene by Heme, as Mediated by the *ROX1* Gene Product. Mol. Cell. Biol. **6**, 4145-4148.

Lowry, C. V., Weiss, J. L., Walthall, D. A., & Zitomer, R. S. (1983) Modulator sequences mediate oxygen regulation of *CYC1* and a neighboring gene in yeast. Proc. Nat. Acad. Sci. USA **80**, 151-155.

McKnight, G. L., Cardillo, T. S., & Sherman, F. (1981) An extensive deletion causing overproduction of yeast iso-2-cytochrome c. Cell **25**, 409-419.

Mehta, K. D., Leung, D., Lefebvre, L., & Smith, M. (1990) The *ANB1* Locus of *Saccharomyces cerevisiae* Encodes the Protein Synthesis Initiation Factor eIF-4D. J. Biol. Chem. **265**, 8802-8807.

Park, M. H., Liberato, D. J., Yergey, A. L., & Folk, J. E. (1984) The Biosynthesis of Hypusine ($N^{\varepsilon}$-(4-amino-2-hydroxybutyl)lysine). J. Biol. Chem. **259**, 12123-12127.

Schnier, J., Schwelberger, H. G., Smit-McBride, Z., Kang, H. A., & Hershey, J. W. B. (1991) Translation Initiation Factor 5A and Its Hypusine Modification Are Essential for Cell Viability in the Yeast *Saccharomyces cerevisiae*. Mol. Cell. Biol. **11**, 3105-3114.

Smit-McBride, Z., Dever, T. E., Hershey, J. W. B., & Merrick, W. C. (1989) Sequence Determination and cDNA Cloning of Eukaryotic Initiation Factor 4D, the Hypusine-Containing Protein. J. Biol. Chem. **264**, 1578-1583.

Smit-McBride, Z., Schnier, J., Kaufman, R. J., & Hershey, J. W. B. (1989) Protein Synthesis Initiation Factor eIF-4D: Functional Comparison of Native and Unhypusinated Forms of the Protein. J. Biol. Chem. **264**, 18527-18530.

# Analysis of the genes encoding eIF-4A from yeast

Stefanie R. Schmid[1], Peter Buser, Raffaella Coppolecchia, Andrea Fischli and Patrick Linder[2]

Dept. of Microbiology, Biozentrum, Klingelbergstrasse 70, 4056 Basel

## Abstract

It has been postulated that the translation initiation factor eIF-4A functions as a RNA helicase in the initiation of protein synthesis. This function has recently been confirmed by *in vitro* helicase assays. We have defined a family of proteins which share several sequence domains with eIF-4A. This family was termed the D-E-A-D-box family according to of one of the motifs (Asp-Glu-Ala-Asp) present in these proteins. Members of this family come from different organisms ranging from bacteria to humans and are also present in some viruses. They are involved in processes as different as transcription, splicing, ribosome assembly or translation. All these processes have some relation to RNA metabolism explaining the requirement of a putative RNA helicase. Thus, the translation initiation factor 4A represents an ideal system to study translation initiation and RNA helicase activity. We summarize here our work on this translation initiation factor from yeast.

## Introduction

Translation initiation depends for most eukaryotic mRNA's on the presence of a 5' cap structure. This structure is recognized by the cap-binding protein eIF-4E present in the cap-binding complex, eIF-4F (Grifo et al., 1983). The other proteins of this complex are a high molecular weight protein (p220) and eIF-4A (Edery et al., 1983). The cap-binding

---

[1] present address: Agouron Institute, La Jolla, California 92037
[2] To whom correspondence should be addressed: Tel +41 61 267 21 35,
FAX +41 61 267 21 18, Email linder@urz.unibas.ch

NATO ASI Series, Vol. H 71
Protein Synthesis and Targeting in Yeast
Edited by A. J. P. Brown, M. F. Tuite, and J. E. G. McCarthy

complex, together with free eIF-4A and eIF-4B are required for the successful binding of the 43S preinitiation complex to the mRNA. The translation initiation factor eIF-4A is a RNA dependent ATPase which can change the conformation of RNA molecules, as assayed by RNAase susceptibility experiments (Ray et al., 1985). Moreover, it has been shown that eIF-4A, together with eIF-4B, can unwind partially double stranded RNA molecules in an ATP dependent process (Rozen et al., 1990). This activity depends strictly on single stranded RNA tails and can be observed in both 5'->3' and 3' -> 5' directions. This strongly supports the translation initiation model in which potential secondary structures or RNA-RNA interactions in the 5' untranslated region have to be unwound by eIF-4A and eIF-4B to allow binding of the 43S preinitiation complex and subsequent scanning for the first AUG on the mRNA (for reviews see Moldave, 1985; Sonenberg, 1988).

In recent years the yeast *Saccharomyces cerevisiae* has been widely used for studying the translation initiation process (Linder, 1992; Linder & Prat, 1990; Tuite, 1989). It has become clear from the protein sequence conservation of different initiation factors and their functional analogy that translation in this single celled eukaryote is very similar to that of higher eukaryotes. We have isolated two genes from *S. cerevisiae*, *TIF1* and *TIF2*, encoding exactly the same protein (Linder & Slonimski, 1989). This protein is 65% identical to the two eIF-4A proteins from mouse (Nielsen et al., 1985; Nielsen & Trachsel, 1988). Gene replacement studies have proven that either *TIF* gene can be inactivated, but that inactivation of both genes is lethal to cell. *In vitro* translation experiments have shown that the *TIF* genes are required for translation. Altogether this clearly demonstrates that they code for the translation initiation factor 4A (Blum et al., 1989).

We have shown that eIF-4A is a member of a large family of proteins which has been termed the D-E-A-D box family according to one of the conserved sequence elements (Asp-Glu-Ala-Asp) present in all the proteins (Linder et al., 1989; Schmid & Linder, 1992). These proteins have several domains in common (Fig. 1; Schmid & Linder, 1992). The corresponding genes are found in organisms ranging from bacteria to human and in viruses (Koonin, 1991; Lain et al., 1990). They are involved in diverse cellular functions such as RNA splicing, ribosome assembly, initiation of translation, spermatogenesis, embryogenesis, and cell growth and division. According to the *in vitro* RNA helicase activities of eIF-4A (together with eIF-4B) and p68, RNA helicase activity has been attributed to all these DEAD-box proteins.

Two of the domains present in the proteins of this family have previously been described in ATPases (Walker et al., 1982). The ATPase A (A--G-GKT) domain of most of the proteins of this family have an alanine instead of a glycine present in other ATPases. The sequence L-D-E-A-D present in this family seems to be a special form of the ATPase B motif, since many other ATPases have only Asp-Glu sequence at the corresponding position (Gorbalenya et al., 1989; Linder et al., 1989; Walker et al., 1982). From the combined

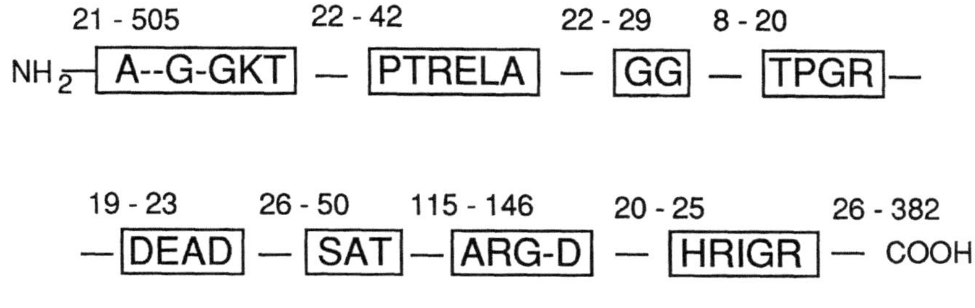


Figure 1: Conserved amino acids in the D-E-A-D family. The members considered in the establishment of this consensus sequence have been described in (Schmid & Linder, 1992),the eIF-4A protein from *S. pombe* (A. F. and P.L., unpublished) and the Ssl2 protein (Gulyas & Donahue, 1992). The motifs A--G-GKT and DEAD represent the ATPase A and ATPase B motifs, respectively.

analysis of X-ray diffraction and nuclear magnetic resonance of adenylate cyclase, it has been suggested that the aspartic acid residue of the ATPase B motif interacts with the Mg-ATP (Fry et al., 1986). No function has so far been determined for the other conserved domains.

More recently, several proteins have been found to form a subfamily (Wassarman & Steitz, 1991). The corresponding genes have been defined by a genetic analysis of splicing in the yeast *Saccharomyces cerevisiae*. The proteins of this subfamily have a D-E-A-H instead of the D-E-A-D motif and differ in the H-R-I-G-R sequence. Some of them have been shown to be RNA dependent ATPases, but so far have not been shown to possess a RNA helicase activity (Kim et al., 1992; Schwer & Guthrie, 1991).

We have made an extensive mutational analysis of the *TIF1* and *TIF2* genes (Schmid & Linder, 1991) using random chemical mutagenesis and site-directed mutagenesis of residues conserved in the family. We could show that most of the mutations abolish *in vivo* function of the protein. Only conserved substitutions retained partial or full activity, as judged from *in vivo* analysis. A temperature sensitive mutation isolated in this study also confirmed the requirement of the *TIF* gene for translation *in vivo*.

Purification of some of the mutant proteins from an *E. coli* expression system allowed us to look more closely at the defects of the mutants (Blum et al., 1992). From this study it is clear that mutations in the ATPase A motif which reduce the rate of ATP hydrolysis activitiy also diminish RNA helicase and translation activity.

## Results and Discussion

Mutagenesis of some of the conserved domains

The comparison of the different proteins from the D-E-A-D protein family has revealed eight conserved domains (Linder et al., 1989; Schmid & Linder, 1992). We have mutated some of these domains by site directed mutagenesis of the *TIF1* or *TIF2* genes (which encode exactly the same protein) of *Saccharomyces cerevisiae* (Schmid & Linder, 1991). Since the eIF-4A protein is essential for translation we used a condititonal system to analyze the phenotypes of the mutant genes. In this conditional system all copies of the *TIF1* and the *TIF2* genes in the genome of a diploid cell were inactivated by gene replacement and the only functional *TIF* gene was under control of a galactose inducible promoter (Blum et al., 1989). We have shown that cells which depend on this conditional system stop growth on glucose medium after a few generations. Thus if a mutant *tif1* gene is introduced in such a background only genes producing active or partially active proteins allow the cells to grow on glucose containing medium (Fig. 2).

The results of this analysis are summarized in Table 1. It shows that the alanine residue (position 66 in the eIF-4A protein from *S. cerevisiae*) in the ATPase A (A--G-GKT) domain can be replaced by glycine without a major effect. This is somewhat surprising as most proteins of the D-E-A-D proteins have an alanine at this position, whereas most other ATPases have a glycine at this position. Replacement of the alanine by an aspartic acid clearly reduces growth and its replacement by a valine is lethal to the cell. Two mutations were also made in the ATPase B (L-D-E-A-D) domain. The substitution of the aspartic acid at position 170 by a glutamic acid abolishes the function of the eIF-4A protein. This is not surprising since, so far, all the proteins of the family contain an aspartic acid at this position as do other ATPases not belonging to the D-E-A-D protein family (Gorbalenya et al., 1989; Walker et al., 1982). It is interesting, however, that a substitution of the aspartic acid at position 173 by a histidine, as is found in the D-E-A-H subfamily of splicing proteins (Schmid & Linder, 1992; Wassarman & Steitz, 1991), also completely abolishes eIF-4A function. This clearly shows that the D-E-A-H proteins form a subfamily. Two mutations affecting two conserved consecutive glycine residues (positions 126 and 127, see also Fig. 1) between the ATPase A and B domains have been analyzed. One affects protein function leading to a lethal phenotype, whereas the other leads to a leaky phenotype. The third domain analyzed is the HRIGR domain in the C-terminal region of the eIF-4A protein. This region is characterized by the presence of many positively charged residues, which lead us to mutagenize the arginine at position 347 of the eIF-4A protein. All mutations made at this position, except the replacement by a lysine, abolish completely the function of the protein, as judged from the *in vivo* assay. Surprisingly, this is also true for the replacement of the arginine by a threonine, which is again found in another member of the family (Prp5, Dalbadie-McFarland & Abelson, 1990).

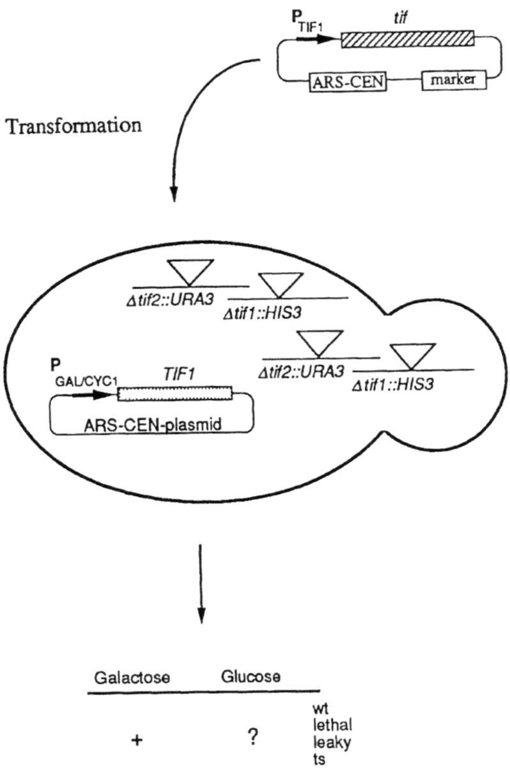


Figure 2: The conditional system. In this system the two copies of *TIF1* and the two copies of *TIF2* were replaced by the marker genes *HIS3* and *URA3*, respectively (Schmid & Linder, 1991). The strain carries the *TIF1* gene under the control of a *CYC1-GAL10* promoter on a plasmid (Blum et al., 1989). Also indicated is a plasmid coding for a mutated *tif* gene. Depending on the mutant plasmid tested, variations of this system were used.

| Position | Name | amino acid change | D-E-A-D | Phenotype | Dominance |
|---|---|---|---|---|---|
| 1 | M1I | Met->Ile | - | - | n.d. |
| 66 | A66D | Ala->Asp | Ala | +/- | - |
| | A66G | Ala->Gly | Ala | + | - |
| | A66V | Ala->Val | Ala | - | d |
| 79 | A79V | Ala->Val | Ala, Pro, Ser | ts | n.d. |
| 126 | G126D | Gly->Asp | Gly | - | d |
| 127 | G127D | Gly->Asp | Gly | +/- | - |
| 145 | G145D | Gly->Asp | - | - | d |
| 145 | G145S | Gly->Ser | Gly, Thr, Ala | +/- | - |
| 170 | D170E | Asp->Glu | Asp | - | d |
| 173 | D173H | Asp->His | Asp | - | d |
| 311 | S311F | Ser->Phe | - | +/- | n.d. |
| 347 | R347G | Arg->Gly | Arg | - | D |
| | R347E | Arg->Glu | Arg | - | D |
| | R347I | Arg->Ile | Arg | - | D |
| | R347K | Arg->Lys | Arg | + | - |
| | R347S | Arg->Ser | Arg | - | D |
| | R347T | Arg->Thr | Arg | - | D |

Table 1: Mutations in the yeast eIF-4A gene. The mutations within the boxes were obtained by site directed mutagenesis. "Position" refers to the amino acid position in the yeast eIF-4A protein. "Name" reflects position and amino acid change. The column "D-E-A-D" indicates the amino acid which is present in other members of the family of putative RNA helicases. A "+" in the column "Phenotype" indicates growth on glucose medium in the system shown in Fig. 2, "+/-" indicates reduced growth and "-" indicates absence of growth, which reflects non-functionality of the eIF-4A protein. If the mutant gene cloned on a multicopy plasmid shows a negative dominant effect it is marked "D" (strong) or "d" (weak) in the column "Dominance" (n.d. = not determined).

Suppressors of the A79V mutation

One goal in the analysis of the mutants is also the isolation of suppressor mutations to learn more about the eIF-4A protein itself and its partners in translation initiation. In the hunt for mutants in the *TIF* genes we have also isolated a temperature sensitive mutant. This mutant yeast eIF-4A protein carries an alanine to valine substitution at position 79 close to the ATPase A motif. Interestingly, the same position is mutated in the *prp16-1* allele (Burgess et al., 1990). We have previously shown by *in vivo* labelling that protein synthesis stops very quickly in this mutant if the cells are shifted to nonpermissive temperature, suggesting that the elevated temperature causes a change in protein conformation (Schmid & Linder, 1991). From the strain SS13-3A (*MATa ade2 his3 leu2 trp1 ura3 Δtif1::HIS3 Δtif2::ADE2*), carrying a plasmid with the temperature sensitive *tif1* allele, we isolated many independent suppressors by selection for growth at non permissive temperature. The candidate suppressors were further analyzed by isolation of the plasmid and retransformation in a conditional system, to determine whether the suppressor mutations mapped on the plasmid or in the genome. All the suppressors so far analyzed are of the intragenic type, mapping at different positions of the *tif1* gene. Despite the fact that all the mutants were independent isolates, five out of the eight analyzed mutations map at position 200 of the eIF-4A protein, changing the leucine residue to a phenylalanine residue. A phenylalanine is found at this position in 17 out of 28 D-E-A-D proteins. The other suppressor mutations are at positions 95 (M->I), 208 (V->F) and 233 (G->C). In 14 out of 28 D-E-A-D proteins an isoleucine is present at position 95 (11 have a valine) whereas the two other mutations (positions 208 and 233) do not affect conserved positions. Thus it is striking that many of the suppressor mutations change an amino acid of the yeast eIF-4A protein towards a residue which is represented within a majority of D-E-A-D proteins. These mutations may restore the conformation of the protein. A larger sample of suppressor mutations must be analyzed to strengthen this observation. Having the information of more suppressor mutations and structural data from other ATPases it will be possible to perform a molecular modelling of the eIF-4A protein. This in turn will provoke predictions for the interaction with ATP and RNA, which can be tested experimentally.

Multiple copies of inactive eIF-4A can be lethal

Although the exact effect of the mutations on the different mutant proteins is not known, it is possible that multiple copies of the mutant *tif* genes either show enough activity to allow survival or, in contrast, show a dominant mutant phenotype. To test this, we cloned all mutant *tif* genes on a multicopy plasmid (YEplac181, Gietz & Sugino, 1988). These plasmids were introduced into a strain in which the *TIF2* gene is replaced by a cassette carrying the *URA3* gene and the *TIF1* gene under control of the hybrid *CYC1-GAL* promoter. Since the expression of the *TIF1* gene is several fold lower than that of the *TIF2* gene (Prat et al., 1990) this strain produces less Tif protein on glucose containing medium than a normal strain but can be induced for Tif production on galactose containing medium.

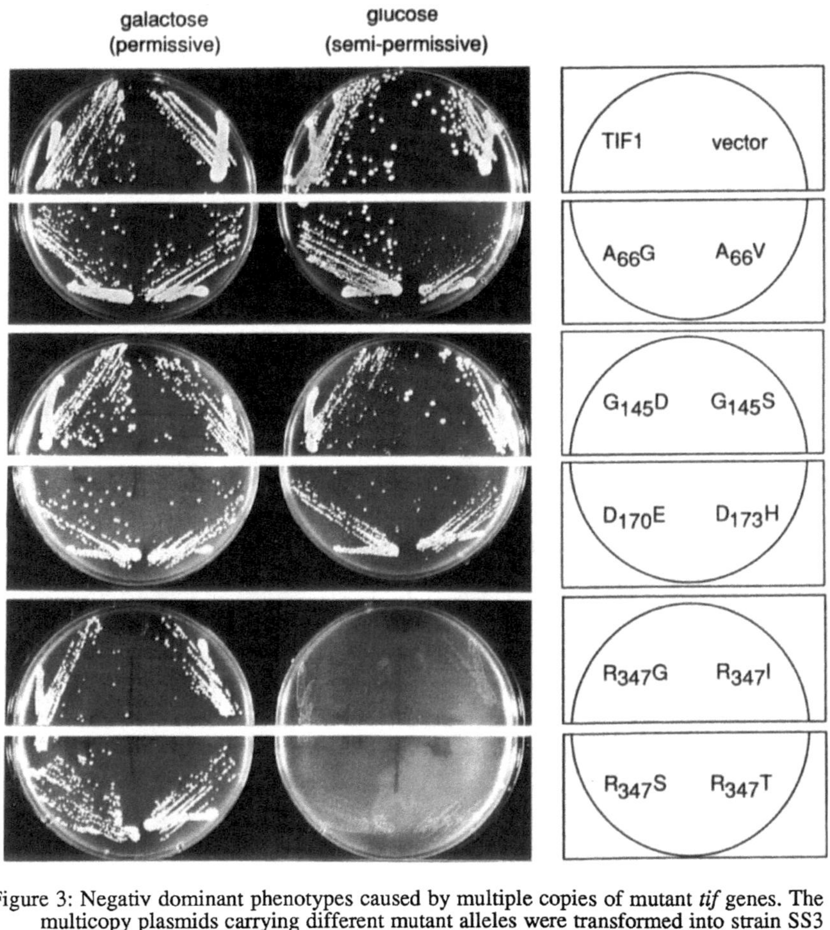


Figure 3: Negativ dominant phenotypes caused by multiple copies of mutant *tif* genes. The multicopy plasmids carrying different mutant alleles were transformed into strain SS3 (*MATa ade2 his3 leu2 trp1 ura3 tif2::URA3-GAL-TIF1*) and the resulting transformants were streaked on galactose (left) and glucose containing medium (middle). The plasmids carrying the different tif alleles are indicated in the schematic presentation on the right.

Transformation of this strain with the multicopy plasmid bearing the mutant *tif* genes showed that for some mutants growth was severely reduced on glucose medium compared to the cells transformed with the vector alone or the plasmid carrying the wild-type *TIF1* gene. The reduction in growth, however, was not observed on galactose containing medium. The dominant phenotype is not the same for all the mutants, ranging from no effect to almost complete growth inhibition (Fig. 3). The most pronounced effect was observed with mutations in the HRIGR motif. A likely explanation for this phenotype is that overexpression of the mutant Tif proteins leads to the formation of an inactive complex with some other translation factor and that mutant Tif proteins titrate out a vital function. We are using this system to search for this factor, which most likely is an essential partner of eIF-4A in the cell.

The eIF-4A protein from *Schizosaccharomyces pombe*

We have previously reported the expression of the mouse eIF-4AI gene in *S. cerevisiae* (Prat et al., 1990). Despite the high conservation between the eIF-4A proteins from *S. cerevisiae* and mouse no complementation could be observed using a temperature sensitive *TIF1* gene (A79V). In order to learn more about the conservations required to obtain a functional eIF-4A in *S. cerevisiae*, we isolated the corresponding gene from the fission yeast *Schizosaccharomyces pombe*. (A. Fischli and P. Linder, in preparation). Using sequence

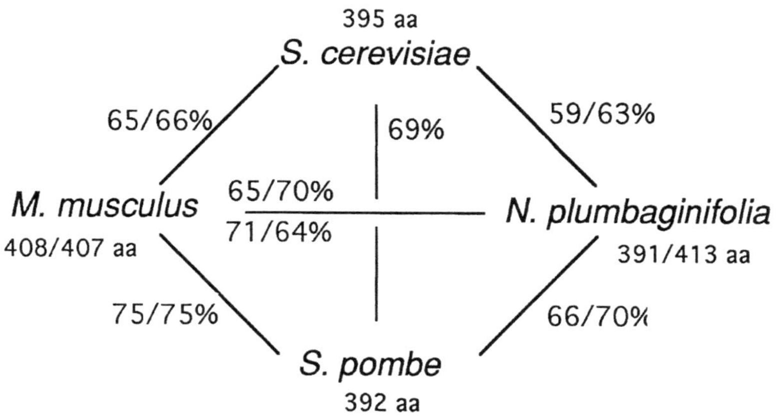


Figure 4: Conservation of eIF-4A proteins from different organisms. The length of the proteins in amino acids and the identity of the proteins are indicated. In the case of the proteins from *Nicotiana pumbaginifolia* and *Mus musculus* both copies have been considered. See text for more information.

information from the two eIF-4A genes from mouse (Nielsen et al., 1985; Nielsen & Trachsel, 1988), the two genes from the plant *Nicotiana plumbaginifolia* (Owttrim et al., 1991), and an unpublished sequence from *Campoletis sonorensis* (B. Webb, S. Dib-Hajj, and M. Summers, personal communication, see in Schmid & Linder, 1992) we designed two oligonucleotides specific for the eIF-4A subgroup of D-E-A-D proteins. Using these oligonucleotides we PCR amplified *S. pombe* genomic DNA. The resulting fragment was sequenced to confirm that it encodes part of an eIF-4A like gene and used to screen a *S. pombe* cDNA library (obtained from F. Lacroute). The protein sequence obtained from two positive clones showed that the *S. pombe* eIF-4A protein is 392 amino acids long. It is 69% identical to the yeast *S. cerevisiae* eIF-4A and 75% identical to the mouse eIF-4A proteins (Fig. 4). It appears therefore that the *S. pombe* eIF-4A protein shows a higher degree of similarity to the mammalian protein than to the protein from budding yeast. Although we have no functional proof that the cDNA we have isolated codes for a *bona fide* translation initiation factor, the high homology with the known eIF-4A genes is a strong argument for this conclusion. Further experiments must be performed to show whether this gene can functionally replace *TIF1* and *TIF2* in *S. cerevisiae*. To our knowledge, this is the first gene encoding a translation initiation factor isolated from *S. pombe*.

In conclusion, we have performed a detailed analysis of the yeast translation initiation factor eIF-4A by extensive mutational analysis of the *S. cerevisiae* genes encoding this factor. The *in vivo* and *in vitro* characterization of the mutants will certainly provide greater insight into the complex system of translation initiation and the RNA helicase activity of the eIF-4A/eIF-4B complex. We have also isolated suppressors of the temperature sensitive *tif* allele. Overexpression of some of the mutant proteins leads to negative dominant effect which can be used to isolate gene dosage dependent suppressors. Analysis of these suppressors will lead to the characterization of other known or unknown factors which interact with eIF-4A.

**Acknowledgements**

We would like to thank George Owttrim for helpful comments on the manuscript. We are most grateful to François Lacroute for making the *S. pombe* cDNA library available to us. This work was supported by grants from the Swiss National Science Foundation.

**References**

Blum S, Mueller M, Schmid SR, Linder P & Trachsel H (1989) Translation in *Saccharomyces cerevisiae*: initiation factor 4A-dependent cell-free system. Proc. Natl. Acad. Sci. USA 86: 6043-6046

Blum S, Schmid SR, Pause A, Buser P, Linder P, Sonenberg N & Trachsel H (1992) ATP hydrolysis by initiation factor 4A is required for translation initiation in *Saccharomyces cerevisiae*. Proc. Natl. Acad. Sci. USA: in press

Burgess S, Couto JR & Guthrie C (1990) A putative ATP binding protein influences the fidelity of branchpoint recognition in yeast splicing. Cell 60: 705-717

Dalbadie-McFarland G & Abelson J (1990) PRP5: a helicase-like protein required for mRNA splicing in yeast. Proc. Natl. Acad. Sci. USA 87: 4236-4240

Edery I, Humbelin M, Darveau A, Lee KAW, Milburn S, Hershey JWB, Trachsel H & Sonenberg N (1983) Involvement of eukaryotic initiation factor 4A in the cap recognition process. J. Biol. Chem. 258: 11398-11403

Fry DC, Kuby SA & Mildvan AS (1986) ATP-binding site of adenylate kinase: mechanistic implications of its homology with *ras*-encoded p21, F1-ATPase, and other nucleotide-binding proteins. Proc. Natl. Acad. Sci. USA 83: 907-911

Gietz RD & Sugino A (1988) New yeast-*Escherichia coli* shuttle vectors constructed with in vitro mutagenized yeast genes lacking six-base pair restriction sites. Gene 74: 527-534

Gorbalenya AE, Koonin EV, Donchenko AP & Blinov VM (1989) Two related superfamilies of putative helicases involved in replication, recombination, repair and expression of DNA and RNA genomes. Nucleic Acids Res. 17: 4713-4730

Grifo JA, Tahara SM, Morgan MA, Shatkin AJ & Merrick WC (1983) New initiation factor activity required for globin mRNA translation. J. Biol. Chem. 258: 5804-5810

Gulyas KD & Donahue TF (1992) *SSL2*, a suppressor of a stem-loop mutation in the *HIS4* leader encodes a yeast homologue of human *ERCC-3*. Cell

Kim S-H, Smith J, Claude A & Lin R-J (1992) The purified yeast pre-MRNA splicing factor PRP2 is an RNA-dependent NTPase. EMBO-J. 11: 2319-2326

Koonin EV (1991) Similarities in RNA helicases. Nature 352: 290

Lain S, Riechmann JL & Garcia JA (1990) RNA helicase: a novel activity associated with a protein encoded by a positive strand RNA virus. Nucleic Acids Res. 18: 7003-7006

Linder P (1992) Molecular biology of translation in yeast. Antonie v. Leeuwenhoeck: in press

Linder P, Lasko PF, Ashburner M, Leroy P, Nielsen PJ, Nishi K, Schnier J & Slonimski PP (1989) Birth of the D-E-A-D box. Nature 337: 121-122

Linder P & Prat A (1990) Baker's yeast, the new work horse in protein synthesis studies: analyzing eukaryotic translation initiation. BioEssays 12: 519-526

Linder P & Slonimski PP (1989) An essential yeast protein, encoded by duplicated genes *TIF1* and *TIF2* and homologous to the mammalian translation initiation factor eIF-4A, can suppress a mitochondrial missense mutation. Proc. Natl. Acad. Sci. USA 86: 2286-2290

Moldave K (1985) Eukaryotic protein synthesis. Ann. Rev. Biochem. 54: 1109-1149

Nielsen PJ, McMaster GK & Trachsel H (1985) Cloning of eukaryotic protein synthesis initiation factor genes: isolation and characterization of cDNA clones encoding factor eIF-4A. Nucleic Acids Res. 13: 6867-6880

Nielsen PJ & Trachsel H (1988) The mouse protein synthesis initiation factor 4A gene family includes two related functional genes which are differentially expressed. EMBO J. 7: 2097-2105

Owttrim GW, Hofmann S & Kuhlemeier C (1991) Divergent genes for translation initiation factor eIF-4A are coordinately expressed in tobacco. Nucleic Acids. Res. 19: 5491-5496

Prat A, Schmid SR, Buser P, Blum S, Trachsel H, Nielsen PJ & Linder P (1990) Expression of translation initiation factor 4A from yeast and mouse in *Saccharomyces cerevisiae*. Biochim. Biophys. Acta 1050: 140-145

Ray BK, Lawson TG, Kramer JC, Cladaras MH, Grifo JA, Abramson RD, Merrick WC & Thach RE (1985) ATP-dependent unwinding of messenger RNA structure by eukaryotic initiation factors. J. Biol. Chem. 260: 7651-7658
Rozen F, Edery I, Meerovitch K, Dever TE, Merrick WC & Sonenberg N (1990) Bidirectional RNA helicase activity of eucaryotic translation initiation factors 4A and 4F. Mol. Cell. Biol. 10: 1134-1144
Schmid SR & Linder P (1991) Translation initiation factor 4A from *Saccharomyces cerevisiae*: Analysis of residues conserved in the D-E-A-D family of RNA helicases. Mol. Cell. Biol. 11: 3463-3471
Schmid SR & Linder P (1992) D-E-A-D protein family of putative RNA helicases. Mol. Microbiol. 6: 283-292
Schwer B & Guthrie C (1991) PRP16 is an RNA-dependent ATPase that interacts transiently with the spliceosome. Nature 349: 494-499
Sonenberg N (1988) Cap-binding proteins of eukaryotic messenger RNA: functions in initiation and control of translation. Prog. Nucleic Acid Res. Mol. Biol. 35: 173-207
Tuite MF (1989) Protein synthesis. In: Rose AH & Harrison JS (Eds) The yeasts (pp 161-204) Academic Press, N. Y.
Walker JE, Saraste M, Runswick MJ & Gay NJ (1982) Distantly related sequences in the alpha- and beta-subunits of ATP synthase, myosin, kinases and other ATP-requiring enzymes and a common nucleotide binding fold. EMBO J. 1: 945-951
Wassarman DA & Steitz JA (1991) Alive with DEAD proteins. Nature 349: 463-464

# UNCHARGED tRNA AND DEREPRESSION OF THE GENERAL AMINO ACID CONTROL: AUTOREGULATION OF YEAST LYSYL-tRNA SYNTHETASE

Stefan Lanker[1], Janet L. Bushman[2], Alan G. Hinnebusch[2], Hans Trachsel[1] and Peter P. Mueller[1]

[1]Institute of Biochemistry and Molecular Biology
University of Berne, Switzerland
[2]National Institute of Child Health and Human Development
National Institutes of Health, Bethesda, MD 20892

## 1. SUMMARY

The yeast *S. cerevisiae* is recognized as an ideal eukaryotic model organism to study basic biological processes due the possibility to combine biochemical methods, recombinant DNA technology and classical genetics and the fact that many biochemical pathways including translation are very similar in lower and higher eukaryotes. One of the best characterized examples of translational control in eukaryotes is the regulation of *GCN4* mRNA translation which is mediated by phosphorylation of eukaryotic initiation factor 2$\alpha$ (eIF-2$\alpha$) (Dever et al., 1992). Expression of GCN4, a transcriptional activator of more than 30 amino acid biosynthetic genes, is modulated according to the availability of amino acids and the amount of uncharged tRNAs. In this chapter, we present an autoregulatory model in which aminoacyl-tRNA synthetases can be autoregulated by making their transcription dependent on GCN4 protein and by signaling translational derepression of *GCN4* through the cognate uncharged tRNAs when aminoacylation activity is reduced (Lanker et al., 1992).

In the following sections we describe the cloning of the *GCD5* gene of *S. cerevisiae* and show that it is identical to *KRS1* encoding lysyl-tRNA synthetase (LysRS). The mutation *gcd5-1* changes a conserved residue in the putative lysine binding domain of LysRS. This leads to a defect in lysine binding and, consequently, to reduced charging of $tRNA^{Lys}$. Mutant *gcd5-1* cells compen-

NATO ASI Series, Vol. H 71
Protein Synthesis and Targeting in Yeast
Edited by A. J. P. Brown, M. F. Tuite, and J. E. G. McCarthy

sate for the defect in LysRS by increasing *GCN4* expression at the translational level. GCN4 protein in turn stimulates transcription of *GCD5*, leading to increased LysRS activity (Lanker et al., 1992).

## 2. INTRODUCTION

Starvation of the yeast *Saccharomyces cerevisiae* for a single amino acid leads to increased synthesis of GCN4, the transcriptional activator of a number of unlinked genes encoding enzymes involved in amino acid biosynthesis (general amino acid control). *GCN4* expression is controlled at the translational level by four short upstream open reading frames (uORFs) in the mRNA leader which reduce translation when amino acids are abundant but allow increased translation of *GCN4* in response to amino acid starvation (for reviews, see Hinnebusch, 1988; Hinnebusch, 1990; Hinnebusch & Klausner, 1991).
A recently proposed model (Abastado et al., 1991) suggests that under nonstarvation conditions, ribosomes scanning from the mRNA cap initiate at the first uORF encountered (uORF1) and following termination, resume scanning and reinitiate translation at one of the downstream uORFs 2, 3 or 4. After translating one of these uORFs, ribosomes are released from the mRNA; therefore, only a small fraction of the ribosomes initiate translation of the *GCN4* open reading frame. Under conditions of amino acid starvation, ribosomes which resume scanning after translating uORF1 become competent for reinitiation more slowly than under nonstarvation conditions. Therefore, a significant fraction of these ribosomes bypass uORFs 2-4 and reinitiate at the *GCN4* open reading frame instead. It was recently demonstrated (Dever et al., 1992) that the protein kinase GCN2, a positive regulator of *GCN4* expression (Wek et al., 1989), phosphorylates translation initation factor 2 (eIF-2) on its $\alpha$ subunit under amino acid starvation conditions. Phosphorylation of mammalian eIF-2$\alpha$ has long been known to inhibit protein synthesis at the initiation step (reviewed in Moldave, 1985; London et al., 1987). Dever et al. proposed that because of reduced amounts of active eIF-2, reinitiation of translation is less efficient under starvation conditions, allowing more ribosomes to scan past upstream ORFs 2-4 and initiate at the *GCN4* start codon. This model is supported by the fact that mutations in serine-51 of eIF-2$\alpha$, the site of phosphorylation on eIF-2$\alpha$ in mammalian cells (Colthurst et al., 1987), impairs translational derepression of *GCN4* expression (Dever et al., 1992).

It is thought that accumulation of uncharged tRNA in amino acid-starved cells stimulates GCN2 to phosphorylate eIF-2$\alpha$ because the *ils1-1* mutation affecting isoleucyl-tRNA synthetase leads to derepression of the general control system in amino acid-complete medium (Delforge et al., 1975). The GCN2 protein contains a large domain related to histidyl-tRNA synthetase (HisRS) adjacent to its protein kinase moiety, and point mutations in this HisRS-like region impair the derepression of *GCN4* translation in response to amino acid starvation (Wek et al., 1989). These observations led to the suggestion that binding of uncharged tRNA to the HisRS-like region increases the activity of the protein kinase domain of GCN2. Presumably, the HisRS-related region of GCN2 has diverged sufficiently from authentic HisRS that it would bind other uncharged tRNAs besides $tRNA^{His}$ (Wek et al., 1989).

Mutants lacking GCN2 kinase function cannot efficiently translate *GCN4* mRNA under amino acid starvation conditions (Gcn$^-$ phenotype, general control non-derepressible). By contrast, mutations in multiple *GCD* genes lead to high-level translation of *GCN4* mRNA in the absence of amino acid starvation (Gcd$^-$ phenotype, general control constitutively derepressed), the same phenotype associated with reduced eIF-2 activity. Most *gcd* mutations are pleiotropic, leading to slow growth or temperature-sensitive growth under nonstarvation conditions besides derepressing *GCN4* expression (Hinnebusch, 1990). This dual phenotype suggests that GCD proteins have an essential function in addition to the regulation of *GCN4* expression. Recent evidence indicates that GCD1 (Tzamarias et al., 1989; Cigan et al., 1991) and GCD2 (Foiani et al, 1991) are required for general translation initiation, perhaps as components of the eukaryotic initiation factor 2B (eIF-2B) that catalyzes GDP-GTP exchange on eIF-2 and thereby stimulates new rounds of translation initiation.

## 3. RESULTS AND DISCUSSION

### a) Cloning of the *GCD5* gene

The *gcd5-1* mutation was isolated on the basis of causing increased resistance to the tryptophan analogue 5-fluorotryptophan (5-FT), a phenotype indicative of constitutive derepression of enzymes subject to general amino acid control. The mutation also leads to slow growth (Slg$^-$) on nutrient-rich or minimal media (Niederberger et al., 1986). We isolated the *GCD5* gene from a yeast genomic library on plasmid pJB12 by its ability to complement the Slg$^-$ and 5-

FT-resistance phenotypes of the *gcd5-1* mutation (Lanker et al., 1992). The *GCD5* gene was localized on the genomic insert of pJB12 and part of it was sequenced. A database search with the resulting sequences revealed 100% identity with the open reading frame or flanking sequences of the *KRS1* gene, encoding the yeast lysyl-tRNA synthetase (LysRS; Mirande and Waller, 1988).

**b) The *gcd5-1* mutation leads to reduced lysyl-tRNA charging activity and lowers the affinity of LysRS for lysine**

Having identified *GCD5* as *KRS1*, we wished to determine whether the *gcd5-1* mutation affects LysRS activity, and whether altered charging activity could result from impaired binding of any one of the three substrates: ATP, lysine or $tRNA^{Lys}$. To address these questions, we performed *in vitro* lysyl-tRNA charging assays. Extracts prepared from wild-type and *gcd5-1* strains were assayed under conditions of limiting extract for incorporation of radioactively labeled lysine into a trichloroacetic acid (TCA)-precipitable species. The *gcd5-1* mutant extract showed a two-fold reduction in the activity of $^{3}H$-lysine incorporation into tRNA compared to wild-type (Figure 1A). We also examined the level of total aminoacyl-tRNA synthetase activities by adding a radioactively labeled mixture of all amino acids to extracts of wild-type and mutant strains (Figure 1B). The overall synthetase activity in the *gcd5-1* mutant was 2.5-fold higher than in the wild-type strain. This may be an indication that many different aminoacyl-tRNA synthetase genes are subject to the general control, as shown for isoleucyl-tRNA synthetase (Meussdoerffer & Fink; 1983). Consequently, the amount of LysRS activity relative to total tRNA-aminoacylation activity in the *gcd5-1* mutant was only approximately 20% of that seen in the wild-type strain (Figure 1C).

Next we conducted *in vitro* lysyl-tRNA charging assays with varying concentrations of each substrate. The results show that the mutant synthetase had fairly similar affinities for ATP (Figure 2A) and tRNA (Figure 2B), showing a 2.5-fold higher $k_M$ for ATP and a two-fold lower $k_M$ for tRNA compared to wild-type. However, the affinity of the mutant synthetase for lysine was more dramatically altered (Figure 2C), with a $k_M$ 9-fold higher than wild-type. These results suggest that the major defect of the *gcd5-1* mutant LysRS is in binding lysine.

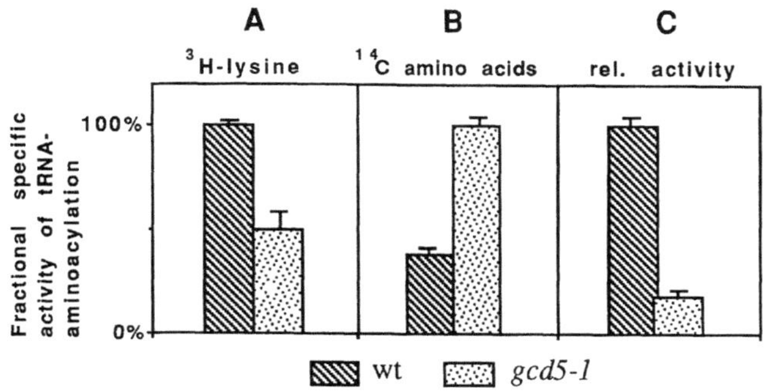


**Fig. 1.** LysRS activity in *gcd5-1* mutant and wild-type extracts.
Yeast extracts of strains T86A (*gcd5-1*) and GRF18 (wt) were prepared and tRNA synthetase activities were measured as described (Lanker et al., 1992). Each reaction mixture (50 µL) contained 3-4 µg protein and 1 µCi $^{3}$H-labeled lysine (**A**) or 1 µCi $^{14}$C-labeled total amino acids (**B**). Specific activities were calculated as TCA-precipitable cpm (charged tRNA) formed per min per mg protein. **A**) Lysyl-tRNA charging activity assayed using [$^{3}$H]-lysine ; the value of the wild-type strain was set to 100%. **B**) Total-tRNA charging activity assayed using a mixture of [$^{14}$C]-labeled amino acids ; the value of the *gcd5-1* mutant strain was set to 100%. **C**) Relative lysyl-tRNA charging activity, calculated as lysyl-tRNA charging activity divided by total-tRNA charging activity.

**c) Excess lysine can reduce the slow growth phenotype of *gcd5-1***

Because the mutant synthetase showed a decreased affinity for lysine, we addressed the possibility that addition of exogenous lysine would diminish the slow growth phenotype of the *gcd5-1* mutant. A *gcd5-1* strain was transformed with plasmid pJB12 containing *GCD5* or with vector alone, and the resulting transformants were grown on minimal essential agar medium and either excess lysine (3 mg/ml; Figure 2E) or, as a control, excess arginine (2 mg/ml; Figure 2D). The lysine supplement appeared to shorten the generation time of a *gcd5-1* mutant strain, increasing the size of colonies formed from single cells after four days of incubation relative to that seen on plates supplemented with arginine. By contrast, *GCD5*-transformants formed equally large colonies on both media. These results were confirmed by measuring the generation times of both strains in liquid minimal medium in the presence or absence of excess lysine.

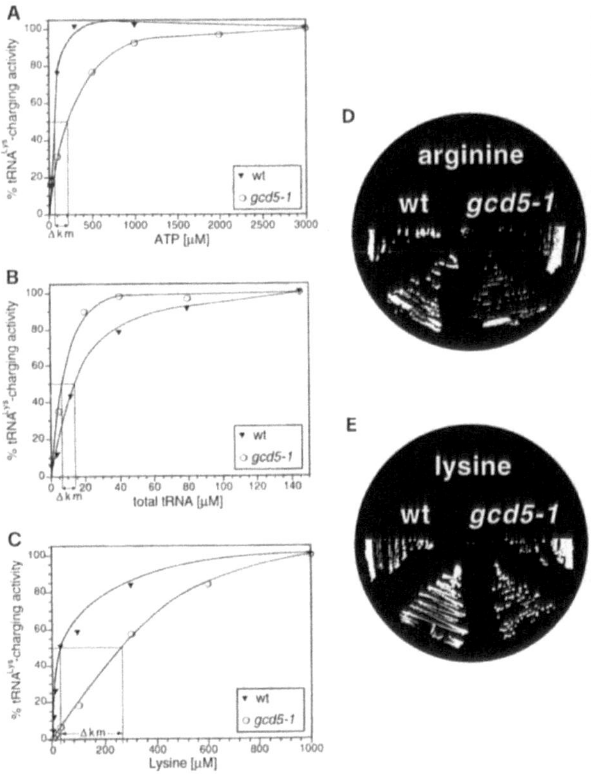


**Fig. 2.** Substrate saturation curves for LysRS activity and effect of lysine on the growth rate. Experiments were performed as described in Fig. 1. In each experiment, the highest values of lysyl-tRNA charging activity for wild-type (▼) and *gcd5-1*(○) mutant strains, respectively, were set to 100%. **A**) ATP dependence; **B**) total yeast tRNA dependence; and **C**) lysine dependence of lysyl-tRNA formation, respectively. $\Delta k_M$: difference of $k_M$ values for wild-type and mutant strains. Panels **D**) and **E**): The *gcd5-1* mutant strain T86A was transformed with pJB12 bearing *GCD5* (wt) or vector alone (*gcd5-1*). Two transformants of each type were grown for 4 days at 30°C on SD plates containing either arginine (2 mg/ml; **D**) or lysine (3 mg/ml; **E**).

## d) The *gcd5-1* allele has a single point mutation in the putative lysine binding domain of LysRS

To investigate the molecular basis of the lysine-binding defect of the mutant enzyme, we decided to sequence the *gcd5-1* allele. A single point mutation was detected in the entire sequence of the mutant 1.94 kb fragment (Figure 3A): a G --> T transversion at position 1463 that changes arginine 488 to leucine. This mutation in LysRS lies in the ß-strand A3 within the proposed amino acid binding domain of class II aminoacyl-tRNA synthetases (Cusack et al., 1991) and affects a residue conserved between *E.coli* and *S. cerevisiae* (Figure 3B). Taken together, our data support the predicition that the ß-strand A3 in class II synthetases forms part of the amino acid binding domain.

**e) *gcd5-1* leads to derepression of *GCN4* expression**

The *gcd5-1* mutation leads to derepression of the *TRP3* gene, expression of which is regulated by the general amino acid control (Niederberger et al., 1986). Because GCN4 is the proximal transcriptional activator in this system, we investigated the effect of *gcd5-1* on *GCN4* expression. A plasmid containing a *GCN4-lacZ* fusion (Hinnebusch, 1984) was introduced into *gcd5-1* mutant and wild-type strains and the transformants were characterized for the levels of ß-galactosidase activity expressed under different conditions. In minimal medium (Figure 4A, unsupplemented), *GCN4-lacZ* expression was more than 12-fold higher in the *gcd5-1* strain than in the wild-type strain. Upon starvation of the strains for histidine by addition of 3-aminotriazole, an inhibitor of the *HIS3* enzyme (Figure 4A, histidine-starved), *GCN4-lacZ* expression was derepressed approximately 9-fold in the wild-type strain, whereas expression in the *gcd5-1* mutant was slightly reduced to about wild-type derepressed levels. The addition of excess lysine to minimal medium, shown above to increase the growth rate of the *gcd5-1* strain, reduced *GCN4-lacZ* expression in the *gcd5-1* mutant by a factor of six (Figure 4A, lysine-supplemented), whereas lysine addition had little or no effect in the wild-type strain. These results show that the *gcd5-1* mutation leads to an increase in *GCN4* expression that can be overcome by addition of excess lysine. The simplest explanation for these findings is that, in the absence of exogenous lysine, the *gcd5-1* mutation leads to a deficiency in charging of $tRNA^{Lys}$ that signals the derepression of *GCN4* .

To verify that the derepression of *GCN4* expression seen in the *gcd5-1* mutant is mediated at the translational level by the uORFs in the *GCN4* mRNA leader (Mueller & Hinnebusch, 1986), we introduced a *GCN4*$^{c}$*-lacZ* fusion lacking all four uORFs into *gcd5-1* mutant and wild-type strains and analyzed the resulting transformants for expression of ß-galactosidase activity. In both strains, *GCN4*$^{c}$*-lacZ* expression was very high (ca. 2000 units) and showed little or no additional derepression upon histidine starvation (data not shown). This result was expected for the *GCD5* strain because the uORFs are required for repression of *GCN4* translation under non-starvation conditions and for its derepression in response to histidine starvation (Mueller & Hinnebusch, 1986). The fact that *gcd5-1* increases expression of ß-galactosidase activity 12-fold from the construct containing the uORFs and only 1.6-fold from the construct lacking uORFs is a strong indication that the mutation derepresses *GCN4* at the translational level (Mueller & Hinnebusch, 1986). Further support for this conclusion was provided by measuring the steady state levels of *GCN4* and

A

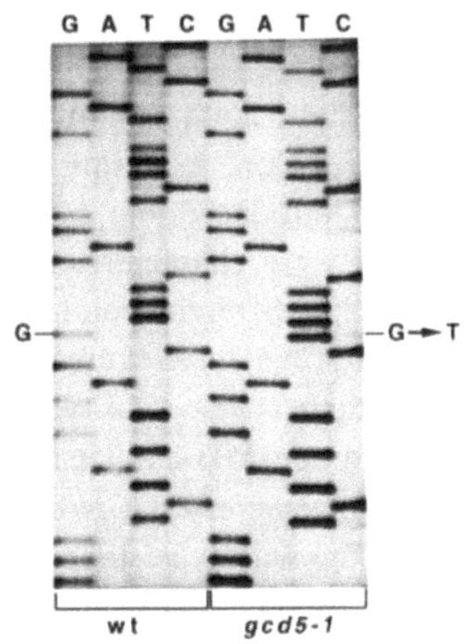


**Fig. 3.** Amplification and sequencing of the mutant *gcd5-1* allele.
Genomic DNA segments from GRF18 (wt) and T86A (*gcd5-1* ) were amplified with the polymerase chain reaction as described (Lanker et al., 1992). **A**) Autoradiogram showing wild-type (wt) and mutant (*gcd5-1*) sequences in the vicinity of the single point mutation G1463 --> T detected in the *gcd5-1* allele. **B**) Alignment of a portion of the sequences of *Saccharomyces cerevisiae* (SC) and *Escherichia coli* (EC) LysRS in the putative lysine binding domain. Amino acids are in italics, conserved residues in bold letters.

B

| | | | | | | | | | | | | | |
|---|---|---|---|---|---|---|---|---|---|---|---|---|---|
| gcd5-1 | | | | | | *Leu* | | | | | | | |
| | | | | | | CTT | | | | | | | |
| SC | 1450 | GGT | CTA | TGT | GAG | **CGT** | **TTC** | **GAG** | GTC | **TTT** | GTA | GCT | ACA |
| | 484 | *Gly* | *Leu* | *Cys* | *Glu* | ***Arg*** | ***Phe*** | ***Glu*** | *Val* | ***Phe*** | *Val* | *Ala* | *Thr* |
| EC | 409 | *Glu* | *Ile* | *Thr* | *Asp* | ***Arg*** | ***Phe*** | ***Glu*** | *Phe* | ***Phe*** | *Ile* | *Gly* | *Gly* |
| | 1486 | AAG | **GAA** | **ATT** | TGT | **AAT** | GCC | TAC | ACT | **GAA** | **TTG** | **AAC** | **GAT** |
| SC | 496 | *Lys* | ***Glu*** | ***Ile*** | *Cys* | ***Asn*** | *Ala* | *Tyr* | *Thr* | ***Glu*** | ***Leu*** | ***Asn*** | ***Asp*** |
| EC | 421 | *Arg* | ***Glu*** | ***Ile*** | *Gly* | ***Asn*** | *Gly* | *Phe* | *Ser* | ***Glu*** | ***Leu*** | ***Asn*** | ***Asp*** |

*GCN4-lacZ* mRNAs in the *gcd5-1* mutant and wild-type strains. The results showed that the levels of these mRNAs differed between the two strains by less than a factor of 2 (data not shown).

### f) *GCD5-lacZ* expression is derepressed in *gcd5-1* strains

Since *GCD5* has a consensus GCN4 binding site (TGACTC-box) in its 5' non-coding sequence (Mirande & Waller, 1988), we reasoned that the elevated *GCN4* expression seen in the *gcd5-1* mutant might lead to an increase in *GCD5* transcription and KRS enzyme levels. To analyze the effects of *gcd5-1* on *GCD5* expression, we constructed a *GCD5-lacZ* fusion and introduced it into *gcd5-1* mutant and wild-type strains. RNA blot hybridization analysis showed that the steady-state level of *GCD5* mRNA was increased substantially (4.4-fold) in the *gcd5-1* mutant relative to wild-type when both strains were grown on minimal medium (compare lanes 1 and 3, Figure 4B). A smaller, but significant (2-fold), increase in *GCD5-lacZ* mRNA was also observed in the mutant (Figure 4B, lane 3). Addition of excess lysine reduced the *GCD5-* and *GCD5-lacZ* transcript levels in the *gcd5-1* mutant strain (lane 4), whereas the *GCD5* mRNA level in the wild-type strain was affected very little by lysine (lane 2).

Measurements of *GCD5-lacZ* fusion enzyme activity in the two strains (Figure 4C) gave results very similar to those just discussed for the *GCD5* transcript (Figure 4B), showing 3.9-fold higher expression in the *gcd5-1* mutant versus the wild-type strain in minimal medium, and a two-fold repression of enzyme activity in the *gcd5-1* mutant upon addition of exogenous lysine. By contrast, in the wild-type strain, GCD5-lacZ enzyme activity was nearly constant in the presence or absence of lysine (Figure 4C).

### g) *gcd5-1 gcn* mutants can be rescued by overproducing GCN4

The *gcd5-1* mutation was previously found to be lethal in combination with *gcn1, gcn2, gcn3* or *gcn4* mutations (Niederberger et al., 1986), each of which impairs derepression of enzymes subject to the general control (Hinnebusch, 1990). The fact that *GCD5* encodes LysRS and that this gene is essential (Martinez et al., 1991), combined with the observation that *gcd5-1* leads to increased transcription of *GCD5,* suggests a model to explain the inviability of *gcd5-1 gcn* double mutants. We propose that derepression of the *gcd5-1* allele is necessary for viability, and that this derepression is dependent on the elevated levels of GCN4 protein that occur under starvation conditions and in response to mutations affecting aminoacyl-tRNA synthetases, such as *gcd5-1*. The experiment shown in Figure 4D confirms one of the tenets of our model, that *GCD5* expression is regulated by GCN4. We compared expression of the *GCD5-lacZ* fusion in strains lacking GCN4 function (*gcn4*) or containing the *GCN4*$^{c}$ allele that expresses high-levels of GCN4 protein constitutively due to the absence of all four uORFs in the *GCN4* mRNA leader (Mueller & Hinnebusch, 1986). The results show that *GCD5-lacZ* expression is substantially higher in the *GCN4*$^{c}$ strain compared to the *gcn4* mutant, whether cells are grown in a medium containing excess lysine, excess arginine (which leads indirectly to lysine limitation (Delforge et al., 1975)), or both amino acids. These data indicate that GCN4 is a positive regulator of *GCD5* transcription.

A second test of our hypothesis was conducted by determining whether expression of GCN4 from a constitutive allele can suppress the lethality of *gcn*

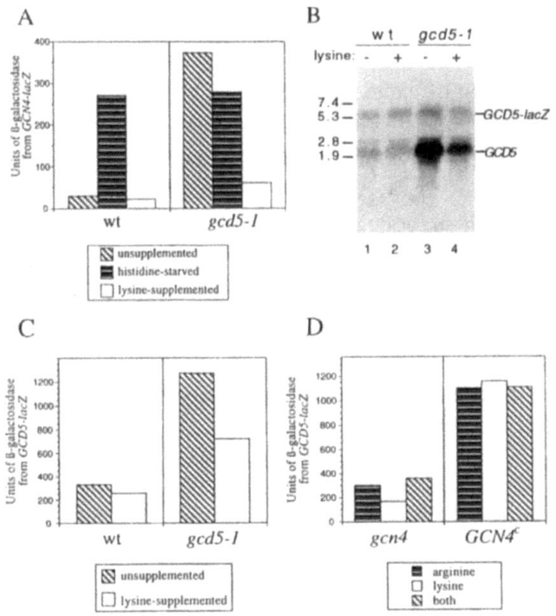


**Fig. 4.** Analysis of *GCN4* expression in *gcd5-1* mutant and wild-type strains and regulation of *GCD5* expression by GCN4. *GCN4-lacZ* (**A**) or *GCD5-lacZ* (**B, C, D**) fusion constructs were introduced into wild-type (wt) and *gcd5-1* mutant strains, respectively (panels **A, B, C**), a *gcn4* mutant (panel **D**, *gcn4* ), or a strain with high-level constitutive expression of *GCN4* (panel **D**, *GCN4*$^{C}$ ).
**A**) Steady-state levels of ß-galactosidase activity expressed from *GCN4-lacZ* in *gcd5-1* or wild-type strains were determined from transformants grown under normal conditions (unsupplemented), histidine-starvation conditions (histidine-starved) or in the presence of excess exogenous lysine (lysine-supplemented).
**B**) Steady-state levels of *GCD5* and *GCD5-lacZ* mRNAs. Strains were grown in minimal medium containing only the required nutrients (lanes 1 and 3) or in medium supplemented with excess lysine (3 mg/ml; lanes 2 and 4).
**C**) Steady-state levels of ß-galactosidase activity expressed from *GCD5-lacZ* in *gcd5-1* mutant or wild-type strains grown in minimal medium (unsupplemented) or minimal medium supplemented with excess lysine (lysine-supplemented).
**D**) Steady-state levels of ß-galactosidase activity expressed from *GCD5-lacZ* in *gcn4* and *GCN4*$^{C}$ strains grown with excess lysine (lysine), excess argine (arginine) or excess lysine and arginine (both). Enzyme activities are expressed in nanomoles ortho-nitrophenyl-ß-D-galactopyranoside cleaved per min per mg of protein.

*gcd5-1* double mutants since constitutive GCN4 alleles express GCN4 independent of GCN1, GCN2 and GCN3 (Thireos et al., 1984; Hinnebusch, 1985; Mueller & Hinnebusch, 1986). To test this, we used genetic crosses to construct *gcn2 gcd5-1* and *gcn3 gcd5-1* mutants bearing plasmid from which GCN4 was expressed at high constitutive levels. When grown for many generations on nonselective medium, none of the *gcn2 gcd5-1* or *gcn3 gcd5-1* cells did loose the plasmid, indicating that GCN4 overexpression from the allele on the plasmid is necessary for survival of these double mutants. These results strongly support the idea that *gcd5-1 gcn* double mutants are lethal because GCN4-mediated derepression of *GCD5* transcription is required for expression of gcd5-1 mutant enzyme activity at a level sufficient for life.

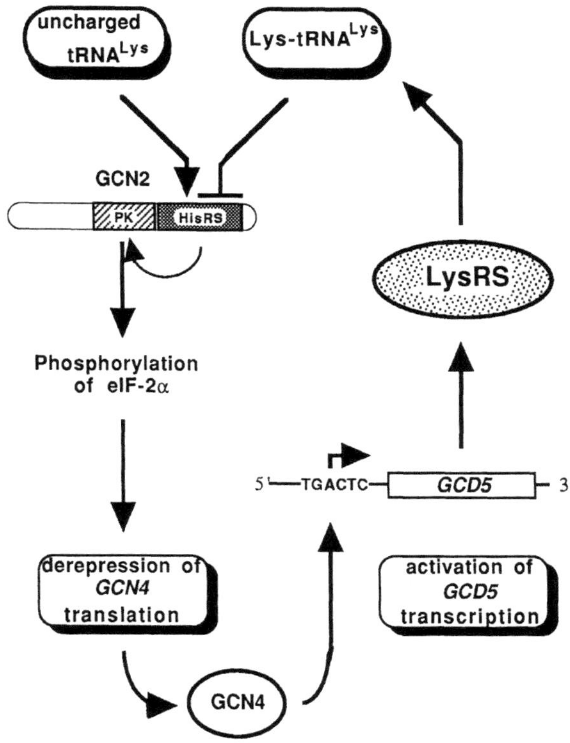


**Fig. 5.** Autoregulatory model of GCD5 expression, involving translational regulation of *GCN4* and transcriptional activation of *GCD5* (Lanker et al., 1992).
Uncharged tRNA$^{Lys}$ accumulation is sensed by the histidyl-tRNA synthetase-like domain (HisRS) of the GCN2 protein kinase, leading to activation of the kinase domain (PK). Activated GCN2 phosphorylates eIF-2α, and phosphorylated eIF-2α causes derepression of *GCN4* translation (Dever et al., 1992). GCN4 protein is a transcriptional activator that binds to the TGACTC DNA sequence present upstream of *GCD5*. This stimulates transcription of *GCD5*, encoding LysRS (LysRS). The increased amount of LysRS actvity leads to a higher level of lysyl-tRNA$^{Lys}$ and a concomitant decrease in uncharged tRNA$^{Lys}$. This eliminates further activation of GCN2 kinase function and both eIF-2α phosphorylation and translation of *GCN4* mRNA return to basal levels.

In summary, we propose the following autoregulatory model for *GCD5* expression to account for our results (Figure 5): i) Under conditions of insufficient charging of tRNA$^{Lys}$ by LysRS, GCN2 kinase is activated through its HisRS-related domain and phosphorylates the α subunit of eIF-2 . This could arise from limiting amounts of lysine or from reduced LysRS activity. ii)

Phosphorylation of eIF-2α leads to increased GCN4 protein synthesis (Dever et al., 1992). iii) GCN4 binds upstream of *GCD5* and stimulates *GCD5* transcription. The additional LysRS thus produced increases the level of charged $tRNA^{Lys}$. Consequently, GCN2 kinase is no longer activated, phosphorylation of eIF-2 returns to basal levels (presumably as the result of dephosphorylation by a protein phosphatase) and translation of *GCN4* is once again repressed. As expected, we found that addition of excess lysine, which should reduce the amount of uncharged $tRNA^{Lys}$, decreased the extent of *GCN4* derepression.

An important corollary of our results is that the GCN2 protein kinase mediates increased translation of *GCN4* in response to increased levels of uncharged $tRNA^{Lys}$. It is known that GCN2 is required for derepression of *GCN4* expression in histidine-starved (Hinnebusch, 1984) and tryptophan-starved cells (Williams et al., 1989). In addition, *gcn2* mutants are sensitive to analogues of various other amino acids (Wolfner et al., 1975). Taken together, these observations support the idea that the HisRS-related domain of GCN2 interacts with multiple tRNA species in addition to $tRNA^{His}$ (Wek et al., 1989).

Our finding that total tRNA-charging activity was elevated in the *gcd5-1* mutant (Figure 2B) may be an indication that many different aminoacyl-tRNA synthetase genes are subject to GCN4 control, as shown for isoleucyl-tRNA synthetase (Meussdoerffer & Fink; 1983), and thus autoregulate their expression in the same fashion described here for LysRS.

## 4. ACKNOWLEDGMENTS

We thank Peter Niederberger and Ralf Huetter for providing us with a *gcd5-1* mutant. This work was supported by grant 31-25565.88 to H.T. and P. P. M. from the Swiss National Foundation.

## 5. REFERENCES

Abastado, J. P., Miller, P. F., Jackson, B. M., and Hinnebusch, A. G. (1991). Suppression of ribosomal reinitation at upstream open reading frames in amino acid-starved cells forms the basis for *GCN4* translational control. **Mol. Cell. Biol.** *11*, 486-496.

Cigan, A. M., Foiani, M., Hannig, E. M., and Hinnebusch, A. G. (1991). Complex formation by positive and negative translational regulators of *GCN4*. **Mol. Cell. Biol.** *11*, 3217-3228.

Colthurst, D. R., Campbell, D. G., and Proud, C. G. (1987). Structure and regulation of eukaryotic initiation factor eIF-2: sequence of the site in the α subunit phosphorylated by the haem-controlled repressor and by the double-stranded RNA-activated inhibitor. **Eur. J. Biochem.** *166*, 357-363.

Cusack, S., Härtlein, M., and Leberman, R. (1991). Sequence, structural and evolutionary relationships between class-2 aminoacyl-transfer RNA synthetases. **Nucleic Acids Res.** *19,* 3489-3498.

Delforge, J., Messenguy, F., and Wiame, J. M. (1975). The regulation of arginine biosynthesis in *Saccharomyces cerevisiae.* **Eur. J. Biochem.** *57,* 231-239.

Dever, T., Feng, L., Wek, R. C., Cigan, A. M., Donahue, T. F., and Hinnebusch, A. G. (1992). Phosphorylation of initiation factor 2α by protein kinase GCN2 mediates gene-specific translational control of *GCN4* in yeast. **Cell** *68,* 585-596.

Foiani, M., Cigan, A. M., Paddon, C. J., Harashima, S., and Hinnebusch, A. G. (1991). GCD2, a translational repressor of the *GCN4* gene, has a general function in the initiation of protein synthesis. **Mol. Cell. Biol.** *11,* 3203-3216.

Hinnebusch, A. G. (1984). Evidence for translational regulation of the activator of general amino acid control in yeast. **Proc. Natl. Acad. Sci. U. S. A.** *81,* 6442-6446.

Hinnebusch, A. G. (1985). A hierarchy of *trans*-acting factors modulate translation of an activator of amino acid biosynthetic genes in yeast. **Mol. Cell. Biol.** *5,* 2349-2360.

Hinnebusch, A. G. (1988). Mechanism of gene regulation in the general control of amino acid biosynthesis in *Saccharomyces cerevisiae.* **Microbiol. Reviews** *52,* 248-273.

Hinnebusch, A. G. (1990). Involvement of an initiation factor and protein phosphorylation in translational control of *GCN4* mRNA. **Trends Biochem. Sci.** *15,* 148-152.

Hinnebusch, A. G., and Klausner, R. D. (1991). Examples of eukaryotic translational control: GCN4 and Ferritin. In: Trachsel, H. (Editor): **Translation in Eukaryotes**. CRC Press, Boca Raton, Florida, 243-272.

Lanker, S., Bushman, J. L., Hinnebusch, A. G., Trachsel, H., and Mueller, P. P. (1992). Autoregulation of the yeast lysyl-tRNA synthetase gene *GCD5/KRS1* by translational and transcriptional control mechanisms. **Cell** *in press.*

London, I. M., Levin, D. H., Matts, R. L., Thomas, N. S. B., Petryshyn, R., and Chen, J. J. (1987). Regulation of Protein Synthesis. In: Boyer, P. D., Krebs, E. G. (Editors): **The enzymes**. New York, Academic Press, 359-380.

Martinez, R., Latreille, M. T., and Mirande, M. (1991). A PMR2 tandem repeat with a modified C-terminus is located downstream from the *KRS1* gene encoding lysyl-tRNA synthetase in *Saccharomyces cerevisiae.* **Mol. Gen. Genet.** *227,* 149-154.

Meussdoerffer, F., and Fink, G. R. (1983). Structure and expression of two aminoacyl-tRNA synthetase genes from *Saccharomyces cerevisiae.* **J. Biol. Chem.** *258,* 6293-6299.

Mirande, M., and Waller, J. P. (1988). The yeast lysyl-tRNA synthetase gene. **J. Biol. Chem.** *263,* 18443-18451.

Moldave, K. (1985). Eukaryotic protein synthesis. **Annu. Rev. Biochem.** *54,* 1109-1149.

Mueller, P. P., and Hinnebusch, A. G. (1986). Multiple upstream AUG codons mediate translational control of *GCN4.* **Cell** *45,* 201-207.

Niederberger, P., Aebi, M., and Huetter, R. (1986). Identification and characterization of four new *GCD* genes in *Saccharomyces cerevisiae.* **Curr. Genet.** *10,* 657-664.

Thireos, G., Driscoll Penn, M., and Greer, H. (1984). 5' untranslated sequences are required for the translational control of a yeast regulatory gene. **Proc. Natl. Acad. Sci. U.S.A.** *81,* 5096-5100.

Tzamarias, D., Roussou, I., and Thireos, G. (1989). Coupling of *GCN4* mRNA translational activation with decreased rates of polypeptide chain initiation. **Cell** *57,* 947-954.

Wek, R. C., Jackson, B. M., and Hinnebusch, A. G. (1989). Juxtaposition of domains homologous to protein kinases and histidyl-tRNA synthetases in GCN2 protein suggests a mechanism for coupling *GCN4* expression to amino acid availability. **Proc. Natl. Acad. Sci. U. S. A.** *86,* 4579-4583.

Williams, N. P., Hinnebusch, A. G., and Donahue, T. F. (1989). Mutations in the structural genes for eukaryotic initiation factors 2α and 2ß of *Saccharomyces cerevisiae* disrupt translational control of *GCN4* mRNA. **Proc. Natl. Acad. Sci. U. S. A.** *86,* 7515-7519.

Wolfner, M., Yep, D., Messenguy, F., and Fink, G. R. (1975). Integration of amino acid biosynthesis into the cell cycle of *Saccharomyces cerevisiae.* **J. Mol. Biol.** *96,* 273-290.

# POSITIVE CONTROL OF TRANSLATION IN ORGANELLAR GENETIC SYSTEMS

Thomas D. Fox and Zonghou Shen*
Section of Genetics and Development
Cornell University
Ithaca, NY 14853-2703
U.S.A.

The expression of genes in the mitochondria and chloroplasts of eucaryotic cells is, in many cases, modulated in response to environmental factors. In the case of yeast mitochondria, gene expression is lowered in the absence of oxygen (Woodrow and Schatz 1979) and/or in the presence of glucose (Falcone et al. 1983). In the case of chloroplasts, gene expression is dependent on the presence of light, in most cases (Mullet 1988). In both of these organellar systems, available genetic and physiological evidence indicates that the expression of many genes is dependent upon mRNA-specific translational activators, and that these activators may play a role in modulating gene expression.

## mRNA-specific activators of yeast mitochondrial translation

Among collections of *Saccharomyces cerevisiae* nuclear mutants exhibiting the Pet$^-$ (non-respiratory) phenotype, a number of strains with defects in the expression of specific mitochondrial genes have been identified (reviewed in Costanzo and Fox 1990). In many cases, these mutants have recessive defects in the translation of single mitochondrially encoded mRNAs. To date, such mRNA-specific positive

* Department of Biology
Wuhan University
Wuhan, People's Republic of China

NATO ASI Series, Vol. H 71
Protein Synthesis and Targeting in Yeast
Edited by A. J. P. Brown, M. F. Tuite, and J. E. G. McCarthy

activators of translation have been identified for five of the eight major mitochondrially coded mRNAs: these are the mRNAs encoding the three largest subunits of cytochrome oxidase, *COX1* (Decoster et al. 1990), *COX2* (Ebner et al. 1973; Cabral and Schatz 1978; Poutre and Fox 1987), *COX3* (Ebner et al. 1973; Cabral and Schatz 1978; Costanzo and Fox 1986; Costanzo et al. 1986; Kloeckener-Gruissem et al. 1987; Fox et al. 1988; Kloeckener-Gruissem et al. 1988; Costanzo et al. 1989), the *COB* mRNA, encoding cytochrome *b* (Dieckmann and Tzagoloff 1985; Rödel et al. 1985; Rödel 1986; Rödel et al. 1986; Rödel and Fox 1987; Michaelis et al. 1988) and the *OLI1* mRNA encoding ATPase subunit 9 (Ackerman et al. 1991; Payne et al. 1991). It seems likely that the translation of the remaining three major mitochondrial mRNAs is similarly controlled.

The mechanism(s) by which these activators work is not understood in detail. However several features are clear. First, in those cases studied [the *COX2*-specific activator PET111 (Strick and Fox 1987), the *COX3*-specific activators PET54, PET122 and PET494 (Costanzo and Fox 1986; Ohmen et al. 1988; Costanzo et al. 1989), and the *COB*-specific activators CBS1 and CBS2 (Michaelis and Rödel 1990; Körte et al. 1991)] the protein products of the nuclear genes are located in mitochondria and thus presumably act directly. Interestingly, available evidence indicates that all of these proteins are are least partially associated with the inner mitochondrial membrane (Michaelis et al. 1991; T.W. McMullin, C.A. Strick & T.D.F., unpublished) suggesting that they may play a role in localizing translation to sites on the membrane. Second, translational activation of the *COX2* (Poutre and Fox 1987; J.J. Mulero & T.D.F., unpublished), *COX3* (Costanzo and Fox 1988) and *COB* (Rödel 1986; Rödel and Fox 1987) mRNAs occurs via sites in their 5'-untranslated leaders whose positions relative to the initiation codon can be changed, at least for *COX3* and *COB*.

Finally, at least one of these translational activators interacts functionally with the small subunit of mitochondrial ribosomes. This was demonstrated for the *COX3*-specific activator PET122 by analyzing allele-specific suppressors of *pet122* mutations that truncated the C-terminal domain of PET122 (Haffter et al. 1990). Suppressor mutations were identified in nuclear genes encoding three different mitochondrial small subunit ribosomal proteins (McMullin et al. 1990; Haffter et al. 1991; Haffter and Fox 1992). [A fourth suppressor gene, encoding a large protein of unidentified function was also found (Haffter and Fox 1992).] Taken together these findings suggest a picture in which mRNA-specific translational activators, bound to the inner mitochondrial membrane, mediate the interaction between mRNAs and the mitochondrial ribosomal small subunit.

**Possible translational modulation of yeast mitochondrial gene expression in response to environmental factors**

The synthesis of mitochondrial gene products is generally studied in cells poisoned with the drug cycloheximide, which does not inhibit mitochondrial translation. While this procedure is not highly quantitative, it appears to provide a reasonably accurate view of mitochondrial protein synthesis. Frontali and co-workers (Falcone et al. 1983) used labeling in the presence of cycloheximide to follow the recovery of mitochondrial gene expression in yeast cells recovering from glucose repression during a period of six hours. While they found that the synthesis of all mitochondrial gene products increased, the degree of induction varied widely. They concluded that these effects were probably not due to a single global regulator of mitochondrial gene expression.

In a subsequent study (Zennaro et al. 1985) the same group examined the steady-state levels of mitochondrially coded mRNAs during release from glucose repression. During the three hour time course examined the level of the *COX2* mRNA increased several fold while the level of the *COX3* mRNA remained constant.

While these two studies are not directly comparable, taken together they argue that the synthesis of the cytochrome oxidase subunit III protein is probably modulated translationally. Although the data concerning *COX2* mRNA levels might indicate that this gene is controlled transcriptionally, they are also consistent with translational control in view of the fact that *COX2* mRNA levels are known to be reduced in *pet111* mutants as a secondary consequence of the block in translation (Poutre and Fox 1987).

The synthesis of mitochondrial proteins has also been examined in anaerobically growing cells by labeling in the presence of cycloheximide

(Woodrow and Schatz 1979). In these experiments, labeling rates were reduced, relative to aerobically grown cells, by 19 to 50 fold. In contrast, the steady-state level of the *COX3* mRNA in anaerobically grown cells was only two to three fold lower than that of aerobically grown cells (Marykwas 1991).

## Do mRNA-specific translational activators modulate yeast mitochondrial gene expression?

If mRNA-specific translational activators play a role in modulating gene expression, then their activity in mitochondria must be rate limiting for translation, at least under some conditions. To date, no direct data are available on this point, since accurate measurements of mitochondrial translation rates are very difficult. To circumvent these problems we are currently attempting to insert reporter genes into mtDNA (Fox et al. 1991), a task made potentially difficult by anomalies in the yeast mitochondrial genetic code (Fox 1987).

If the level of a single mitochondrial gene product, for example coxIII, were limiting for respiratory growth, then one might might ask whether its translational activators were rate limiting by examining the growth phenotype of strains in which the gene dosage of the activators had been altered. While coxIII is apparently not rate limiting for respiratory growth in wild-type cells, it can be made artificially limiting by changing its AUG initiation codon to AUA (Folley and Fox 1991). This mitochondrial initiation codon mutation causes a roughly seven fold decrease in coxIII synthesis and thereby results in a "leaky" non-respiratory growth phenotype. Interestingly, diploid cells that carry these mutant mitochondria, and only a single functional copy of either the *PET122* or *PET494* genes in their nuclei, grow substantially slower on respiratory media than the corresponding diploid strains with a single functional *PET54* gene or with homozygous wild-type nuclei (Folley and Fox 1991). In contrast, the growth of diploids carrying wild-type mitochondria is unaffected by heterozygous null mutations in these nuclear genes, presumably because coxIII levels already exceed levels limiting for growth. These results can be interpreted to suggest that the levels of PET122 and PET494 protein in mitochondria are limiting, or nearly so, for translation of the *COX3* mRNA. However, only when coxIII levels are artificially low and limiting for growth, as a result of the initiation codon mutation, do changes in the levels of these two activators produce growth phenotypes.

We have previously reported that *PET494* expression, monitored with a *pet494::lacZ* fusion, is modulated by oxygen and carbon source in a manner consistent with observed cytochrome oxidase levels (Marykwas and Fox 1989). (Interestingly, oxygen regulation of *PET494* expression is at the level of translation!) Similar experiments in which expression of a *pet54::lacZ* fusion was monitored in a respiring diploid strain revealed that while *PET54* expression is approximately ten fold greater that *PET494* expression, *PET54* responds in parallel to environmental conditions over an approximately 20 fold range (Table 1). The level of this regulation remains to be determined.

Table 1. Expression of a *pet54::lacZ* fusion in a respiring diploid strain as a function of carbon source and oxygen availability.

| Carbon source | $O_2$ or $N_2$ | β-galactosidase specific activity |
|---|---|---|
| Glucose (2%) | $N_2$ | 0.14 |
| Glucose (2%) | $O_2$ | 0.6 |
| Galactose (2%) | $O_2$ | 1.2 |
| Ethanol (3%) +Glycerol (3%) | $O_2$ | 3.0 |

These observations are consistent with the idea that mRNA-specific activators could modulate mitochondrial gene expression. However it is difficult to rationalize, solely on regulatory grounds, the need for separate activators for each mRNA (particularly in the case of the three subunits of cytochrome oxidase which are presumably required in equal amounts). If this complex system confers a functional advantage, it may be in the localization of translation at sites on the membrane (see above) which might promote optimal assembly of enzyme complexes.

## Chloroplast translational control

While mRNA-specific translational activation may be difficult to rationalize, its occurrence does not appear to be restricted to yeast mitochondria. Studies on light regulation of chloroplast gene expression in several species clearly indicate the importance of translational control in this organelle as well (reviewed in Mullet 1988).

Genetic studies in *Chlamydomonas* have identified several nuclear genes that appear to function as activators for translation of specific chloroplast mRNAs (Jensen et al. 1986; Kuchka et al. 1988; Rochaix et al. 1989) and in at least one case, a nuclear mutation can be suppressed by a chloroplast mutation in the 5'-leader of the affected mRNA, *psbC* (Rochaix et al. 1989). Nuclear mutations in maize that might also affect chloroplast gene expression in a similar fashion have also been reported (Metz and Miles 1982).

Biochemical studies have identified proteins from whole cell extracts of *Chlamydomonas* that bind specifically to a stem-loop structure just upstream of the initiation codon of the chloroplast *psbA* mRNA (Danon and Mayfield 1991). While none of these proteins has been shown to be the product of a genetically identified regulatory gene, the binding activity is absent in dark grown cells as well as in a nuclear mutant blocked in *psbA* translation. Thus they appear to play an important role in regulation. The position of the stem-loop structure to which they bind suggests a mechanism similar to that proposed for examples of positive control of translation in bacteria (Altuvia et al. 1987; Wulczyn and Kahmann 1991).

## Acknowledgements

Work in the authors' laboratory is supported by a grant (GM29362) from the US National Institutes of Health.

## References

Ackerman SH, Gatti DL, Gellefors P, Douglas MG, Tzagoloff A (1991) ATP13, a nuclear gene of Saccharomyces cerevisiae essential for the expression of subunit 9 of the mitochondrial ATPase. FEBS Lett. 278:234-238

Altuvia S, Locker-Giladi H, Koby S, Ben-Nun O, Oppenheim AG (1987) RNase III stimulates the translation of the cIII gene of bacteriophage λ. Proc. Natl. Acad. Sci. USA 84:6511-6515

Cabral F, Schatz G (1978) Identification of cytochrome *c* oxidase subunits in nuclear yeast mutants lacking the functional enzyme. J. Biol. Chem. 253:4396-4401

Costanzo MC, Fox TD (1986) Product of Saccharomyces cerevisiae nuclear gene PET494 activates translation of a specific mitochondrial mRNA. Mol. Cell. Biol. 6:3694-3703

Costanzo MC, Fox TD (1988) Specific translational activation by nuclear gene products occurs in the 5' untranslated leader of a yeast mitochondrial mRNA. Proc. Natl. Acad. Sci. USA 85:2677-2681

Costanzo MC, Fox TD (1990) Control of mitochondrial gene expression in Saccharomyces cerevisiae. Annu. Rev. Genet. 24:91-113

Costanzo MC, Seaver EC, Fox TD (1986) At least two nuclear gene products are specifically required for translation of a single yeast mitochondrial mRNA. EMBO J. 5:3637-3641

Costanzo MC, Seaver EC, Fox TD (1989) The PET54 gene of Saccharomyces cerevisiae: Characterization of a nuclear gene encoding a mitochondrial translational activator and subcellular localization of its product. Genetics 122:297-305

Danon A, Mayfield SPY (1991) Light regulated translational activators: identification of chloroplast gene specific mRNA binding proteins. EMBO J. 10:3993-4001

Decoster E, Simon M, Hatat D, Faye G (1990) The MSS51 gene product is required for the translation of the COX1 mRNA in yeast mitochondria. Mol. Gen. Genet. 224:111-118

Dieckmann CL, Tzagoloff A (1985) Assembly of the mitochondrial membrane system: CBP6, a yeast nuclear gene necessary for the synthesis of cytochrome b. J. Biol. Chem. 260:1513-1520

Ebner E, Mason TL, Schatz G (1973) Mitochondrial assembly in respiration-deficient mutants of Saccharomyces cerevisiae. II. Effect of nuclear and extrachromosomal mutations on the formation of cytochrome c oxidase. J. Biol. Chem. 248:5369-5378

Falcone C, Agostinelli M, Frontali L (1983) Mitochondrial translation products during release from glucose repression in Saccharomyces cerevisiae. J. Bacteriol. 153:1125-1132

Folley LS, Fox TD (1991) Site-directed mutagenesis of a Saccharomyces cerevisiae mitochondrial translation initiation codon. Genetics 129:659-668

Fox TD (1987) Natural variation in the genetic code. Annu. Rev. Genet. 21:67-91

Fox TD, Costanzo MC, Strick CA, Marykwas DL, Seaver EC, Rosenthal JK (1988) Translational regulation of mitochondrial gene expression by nuclear genes of Saccharomyces cerevisiae. Phil. Trans. Royal Soc. Lond. B 319:97-105

Fox TD, Folley LS, Mulero JJ, McMullin TW, Thorsness PE, Hedin LO, Costanzo MC (1991) Analysis and manipulation of yeast mitochondrial genes. Methods Enzymol. 194:149-165

Haffter P, Fox TD (1992) Suppression of carboxy-terminal truncations of the yeast mitochondrial mRNA-specific translational activator PET122 by mutations in two new genes, MRP17 and PET127. Mol. Gen. Genet. in press

Haffter P, McMullin TW, Fox TD (1990) A genetic link between an mRNA-specific translational activator and the translation system in yeast mitochondria. Genetics 125:495-503

Haffter P, McMullin TW, Fox TD (1991) Functional interactions among two yeast mitochondrial ribosomal proteins and an mRNA-specific translational activator. Genetics 127:319-326

Jensen KH, Herrin DL, Plumley FG, Schmidt GW (1986) Biogenesis of photosystem II complexes: transcriptional, translational, and posttranslational regulation. J. Cell Biol. 103:1315-1325

Kloeckener-Gruissem B, McEwen JE, Poyton RO (1987) Nuclear functions required for cytochrome c oxidase biogenesis in Saccharomyces cerevisiae: multiple trans-acting nuclear genes exert specific effects on the expression of each of the cytochrome c oxidase subunits encoded on mitochondrial DNA. Curr. Genet. 12:311-322

Kloeckener-Gruissem B, McEwen JE, Poyton RO (1988) Identification of a third nuclear protein-coding gene required specifically for posttranscriptional expression of the mitochondrial COX3 gene in Saccharomyces cerevisiae. J. Bacteriol. 170:1399-1402

Körte A, Michaelis U, Lottspeich F, Rödel G (1991) Over-expression, purification and determination of the proteolytic processing site of the yeast mitochondrial CBS1 protein. Curr. Genet. 20:87-90

Kuchka MR, Mayfield SP, Rochaix J-D (1988) Nuclear mutations specifically affect the synthesis and/or degradation of chloroplast-encoded D2 polypeptide of photosystem II in Chlamydomonas reinhardtii. EMBO J. 7:319-324

Marykwas DL (1991) Environmental control of the yeast nuclear gene PET494, a specific activator of mitochondrial translation. Ph.D. Thesis, Cornell University

Marykwas DL, Fox TD (1989) Control of the Saccharomyces cerevisiae regulatory gene PET494: transcriptional repression by glucose and translational induction by oxygen. Mol. Cell. Biol. 9:484-491

McMullin TW, Haffter P, Fox TD (1990) A novel small subunit ribosomal protein of yeast mitochondria that interacts functionally with an mRNA-specific translational activator. Mol. Cell. Biol. 10:4590-4595

Metz JG, Miles D (1982) Use of a nuclear mutant of maize to identify components of photosystem II. Biochim. Biophys. Acta 681:95-102

Michaelis U, Körte A, Rödel G (1991) Association of cytochrome b translational activator proteins with the mitochondrial membrane: implications for cytochrome b expression in yeast. Mol. Gen. Genet. 230:177-185

Michaelis U, Rödel G (1990) Identification of CBS2 as a mitochondrial protein in Saccharomyces cerevisiae. Mol. Gen. Genet. 223:394-400

Michaelis U, Schlapp T, Rödel G (1988) Yeast nuclear gene CBS2, required for translational activation of cytochrome b, encodes a basic protein of 45 kDa. Mol. Gen. Genet. 214:263-270

Mullet JE (1988) Chloroplast development and gene expression. Annu. Rev. Plant Physiol. Plant Mol. Biol. 39:475-502

Ohmen JD, Kloeckener-Gruissem B, McEwen JE (1988) Molecular cloning and nucleotide sequence of the nuclear PET122 gene required for expression of the mitochondrial COX3 gene in S. cerevisiae. Nucleic Acids Res. 16:10783-10802

Payne MJ, Schweizer E, Lukins HB (1991) Properties of two nuclear pet mutants affecting expression of the mitochondrial oli1 gene of Saccharomyces cerevisiae. Curr. Genet. 19:343-351

Poutre CG, Fox TD (1987) PET111, a Saccharomyces cerevisiae nuclear gene required for translation of the mitochondrial mRNA encoding cytochrome c oxidase subunit II. Genetics 115:637-647

Rochaix J-D, Kuchka M, Mayfield S, Schirmer-Rahire M, Girard-Bascou J, Bennoun P (1989) Nuclear and chloroplast mutations affect the synthesis or stability of the chloroplast psbC gene product in Chlamydomonas reinhardtii. EMBO J. 8:1013-1021

Rödel G (1986) Two yeast nuclear genes, CBS1 and CBS2, are required for translation of mitochondrial transcripts bearing the 5'-untranslated COB leader. Curr. Genet. 11:41-45

Rödel G, Fox TD (1987) The yeast nuclear gene CBS1 is required for translation of mitochondrial mRNAs bearing the cob 5'-untranslated leader. Mol. Gen. Genet. 206:45-50

Rödel G, Körte A, Kaudewitz F (1985) Mitochondrial suppression of a yeast nuclear mutation which affects the translation of the mitochondrial apocytochrome b transcript. Curr. Genet. 9:641-648

Rödel G, Michaelis U, Forsbach V, Kreike J, Kaudewitz F (1986) Molecular cloning of the yeast nuclear genes CBS1 and CBS2. Curr. Genet. 11:47-53

Strick CA, Fox TD (1987) Saccharomyces cerevisiae positive regulatory gene PET111 encodes a mitochondrial protein that is translated from an mRNA with a long 5' leader. Mol. Cell. Biol. 7:2728-2734

Woodrow G, Schatz G (1979) The role of oxygen in the biosynthesis of cytochrome c oxidase of yeast mitochondria. J. Biol. Chem. 254:6088-6093

Wulczyn FG, Kahmann R (1991) Translational stimulation: RNA sequence and structure requirements for binding of com protein. Cell 65:259-269

Zennaro E, Grimaldi L, Baldacci G, Frontali L (1985) Mitochondrial transcription and processing of transcripts during release from glucose repression in "resting cells" of Saccharomyces cerevisiae. Eur. J. Biochem. 147:191-196

**TRANSLATION IN YEAST MITOCHONDRIA**

L.A. Grivell, P.J.T. Dekker and H.J. Pel
Section for Molecular Biology
Department of Molecular Cell Biology
University of Amsterdam
Kruislaan 318
1098 SM Amsterdam
The Netherlands

## Introduction

Mitochondria make a small number of their own proteins. They do this employing a translational apparatus that is clearly unusual (see Benne and Sloof, 1987 for review). A different genetic code; a low number of tRNAs, many of them unconventional; diminutive rRNAs; a lack of a typical 5S rRNA; mRNAs with shrunken, or no leader sequences (mammals); mRNAs with long leader and trailer sequences (yeast) and a large number of what appear to be mRNA-specific translation factors (yeast) are just some of the hallmarks that distinguish this machinery from typical pro- or eukaryotic cell sap systems. These features have both livened up the study of mitochondrial translation in the past few years and have contributed to our understanding of such diverse biological questions as decoding mechanisms, the definition of the minimal requirements for rRNA and tRNA function and the evolution of ribosomes. This knowledge has been acquired despite the undoubted disadvantage that no one has yet managed to develop an mRNA-dependent translation system from mitochondria. There are two reasons for this fortunate situation:

1. Much has been learnt from comparisons of mitochondrially-coded translational components (mostly rRNAs and tRNAs) in various organisms. These reveal a remarkable and informative diversity that is perhaps best explained in terms of the low demands the cell puts on these components. Generally speaking, mitochondrial translational systems produce a limited number of proteins at relatively constant rate and consequently may be able to evolve more rapidly than systems subject to complex translational controls and producing a large number of proteins with maximal efficiency (Borst, 1981).

2. Mitochondrial translation is very amenable to genetic analysis. This is especially so in a facultative anaerobe, such as *Saccharomyces cerevisiae*, which when growing on fermentable carbon sources, can dispense with respiratory function, thus allowing the ready selection of mutations affecting various steps in mitochondrial

NATO ASI Series, Vol. H 71
Protein Synthesis and Targeting in Yeast
Edited by A. J. P. Brown, M. F. Tuite, and J. E. G. McCarthy

gene expression. Analysis is further facilitated by the fact that, in contrast to other genetic systems, genes for rRNAs and tRNAs each occur in only one copy per mitochondrial genome. Thus, mutations conferring antibiotic resistance, or heat/cold sensitivity to mitochondrial protein synthesis have turned out to lie within the rRNA genes and identify rRNA domains involved in the peptidyl transferase centre and in translational fidelity (Sor and Fukuhara, 1982, 1984; Kruszewska and Slonimski, 1984; Weiss-Brummer et al., 1987; Daignan-Fornier et al., 1988; Sakai et al., 1991). Second-site nuclear or mitochondrial suppressors of some of these mutations have also been isolated (Julou et al., 1984; Daignan-Fornier and Bolotin-Fukuhara, 1988; Valens et al., 1991) and these are likely to yield information on RNA-protein interactions within the ribosome and possibly also on long-range folding interactions within the rRNAs themselves.

Current points of interest in mitochondrial translation concern the structure, function and evolutionary conservation of individual components of the translational machinery, details of mRNA selection and subsequent steps in translational initiation and the mechanism(s) of translational regulation. Results relevant to a number of these points are touched on in this brief report.

**Getting started: translational initiation**

Of all the steps in mitochondrial gene expression, probably least is known about the initiation of translation. Unlike their cell sap counterparts, for which cap recognition and leader scanning are important steps in initiation (Kozak, 1989), most mRNAs in mammalian mitochondria lack leaders altogether (Attardi, 1985; Denslow et al., 1989), whilst in yeast, mitochondrial mRNAs are uncapped and most possess extremely long untranslated leader and trailer sequences of high average A+U content (>95%). These leader sequences may be difficult for a ribosome to scan, since they often contain both short open reading frames and short G+C rich clusters capable of forming highly stable secondary structures. They are, nevertheless, important for translation. First, mutations within them can affect translatability (Ooi et al., 1981) and second, parts of the leaders are absolutely required for interaction with nuclear-encoded activator proteins that govern translation of specific mRNAs (Fox, 1986; Costanzo and Fox, 1990).

As part of a project aimed at an understanding of the mechanism of mRNA selection in yeast mitochondria, we have set out to identify trans-acting factors that specifically interact with the 5'-leaders of mitochondrial transcripts and to study their role in regulating translational initiation. This work has led to the identification of a 40 kDa protein (p40) that forms a protein-RNA complex with the 5'-untranslated leaders of all major yeast mitochondrial mRNAs (Papadopoulou et al., 1990; Dekker et al. 1992). p40 is nuclear-coded, but is distinct from the specific translational activators characterized so far. It is exclusively present in mitochondria, where it is predominantly matrix-localised and is so abundant (about 0.4% total mitochondrial protein) that all mitochondrial mRNAs may be permanently complexed with the protein in vivo (Dekker et al., 1991).

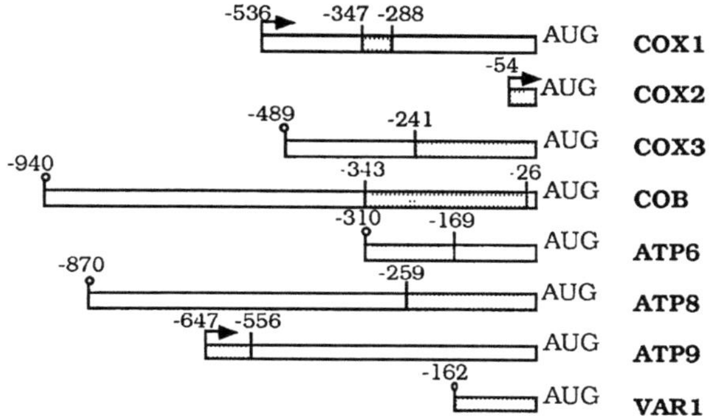


Figure 1: Localization of p40 binding sites in the 5'-leaders of mitochondrial mRNAs. Shaded areas define sequences sufficient for binding of p40. Arrows and open circles indicate the 5'-termini of primary and processed transcripts respectively.

What does p40 see in the leaders it binds to? In COX2, the regions protected against DEPC modification by this protein form a bulged stem-loop structure (Papadopoulou et al., 1990), which overlaps both a putative ribosome binding site (Li et al., 1982) and a UCUAA motif that is conserved in the leader of *COX2* messengers in different yeast species (Hardy and Clark-Walker, 1990). Site-directed mutagenesis has confirmed that the single-stranded area is essential for protein binding. However, subsequent analysis of other mitochondrial mRNAs shows that the site of p40 binding varies widely with respect to the initiator codon (Fig. 1). Footprinting and deletion analysis shows that single-stranded regions are required for protein binding in the COX1 and ATP9 leaders. These regions do not show an exact sequence match to the COX2 binding site, but in all cases, a nearby helical region is conserved (Fig. 2).

What does p40 do? Three main possibilities are worth consideration:

1. The protein may act as a translational activator. Dekker et al. (1992) have consistently found that binding of p40 stabilizes structures in the 5'-leader. The protein could thereby facilitate ribosome binding and thus promote translational initiation. Such a mechanism would resemble the process of internal translation initiation found with picornaviral mRNAs, in which a specific structure in the RNA,

the internal ribosome entry site ( IRES ), is thought to be involved in guiding the ribosome to the correct start site (Jackson et al., 1990).

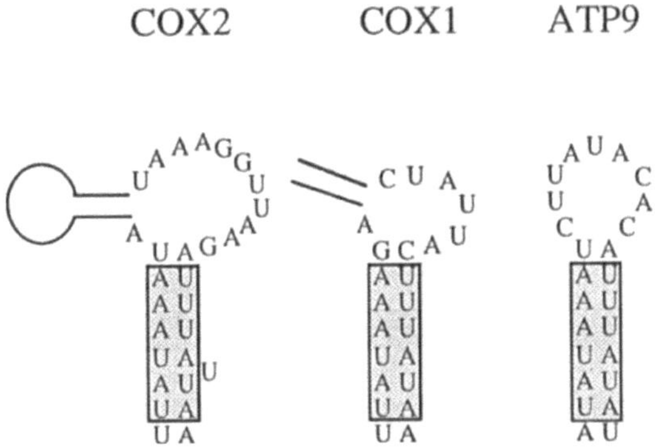


Figure 2: Structural similarities in sites recognized by p40 in the 5'-leaders of mitochondrial mRNAs.

2. Alternatively, p40 might act as a translational repressor. In this model, the observed stabilization of secondary structure within the leader could in fact specifically regulate initiation codon recognition by preventing ribosome access to the start site.

3. Finally, p40 could influence mRNA stability. At least two instances of targeted mRNA instability in yeast mitochondria have been documented and in both cases, nuclear-encoded, mRNA-specific factors are required to stabilize the transcript (Dieckman and Mittelmeier, 1987; Finnegan et al., 1991) By interacting with the 5'-leaders of mitochondrial mRNAs, and forcing the RNA into a stable structure, p40 could act in conjunction with more specific factors to maintain stability.

As yet, it is not possible to rule out any of these possibilities, although both the first and third are made less attractive by the finding that synthesis of at least two proteins (cytochrome b and ATPase 9) from chimaeric constructs containing either part of the COX3 leader fused to the COB coding region, or part of the COB leader fused to the ATP9 coding region, is not absolutely dependent on binding of p40 (Costanzo and Fox, 1988; Dieckmann and Mittelmeier, 1987; Mittelmeier and Dieckmann, 1990). Translation of these chimeric RNAs is still possible, despite the fact that they lost their p40 binding sites in the course of the rearrangements that generated them. Of course, it is conceivable that minor binding sites could have been missed in the study summarized in Fig. 1, or that sites might have been adventitiously created in the chimeras. However, the simplest outcome of these considerations is that p40 may fulfil a somewhat conventional role: as a translational repressor. In both pro- and eukarotic mRNAs, secondary structures normally have an inhibitory effect on translation, either by shielding the ribosome binding site and/or the initiation codon, or by hampering scanning. Regulatory proteins that bind to and stabilize such structures will thus usually repress translation (Gold, 1988; Kozak, 1991). Why should yeast mitochondria wish to repress translation of all mRNAs simultaneously and uniformly? One speculation, advanced by Dekker et al. (1992), is that p40 binding may serve to prevent premature translation of mRNAs in the mitochondrial matrix, thus preventing possible misfolding and subsequent degradation of what are predominantly hydrophobic translation products. This block on translation might only be reversible by interaction of a p40-mRNA complex with mRNA-specific activators. Several of which are known to be associated with either the mitochondrial membrane (Costanzo and Fox, 1990; Michaelis et al., 1991), or the small ribosomal subunit (Haffter et al., 1991; Michaelis et al., 1991), so the net effect of the various interactions would be to limit protein synthesis to sites on the inner membrane, where the products can directly assemble into respiratory chain complexes.

**Putting an end to it all: the yeast mitochondrial polypeptide chain release factor mRF-1**

Our interest in translational termination, a still poorly understood step in protein synthesis in any system, was awakened by our recent discovery of the nuclear gene coding for a polypeptide chain release factor in yeast mitochondria (Pel et al., 1992a). To our initial surprise, the gene for this protein (MRF1) turned up in a screen for DNA sequences capable of correcting respiratory deficiency in a mutant disturbed in its ability to excise the first two introns from transcripts of the gene coding for subunit I of cytochome c oxidase. That we had indeed cloned the gene for a release factor was, however, suggested by the results of database searches, which uncovered a remarkable resemblance of the predicted protein sequence to release factors RF-1 and RF-2 from *E. coli* (38.4 and 29.8% identical residues respectively). Two further observations clinched the matter:

1. Yeast strains with a disrupted copy of the MRF1 gene convert rapidly to cytoplasmic petites ($rho^-/rho^0$). Loss of wild type mtDNA is a familiar feature of mutants with a defective mitochondrial translational apparatus (Myers et al., 1985) and this phenotype therefore suggests not only that mRF-1 is required for mitochondrial translation, but also that it is in all probability the only release factor present. This is not really surprising. In bacteria, two release factors are required since RF-1 recognizes UAA and UAG, whereas RF-2 recognizes UAA and UGA. However, in the yeast mitochondrial genetic code UGA encodes tryptophan and only

UAA and UAG are available as stop codons. In other words, only one release factor of the RF-1 type is necessary. It may therefore be significant that mRF-1 is slightly more similar to *E. coli* RF-1 than RF-2.

2. Increased gene dosage of MRF1 can efficiently reduce nonsense suppression in a strain containing a point mutation (MSU1) in the so-called 530-loop of the mitochondrial small subunit rRNA (Shen and Fox, 1989). This reduction is entirely consistent with a model for nonsense suppresssion involving competition between aminoacyl-tRNA and release factor at a termination codon.

Table 1: Stop codons and flanking sequences in yeast mitochondrial mRNAs

| mRNA | Stop codon | Stop codon context |
|---|---|---|
| coxI | UAA | CAA UCU UAA GUU<br>gln ser |
| aI1 | UAA | GGU AUA UAA AAU<br>gly ile |
| aI2 | UAA | GGU UUA UAA UAA<br>gly leu |
| aI3 | UAA | AAU AAA UAA AUA<br>asn lys |
| aI4 | UAA | AAU AAA UAA AUU<br>asn lys |
| coxII | UAA | GAA CAA UAA UUA<br>glu gln |
| coxIII | UAA | GGU GUU UAA UAA<br>gly val |
| cob | UAA | AAU AAA UAA UAU<br>asn lys |
| ATPase6 | UAA | UUA CAU UAA AUU<br>leu his |
| ATPase8 | UAA | AAA UUA UAA UAU<br>lys leu |
| ATPAse9 | UAA | GGU GUA UAA UAU<br>gly val |
| VAR1 | UAA | UUU AUU UAA UAU<br>phe ile |

What is the connection between mRF-1 and RNA splicing? An intriguing possibility is that the protein, like a number of others required for mitochondrial RNA splicing, has a dual function with roles in both translational termination and RNA splicing (Lambowitz and Perlman, 1990). However, the simplest explanation and one which we currently prefer is that the defective intron excision seen in mrf1 mutants arises from an impaired synthesis of the aI1 and aI2 RNA maturases - intron-encoded proteins whose action has been shown to be necessary for intron excision (Carignani et al., 1983). Despite its simplicity, this explanation has some unresolved aspects. The most puzzling of these is the fact that mrf1 mutants are defective in the synthesis of only two mitochondrial translation products, namely COX1 and COX2. Such a specificity contrasts with the overall reduction of mitochondrial protein synthesis previously reported for mutants in other general mitochondrial translation factors like IF-2 and EF-G (Vambutas et al., 1991) and is strongly suggestive of stop codon specificity of mutant mRF-1. How this might be achieved is so far unclear, as the only naturally occurring stop codon in yeast mitochondria is UAA. Codon context must therefore supply additional determinants and indeed this seems likely, given the fact that in *E. coli*, efficient termination is signalled by a tetranucleotide (Pedersen and Curran, 1991). Context effects have also been described for ribosomal selection rates of suppressor tRNA and release factor (Parker, 1989; Pedersen and Curran, 1991). However, inspection of sequences flanking the stop codons in yeast mitochondrial genes (Table 1) has not so far identified any features that would account for sensitivity to mutation in mRF-1.

The interesting phenotype of mrf1 mutants prompted us to clone and sequence two of the mutant alleles. As Fig. 3 shows, each contains a different single amino acid change within the same region of the protein. This is quite remarkable: only two other RF-1 mutations (in *Salmonella typhimurium*) have been sequenced to date and both map to the same region (Fig. 3). These mutants grow indistinguishably from wild type strains, yet cause a 14 to 34-fold increase of readthrough at certain UAG codons in reporter genes (Elliott and Wang, 1991), suggesting that here too codon context effects play a role in release factor action.

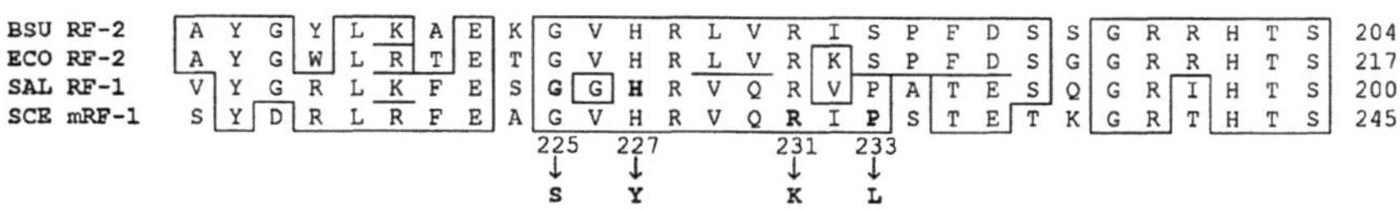


Figure 3: Sequence comparison of prokaryotic and mitochondrial release factors.

What do these mutations tell us about structure-function relationships in release factors? The common feature of codon-context dependence could be taken to imply that the mutated domain is directly involved in stop codon recognition and that it interacts directly with the decoding site. On the other hand, detailed comparison of available prokaryotic and mitochondrial release factor sequences (Pel et al., 1992b) suggests that this may be an oversimplification. This comparison predicts a five-domain model for release factor structure, in which the C-terminal region, with the highest number of RF-type specific residues, interacts with the decoding site. The N-terminal half of the protein, including the region containing the mutations, may in contrast be involved in the initial interaction with the ribosome at the entrance to the cleft between the two subunits (Fig. 4). Two explanations may thus be forwarded for the context-dependent recognition of stop codons by the mutant release factors. In the first, the mutations may reduce binding affinity of the release factor for the ribosome. This may encourage pausing at the stop codon, whereby the likelihood of alternative events such as readthrough, or frameshifting is increased. Alternatively, the mutations may slightly alter positioning of the release factor on the ribosome, in such a way that fidelity of stop codon recognition is decreased and dependence on codon context increased.

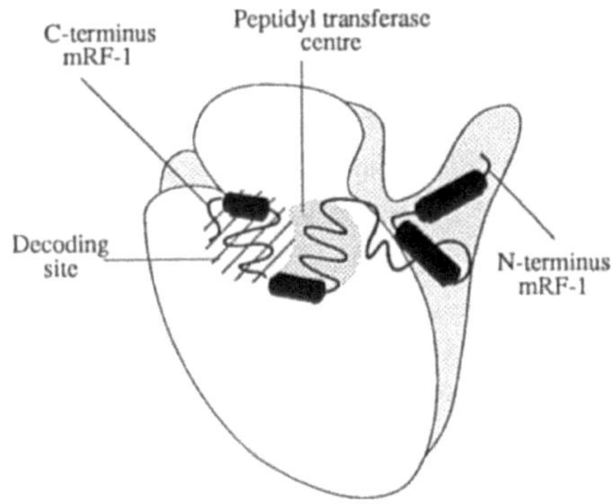


Figure 4: Model for the interaction of mRF-1 with the ribosome.

### Conclusions and Prospects

The retention of a separate genetic system by mitochondria is commonly regarded as an evolutionary accident (Borst, 1977). For the cell, this accident is a costly one in terms of the tradeoff between the number of proteins produced as opposed to the number of proteins required to produce them. For the molecular biologist, however, the situation has amply repaid the effort of study in terms of information on the evolution and function of individual components and is likely to continue to do so. This brief report has highlighted events occurring during translational initiation and termination, two areas of protein synthesis in which current knowledge is clearly incomplete. Results presented suggest that interesting principles are involved both in the roles played by 5'-leaders and leader-binding proteins in initiation and in the role of release factor in determining events in termination. The ease with which yeast mitochondria can be manipulated genetically and biochemically will no doubt facilitate their elucidation.

### Acknowledgements

Our thanks are due to Amy Bednarz, Jauke Diekman, EricJan Dubbink, Katinka van Oosterum, Barbara Papadopoulou, Martijn Rep, Maria Simitsopoulou and Jeroen Stuurman for their contributions to the various phases of the experimental work described and for valuable discussion. This research was supported in part by grants from the Netherlands Foundation for Chemical Research (SON) with financial aid from the Netherlands Organization for the Advancement of Research (NWO).

**References**

Attardi, G. (1985) Animal mitochondrial DNA: an extreme example of genetic economy. Int. Rev. Cytol. 93: 93-145.
Benne, R. and Sloof, P. (1987) Evolution of the mitochondrial protein synthetic machinery. Biosystems 21: 51-68.
Borst, P. (1977) Structure and function of mitochondrial DNA. Trends in Biochem. Sci. 2: 31-34.
Borst, P. (1981) Control of mitochondrial biosynthesis. In: Cellular controls in differentiation (C.W. Lloyd and D.A. Rees, eds) Academic Press pp231-252.
Carignani, G., Groudinsky, O., Frezza, D., Schiavon, E., Bergantino, E. and Slonimski, P.P. (1983) An mRNA maturase is encoded by the first intron of the mitochondrial gene for the subunit I of cytochrome oxidase in *S. cerevisiae*. Cell 35: 733-742.
Costanzo, M.C. and Fox, T.D. (1988) Specific translational activation by nuclear gene products occurs in the 5' untranslated leader of a yeast mitochondrial mRNA. Proc.Natl.Acad.Sci., USA 85: 2677-2681.
Costanzo, M.C. and Fox, T.D. (1990) Control of mitochondrial gene expression in *Saccharomyces cerevisiae*. Ann. Rev. Genet. 24: 91-113.
Daignan-Fornier, B. and Bolotin-Fukuhara, B. (1988) Mutational study of the rRNA in yeast mitochondria: the functional importance of T1696 in the large rRNA gene. Nucl. Acids Res. 16: 9299-9306.
Daignan-Fornier, B., Su, I., Contamine, V. and Bolotin-Fukuhara, B. (1988) A genetic approach to molecular interactions in the yeast mitochondrial ribosome. In: NATO-ASI series, vol. H14 Genetics of Translation (Tuite, M., Picard, M. and Bolotin-Fukuhara, M., eds), Springer Verlag Berlin, Heidelberg pp. 159-169.
Dekker, P.J.T., Papadopoulou, B. and Grivell, L.A. (1991) Properties of an abundant RNA-binding protein in yeast mitochondria. Biochimie 73: 1487-1492.
Dekker, P.J.T., Stuurman, J., van Oosterum, K. and Grivell, L.A. (1992) Determinants for binding of a 40 kDa protein to the leaders of yeast mitochondrial mRNAs. Nucl. Acids Res. 20: 2647-2655.
Denslow, N.D., Michaels, G.S., Montoya, J., Attardi, G. and O'Brien, T.W. (1989) Mechanism of mRNA binding to bovine mitochondrial ribosomes. J. Biol. Chem. 264: 8328-8338.
Dieckmann, C.L. and Mittelmeier, T.M. (1987) Nuclearly-encoded CBP1 interacts with the 5'-end of mitochondrial cytochrome *b* pre-mRNA. Curr. Genet. 12: 391-397.
Elliott, T. and Wang, X. (1991) *Salmonella typhimurium* prfA mutants defective in release factor 1. J. Bacteriol. 173: 4144-4154.
Finnegan, P.M., Payne, M.J., Keramidaris, E. and Lukins, H.B. (1991) Characterization of a yeast nuclear gene, AEP2, required for accumulation of mitochondrial mRNA encoding subunit 9 of the ATP synthase. Curr. Genet. 20: 53-61.
Fox, T.D. (1986) Nuclear gene products required for translation of specific mitochondrially coded mRNAs in yeast. Trends in Genet. 2: 97-100.
Gold,L. (1988) Posttranscriptional regulatory mechanisms in *Escherichia coli*. Ann. Rev. Biochem. 57: 199-233.
Haffter, P., McMullin, T.W. and Fox, T.D. (1991) Functional interactions among two yeast mitochondrial ribosomal proteins and an mRNA-specific translational activator. Genetics 127: 319-326.
Hardy, C.M. and Clark-Walker, G.D. (1990) Nucleotide sequence of the cytochrome oxidase subunit 2 and val-tRNA genes and surrounding sequences from *Kluyveromyces lactis* K8 mitochondrial DNA. Yeast 6: 403-410.
Jackson, R.J., Howell, M.T. and Kaminski, A. (1990) The novel mechanism of initiation of picornavirus RNA translation. Trends in Biochem. Sci. 15: 477-483.
Julou, C., Contamine, V., Sor, F. and Bolotin-Fukuhara, M. (1984) Mitochondrial ribosomal RNA genes of yeast: their mutations and a common nuclear suppressor. Mol. Gen. Genet. 193: 275-279.
Kozak, M. (1989) The scanning model for translation: an update. J. Cell Biol. 108: 229-241.
Kozak, M. (1991) Structural features in eukaryotic mRNAs that modulate the initiation of translation. J. Biol. Chem. 266: 19867-19870.
Kruszewska, A. and Slonimski, P.P. (1984) Mitochondrial and nuclear mitoribosomal suppressors that enable misreading of ochre codons in yeast mitochondria. I. Isolation, localization and allelism of suppressors. Curr. Genet. 9: 1-10.

Lambowitz, A.M. and Perlman, P.S. (1990) Involvement of aminoacyl-tRNA synthetases and other proteins in group I and group II intron splicing. Trends in Biochem. Sci. 15: 440-444.
Li, M., Tzagoloff, A., Underbrink-Lyon, K. and Martin, N.C. (1982) Identification of the paromomycin-resistance mutation in the 15S rRNA gene of yeast mitochondria. J. Biol. Chem. 257: 5921-5928.
Michaelis, U., Körte, A. and Rödel, G. (1991) Association of cytochrome b translational activator proteins with the mitochondrial membrane: implications for cytochrome b expression in yeast. Mol. Gen. Genet. 230: 177-185.
Mittelmeier, T.M. and Dieckmann, C.L. (1990) CBP1 function is required for stability of a hybrid cob-oli1 transcript in yeast mitochondria. Curr. Genet. 18: 421-428.
Myers, A.M., Pape, L.K. and Tzagoloff, A. (1985) Mitochondrial protein synthesis is required for maintenance of intact mitochondrial genomes in *Saccharomyces cerevisiae*. EMBO J. 4: 2087-2092.
Ooi, B.G., Lukins, H.B., Linnane, A.W. and Nagley, P. (1987) Biogenesis of mitochondria: a mutation in the 5'-untranslated region of yeast mitochondrial oli1 mRNA leading to impairment in translation of subunit 9 of the mitochondrial ATPase complex. Nucl. Acids Res. 15: 1965-1977.
Papadopoulou, B., Dekker, P., Blom, J. and Grivell, L.A. (1990) A 40 kd protein binds specifically to the 5'-untranslated regions of yeast mitochondrial mRNAs. EMBO J. 9: 4135-4143.
Parker, J. (1989) Errors and alternatives in reading the universal genetic code. Microbiol. Rev. 53: 273-298.
Pedersen, W.T. and Curran, J.F. (1991)Effects of the nucleotide-3' to an amber codon on ribosomal selection rates of suppressor transfer RNA and release factor-1. J. Mol. Biol. 219: 231-241.
Pel, H.J., Maat, C., Rep, M. and Grivell, L.A. (1992a) The yeast nuclear gene MRF1 encodes a mitochondrial peptide chain release factor that is also involved in mitochondrial RNA splicing. Submitted for publication.
Pel, H.J., Rep, M. and Grivell, L.A. (1992b) Sequence comparison of new prokaryotic and mitochondrial members of the polypeptide chain release factor family predicts a five-domain model for release factor structure. Submitted for publication.
Sakai, H., Stiess, R. and Weiss-Brummer, B. (1991) Mitochondrial mutations restricting spontaneous translational frameshift suppression in the yeast *Saccharomyces cerevisiae*. Mol. Gen. Genet. 227: 306-317.
Shen, Z. and Fox, T.D. (1989) Substitution of an invariant nucleotide at the base of the highly conserved '530-loop' of 15S rRNA causes suppression of yeast mitochondrial ochre mutations. Nucl. Acids Res. 17: 4535-4539.
Sor, F. and Fukuhara, H. (1982) Identification of two erythromycin resistance mutations in the mitochondrial gene coding for the large ribosomal RNA in yeast. Nucl. Acids Res. 21: 6571-6577.
Sor,F. and Fukuhara,H. (1984) Erythromycin and spiramycin resistance mutations of yeast mitochondria: nature of the rib2 locus in the large ribosomal RNA gene. Nucl. Acids Res. 12: 8313-8318.
Valens, M., Rinaldi, T., Daignan-Fornier, B. and Bolotin-Fukuhara, M. (1991) Identification of nuclear genes which participate to mitochondrial translation in S. *cerevisiae*. Biochimie 73: 1525-1532.
Vambutas, A., Ackerman, S.H. and Tzagoloff, A. (1991) Mitochondrial translational-initiation and elongation factors in *Saccharomyces cerevisiae*. Eur. J. Biochem. 201: 643-652.
Weiss-Brummer, B., Sakai, H. and Kaudewitz, F. (1987) A mitochondrial frameshift suppressor (+) of the yeast *S. cerevisiae* maps in the mitochondrial 15S rRNA locus. Curr. Genet. 11: 295-301.

# YEAST MITOCHONDRIAL TRANSLATION: NUCLEAR GENES INVOLVED IN THE EXPRESSION OF THE MITOCHONDRIAL GENOME

T. Rinaldi, M. Valens and M. Bolotin-Fukuhara
Laboratoire de Génétique Moléculaire.
I G M ( URA D 1354 du CNRS)
Université Paris-Sud.
91405 Orsay Cedex
France.

## Abstract

A search for nuclear genes involved in mitochondrial translation has been made using genetic suppressors and multicopy suppressors of mitochondrial mutations which affect protein synthesis.

Three different mitochondrial mutations ( two localized in the 21SrRNA gene and one in the Asp tRNA gene) have been chosen for this purpose. In all cases, nuclear suppressors have been characterized which have biochemical and physiological properties compatible with components of the translational apparatus. Cloning one of those led us to the identification of a new yeast gene called *SMT1*.

Several multicopy suppressors have also been isolated. With this method, we uncovered a new pathway related to mitochondrial translation.We also found that overexpression of both mtEF-Tu and mtAspartyl tRNA synthetase can correct a deficiency in the 3' end maturation of the mitochondrial Asp tRNA .

## Introduction

Mitochondrial translation presents several properties which distinguish it from the same process in other biological systems. Most of the differences are probably related to the mechanism of initiation and the problem of selection of mRNAs. Long leaders, without Shine and Dalgarno sequences as well as several specific proteins are necessary to start translation of mitochondrial genes. However, some unexplored research fields concern also the role of the mitochondrial membrane and the functional relation between the chromosomal genes which encode almost all components of the mitochondrial translational machinery and their mitochondrial partners. The latter code only for the two rRNAs

NATO ASI Series, Vol. H 71
Protein Synthesis and Targeting in Yeast
Edited by A. J. P. Brown, M. F. Tuite, and J. E. G. McCarthy

the mitochondrial tRNAs and the VarI ribosomal protein (see Grivell 1989 for a review).

Efficient translation requires indeed the correct and coordinated expression of many genes encoded by these two different cellular compartments and the correct assembly of their products.Therefore steps like transport from nucleus to mitochondria via the cytosol and import into the mitochondria are part of the process as well as the expression and regulation of the nuclear genes which participate . Furthermore, there is an apparent coupling between translation and correct replication of the mitDNA since a block in mitochondrial translation leads to the loss of the mitochondrial DNA (Myers et al, 1985) .

The number of predictible genes which participate in mitochondrial translation (r-proteins, tRNA synthetases and maturation enzymes, various factors) amounts to about a hundred .The genes really involved are probably much more and only twenty to thirty of them are cloned.

To understand this complex process, it is necessary to identify and characterize all the genes which are directly or undirectly involved in this process, to study their interrelations and their regulation. Finding unexpected genes product linked to the translation process might help to decipher the specific characteristics of the mitochondrial translation.

Cloning the genes is simple as long as the function is easily predictable: in some cases, the protein can be easily characterized (r-proteins for example); in others , a mutant phenotype can be easily defined. However, if the function of the gene is not predictable a priori or its involvment in translation not suspected , no specific phenotype can be looked for, nor a protein purified. Moreover, these mutated genes might not exhibit any proper phenotype and be revealed only by their interactions with other gene products.

One way to overcome these difficulties is to look for second-site mutations (suppressors). In the case of the mitochondrial biogenesis and because of the dispersion of the genes between two different genomes, the search for nuclear suppressors of mitochondrial mutations should be fruitful. Our laboratory has been working in this direction to identify new yeast nuclear genes which are related to the translational process.

We will briefly review here some of the most recent developments, detailed data could be found in several other publications ( Bolotin-Fukuhara, 1979; Julou and Bolotin-Fukuhara, 1982; Bolotin-Fukuhara et al, 1983; Contamine and Bolotin-Fukuhara, 1984; Julou et al, 1984; Guetari et al, 1987; Daignan-Fornier et al, 1988;

Daignan-Fornier and Bolotin-Fukuhara, 1988; Zennaro et al, 1989; ElElj-Fridhi et al, 1991; Valens et al, 1991 ).

## Results and discussion

Because mitochondrial mutations can be easily and precisely localized on the mitochondrial genome and because the genes involved in translation and encoded by it are known, it is easy to select mutations defective in mitochondrial translation. Isolating conditional mutants improves the specific screening since it is a way to prevent the loss of the mitochondrial DNA in the mutated strain (Bolotin-Fukuhara et al, 1977).

Two methods can be applied to search for interacting components:(i) The first one uses the classical selection of second-site mutations.(ii) A second method takes advantages of libraries made on multicopy plasmids to correct the deficient phenotype since wild-type genes are overexpressed in this context and the normal balance of gene expression modified.

If the first method helps to identify functionally (or even physically) interacting components, the second one gives information as for alternative pathways (changing the regulation of the cell, bypassing the defect via another pathway or optimization of some steps between translation of these nuclear genes and the final assembly of macromolecules into the mitochondria) .

Several mutations of the rRNAs ( Daignan-Fornier et al,1988; Daignan-Fornier and Bolotin-Fukuhara,1988 ) , tRNAs (Faye et al, 1976; Zennaro et al, 1989; ElElj-Fridhi et al, 1991) and of the VarI genes (Guetari et al, 1987) have been identified and characterized in our laboratory. We will describe here some of the results obtained when screening three of those mutations for suppression. These mutations are *cs 909*, a cold-sensitive mutation of the 21S rRNA gene ( Daignan-Fornier and Bolotin-Fukuhara, 1988), $C^{R}_{321}$ a chloramphenicol-resistant mutation of the same gene( Dujon, 1980) and *ts 932,* a heat-sensitive mutation of the Asp tRNA (Zennaro et al, 1989).

## I) First approach: selection of second-site mutations

### I-A) Genetic and physiological characteristics

**cs 909:**

*Cs 909* is a mutation in domain V of the large rRNA. Revertants have been obtained , and their genetic characterization described (Contamine and Bolotin-Fukuhara (1984).

We obtained several recessive weak suppressors which were shown to be alleles of the same gene and one of them called *nsu100* was further studied. When separated from the mitochondrial mutation and put in a wild-type context, it was shown to exhibit a cold-sensitive phenotype on respiratory substrate (glycerol) and abnormal mitochondrial ribosomal profiles. This phenotype led us to postulate that this nuclear gene interferes with the ribosome assembly. A second class of revertants which was difficult to characterize genetically was finally shown to be due to mitochondrial suppressors. Nucleotide sequence studies of these second-site mutations showed the importance of the nucleotide T1696 in the large rRNA gene for a functional ribosome (Daignan-Fornier and Bolotin-Fukuhara, 1988). In E. coli , the equivalent primary sequence and secondary structure are identical, this structure being involved in the binding of the r-protein L2.

This observation as well as the fact that the nuclear gene,when mutated, presents also a cold-sensitive phenotype and abnormal ribosomes, suggest that this nuclear gene may encode the mitochondrial equivalent of the L2 ribosomal protein.

**ts 932:**

*Ts 932* is a mutation which impairs the correct 3' end maturation of the Asp tRNA precursor, no mature tRNA being detected by northern experiments (Zennaro et al, 1989). This mutation has been therefore especially interesting to study since one expects a nuclear suppressor gene to encode an enzyme responsible for the 3' end maturation.

However, selection of revertants in a haploid strain carrying the mutation only provided mitochondrial revertants despite the fact that a great number (more than 70) independant revertants were selected. The fact that no nuclear suppressors have been found can be explained. The mutated endonuclease ( the putative product of a nuclear suppressor) is now able to recognize and mature the deficient Asp tRNA but this very same mutation probably prevents the correct maturation of all others. If such is the case, the strain which harbours such a mutated gene ( nuclear suppressor of the mitochondrial *ts 932* mutation) should not grow on glycerol. The only way to select this mutation is to use a diploid context where the mutated copy of the gene will be balanced by the wild-type copy. For this reason, a search for weakly growing strains at 36°C, in diploids homoplasmic for the mitochondrial *ts 932* mutation was undertaken and provided several candidates. One of them -*R117*- was kept for further studies since in the random haploid progeny we found two types of phenotypes with a ratio close to the 2:2 segregation indicative of a nuclear segregation. The phenotype of one class of spores was heat-sensitive (as is the original ts 932 phenotype) while the other exhibited a new phenotype with practically no growth on glycerol (Valens et al, 1991). Phenotype and genotype of the R117 revertant and its segregants were therefore in good agreement with what was expected from a mutation in a 3' endonuclease able to correct the 3' maturation deficiency of *ts 932*.

**CR321**

The *CR321* mutation of the rRNA21S was sequenced by Dujon (1980) and shown to be localized in the peptidyl transferase center of the ribosome.

To identify the components involved in the peptidyl transferase center of the mitochondrial ribosome, we looked for suppressors of the chloramphenicol-resistant phenotype , i.e. strains which are genetically chloramphenicol resistant (they can transmit the resistant allele to the progeny), but are phenotypically sensitive to chloramphenicol. Since the resistance[$C^R$] / sensitivity[$C^S$] phenotype can only be checked on glycerol medium, we introduced as an additional requirement the temperature-sensitive growth of the suppressor on this same medium. This restriction is also helpful to avoid the accumulation of $rho^-$ cells which appear in protein synthesis deficient mutants (Myers et al, 1985). Thus, at 28°C, the suppressor should grow on glycerol and not express the [$C^R$] phenotype any more as compared to the wild-type, while at restrictive temperature there should be no growth at all on glycerol. Two strains having this phenotype has been selected and shown to contain a nuclear suppressor ( Bolotin-Fukuhara,1979) : one is called *cs 1370* -the restrictive temperature is 20°C- and the other *ts 1398* (the restrictive temperature is 36°C). They both exhibit also abnormal ribosomes (Sor and Faye, 1979).

**I-B) Molecular cloning**

The ultimate goal of these genetic studies is to identify in each case the corresponding gene by nucleotide sequence and expression studies.

Cloning a recessive suppressor can be difficult even if one takes care to verify that a ratio of two copies of the recessive mutated allele (active suppressor) versus one copy of the wild-type allele is enough to restore the growth on glycerol. However, if the suppressor has a phenotype of its own, independantly of the mitochondrial allele, the task should be easier since the gene can be cloned by complementation. Such was the case for the three mutations just described. Cloning the gene for the putative endonuclease is in progress. As for the L2-like ribosomal protein, our efforts failed probably because the mitochondrial L2 protein is immunologically too close to the corresponding E.coli protein (discussed in Valens et al, 1991). Finally, suppressors of the $C^{R}_{321}$ mutation have been cloned and allowed the identification of at least one new yeast gene.

Strains carrying the *ts 1398* and *ts 1370* mutations were transformed with a multicopy library constructed in our laboratory ( 8kb average size fragments , prepared by partial Sau3A digestion of chromosomal DNA from the FF1210-6C/R101 strain , were inserted into the BamHI site of the YEP 13 vector). In each case, several overlapping plasmids able to restore growth on glycerol at restrictive temperature were isolated and further analyzed by subcloning . In both cases, the smallest fragment able to restore the growth was transfered to multicopy (YEpLac 181) and centromeric (YCpLac 111) vectors ( Gietz and Sugino, 1988 ). Restoration of growth was achieved as well with the centromeric vector, a good indication that we really cloned the genes corresponding to those mutations.`

Sequencing the DNA fragment which complements the *ts 1398* mutation revealed an open reading frame of 1929 nucleotides which can be translated into a putative protein of 643 aa ( Fig 1). We call this gene *SMT1* (suppression of mitochondrial translation mutations). The sequence at the 5' end of the gene is compatible with the aminoacids composition required for a signal sequence which targets gene products into mitochondria ( review in Pfanner and Neupert, 1990). Such a result is in good agreement with the fact that this gene product has to interact with the mitochondrial rRNA. Although one may suspect this gene to encode a ribosomal protein, no similarities with such proteins or any known yeast genes were revealed by comparison to data bank. However, destruction of the chromosomal copy of the *SMT1* gene showed this gene to be essential for the mitochondria and the null allele is converted extremely rapidly to rho⁻ cells as expected for mitochondrial translation mutants ( see Valens et al, 1991 for details). Epitope tagging experiments have to be performed in order to precisely localize

Fig 1.Nucleotide sequence of the *SMT1* gene

```
   1                                                                          GAATTCCTCGACCTCCTTA
  20 TTGTACAATTCGACGTACTTCGCAATTGTTGGTGCTCCATCCTAGGTGACCGTCGTATGCGCTGTGCTTTCCTAGCTCTTCTT
 103 CCACTTCTGTTTCTCTATACTTTTTTTTTTGTTTTCACCTCGCCCGCCTTTTCTTATTTCAGTGAACTGAAAAAAAAAAAAA
 186 AAGTTATGTATCTTTAATGTAATAAAACTAGTACATAATATATACGGAGCTTGAAGGCACTGGAGAAAATATAGAGTAAGCAT
     Met Trp Lys Phe Asn Lys Lys Leu Ala Arg Leu Thr Tyr Arg Leu Tyr Ser Ser Ser Gly Pro    21
 269 ATG TGG AAA TTC AAC AAA AAG CTC GCC CGT CTT ACT TAC CGT CTT TAT TCA AGC TCA GGC CCA
     Ser Ser Pro Leu His Gly Lys Lys Lys Leu Pro Gln Asn Leu Lys Phe Val Val Leu Asn Pro    42
 333 AGC TCG CCC CTT CAT GGG AAG AAG AAA CTG CCG CAA AAT CTG AAA TTC GTC GTC TTG AAC CCT
     Thr Gln Ser Gly Leu Val Lys Asn Asp Gln Lys Gln Pro Arg His Arg Pro Ser Lys Lys Arg    63
 396 ACC CAA TCT GGG CTA GTG AAA AAT GAT CAG AAG CAG CCT AGA CAC AGG CCG TCT AAG AAA CGT
     Ser His Lys Glu Thr Gly Asp Asn Asn Leu Asp Phe Gly Ser Lys Leu Leu Val Phe Glu Lys    84
 459 AGC CAT AAG GAA ACT GGA GAT AAC AAC CTC GAT TTC GGC TCG AAA CTC CTT GTC TTC GAA AAG
     Gln Asn Ser Leu Asp Ser Ala Leu Asn Ser Ile Arg Leu Lys Lys Pro Thr Ser Ala Ser Leu   105
 522 CAG AAT TCA CTA GAT TCC GCA TTG AAC TCC ATC CGA TTG AAG AAA CCG ACA AGC GCC AGT CTG
     Pro Ser Leu Glu Tyr Asn Ala Leu Leu Gln Ser Leu Thr Ser Ser Tyr Asn Arg Tyr Gln Leu   126
 585 CCC TCC CTG GAA TAC AAT GCC CTT CTC CAA TCG CTT ACA TCA AGC TAC AAT CGC TAC CAA CTG
     Arg Glu Phe Ile Ser Thr His Gln Pro Asp Ser Ser Ser His Leu Thr His Trp Lys Lys Ser   147
 648 CGC GAG TTC ATC TCC ACA CAT CAG CCA GAC TCC TCT TCA CAT CTG ACG CAC TGG AAG AAA AGC
     Lys Leu Ser Gln Tyr Ile Ile Glu Lys Ile Trp Asn Cys Gln Pro Ile Ser Thr Pro Thr Thr   168
 711 AAG CTG TCC CAA TAT ATA ATT GAA AAA ATC TGG AAT TGC CAA CCA ATT TCA ACA CCC ACC ACT
     Pro Thr Gly Ile Lys Ser Thr Ser Leu Thr Phe Gln Phe Asp Ser Pro Arg Glu Ile Phe Leu   189
 774 CCC ACA GGC ATC AAG TCC ACA TCA TTG ACT TTC CAG TTC GAT TCT CCT AGA GAA ATT TTC CTT
     Leu Leu Ile Thr Gln Asn Gly Lys Ile Leu Thr Asn Phe Asn Lys Leu Gly Leu Thr Phe Ile   210
 837 TTA CTC ATC ACT CAA AAT GGT AAA ATA CTC ACC AAC TTC AAT AAG CTC GGG CTG ACA TTC ATC
     Ile Ser Ile Gln Asp Asn Glu Leu Thr Val Lys Gly Ser Pro Ser Leu Leu Lys Tyr Ala Glu   231
 900 ATC TCC ATT CAG GAT AAC GAA TTG ACC GTC AAG GGT TCG CCT AGC TTG CTC AAA TAC GCA GAA
     Ile Ser Leu Asn Lys Ile Trp Ser Asn Ile Thr His Glu Asn Val Arg Met Tyr Ser Leu Met   252
 963 ATA TCC TTA AAC AAA ATA TGG TCA AAC ATC ACT CAT GAA AAT GTG CGT ATG TAT TCC CTC ATG
     Pro Ser Lys Asp Val Ile Asn Leu Ile Gln Lys Glu Thr His Thr Phe Phe Glu Tyr Leu Pro   273
1026 CCA TCA AAA GAC GTC ATC AAC CTT ATT CAG AAG GAA ACT CAC ACT TTT TTT GAA TAC CTC CCG
     Asp Leu Gln Met Tyr Lys Ile Ser Ala Leu Ser Thr Lys Lys Ile Ser Met Ala Lys Val Phe   294
1089 GAC TTG CAA ATG TAC AAG ATT TCA GCC CTA AGC ACC AAG AAA ATA TCC ATG GCC AAA GTC TTT
     Leu Leu Asn Ala Val Ala Ser Asn Pro Asn Thr Thr Gln His His His Thr Ile Ala Ser Pro   315
1152 TTA TTA AAT GCC GTC GCT TCT AAT CCT AAC ACA ACT CAA CAC CAC CAC ACA ATA GCC TCA CCA
     Ala Leu Lys Thr Glu Leu Tyr Pro Phe Asn Asn Thr Leu Glu Asn Leu Asp Trp Leu Asn Lys   336
1215 GCA CTC AAG ACA GAA CTC TAT CCG TTC AAC AAT ACT CTA GAG AAC TTG GAC TGG CTA AAC AAG
     Ser Gln Asp Trp Ala Arg Leu Gln Ser Val Val Pro Lys Asn Cys Thr Asp Leu Met Thr Pro   357
1278 TCA CAA GAT TGG GCA AGA TTA CAG TCA GTT GTC CCT AAA AAT TGC ACA GAT CTG ATG ACT CCC
     Thr Glu Asn Ala Thr Pro Glu Leu Thr Asp Ala Gln Val Ser Gln Phe Glu Ser Ser Leu Ser   378
1341 ACT GAA AAC GCC ACA CCA GAA TTG ACA GAT GCG CAG GTC AGC CAA TTC GAA TCT TCC TTG TCT
     Lys Asn Ile Pro Ser Leu Ser Pro Ser Asp Ser Ile Ser Gln Ser Leu Ser Ile Thr Leu Gly   399
1404 AAA AAT ATA CCA TCT CTC TCC CCT TCC GAC TCT ATT TCC CAA TCT TTA TCA ATA ACG CTA GGT
     His Ser Leu Gln Ser Ala Ser Phe Ser Ser Ile Phe Gln Pro Leu Ile His Lys Ser Phe Ile   420
1467 CAC TCC TTA CAG TCC GCG TCT TTT TCT AGT ATT TTC CAA CCT CTA ATA CAT AAG AGT TTT ATT
     Ser Lys Leu Leu Asn Leu Pro Met Tyr Lys Glu Ser Ser Ser Ser Ala Val Pro Val Pro Val   441
1530 TCA AAA TTA CTG AAT CTA CCA ATG TAC AAA GAA TCA TCA TCA TCA GCC GTA CCT GTG CCT GTG
     Pro Leu Asp Gln His Leu Ile Thr Asn Ala His Gln Ser Phe Ile Gln Leu Asn Phe Thr Pro   462
1593 CCT CTT GAT CAA CAT CTA ATA ACA AAT GCG CAT CAA TCC TTC ATA CAG CTG AAC TTC ACT CCT
     Val Pro Pro Thr Cys Gly Ser Ser Ser Ser Pro Phe Leu Gln Ile Trp Phe Glu Ile Asp Glu   483
1656 GTG CCG CCC ACC TGT GGC TCC TCC TCG TCC CCT TTT TTG CAG ATT TGG TTC GAG ATT GAC GAA
     Phe Asp Asn Ile Val Thr Thr Ser Met Arg Pro Leu Leu Lys Leu Gln Glu Asn Ser Val Ile   504
1719 TTC GAT AAC ATT GTT ACC ACC TCA ATG AGG CCT CTT TTG AAA TTA CAA GAA AAT TCT GTT ATC
     Leu Arg Thr Pro Gln Cys Gln Thr Asp Tyr Lys Ile Thr Ser Asp Tyr Ile Gln Asp Leu Leu   525
1782 TTA AGA ACT CCA CAA TGC CAG ACA GAC TAT AAG ATC ACG TCA GAT TAC ATT CAA GAT CTT TTA
     Pro Asp Phe Asp Gln Thr Asn Pro Asp Ala Trp Leu Ser Glu Gln Lys Gly Leu Gln Glu Phe   546
1845 CCT GAT TTC GAC CAG ACC AAT CCG GAT GCT TGG TTA TCT GAA CAA AAA GGC CTC CAA GAA TTT
     Leu Leu Lys Ser His Trp Lys Leu Asn Lys Tyr Gln Asn Leu Met Lys Lys Ile Asn Ile Ser   567
1908 CTT CTC AAG TCA CAC TGG AAG TTA AAC AAA TAT CAG AAT CTT ATG AAA AAA ATC AAC ATT TCT
     Leu Pro Asp Asn Leu Ile Gln Gln Tyr Gln Leu Thr Asp Val Leu Thr His Arg Val Leu Asn   588
1971 TTA CCA GAT AAC CTA ATA CAG CAA TAC CAA TTG ACT GAT GTT TTG ACC CAT CGT GTC TTG AAT
     Leu Arg Phe Pro Thr Asn Thr Ala Gln Asp Asp Lys Tyr Ile Ile Gln Tyr Ser Asp Ile Ser   609
2034 TTA CGA TTT CCT ACG AAC ACT GCC CAA GAT GAT AAA TAC ATC ATT CAA TAT TCA GAT ATC AGC
     Arg Gly Phe Leu Asn Asn Gly Ser Tyr Arg Gln Leu Asp Phe Ile Asn Val Asn Pro Ser Glu   630
2097 CGC GGA TTT TTG AAC AAT GGC TCC TAC AGA CAA TTA GAT TTC ATC AAT GTG AAC CCC AGT GAA
     Thr Ser Leu Lys Thr Phe Ile Asn Asp Val Leu Ser Phe Stop                              643
2160 ACC TCA CTA AAA ACC TTT ATC AAT GAT GTT TTA AGC TTT TAA GTTTAACGGAAAAGAGTAATGACAAAA
2229 CCGCATACCCCCCCCCCTCCCCCTCTTTACCGCATCTACTCTATTTCCGACACATGTAAATAAAAAAGGCATTTCGTTTATA
2312 TATATACTAAAATAATATATCTTTGTGTATTAATAAATTACTTATTGAGACCAAGCGTTTTTGATAGCGCCGACTTAACAGCA
2395 GCACTCTTTTGCTTTTTTTTT
```

the SMT1 gene product and verify a possible association with the mitochondrial ribosome.
Sequence of the gene which complements the *cs 1370* mutation is in progress.

**I-C) Suppressors as a tool to identify more interacting components.**

We have seen that direct selection of second-site mutations bring informations on interacting components of the mitochondrial translation machinery.

For each of the situations previously described, nuclear suppressors were identified. These second-site mutations, once separated from the corresponding mitochondrial mutations and associated with wild-type mitochondria, presented also defective phenotypes. These defective phenotypes are useful to clone the corresponding genes, but should also help to identify more interacting components.

**Table I . Revertants of cs 1370: phenotypic and genetic analysis**

| | | crosses with 1370-9 | | | |
|---|---|---|---|---|---|
| | | [cr] | [cs] | T | Conclusion |
| **R73** | **[CamR]** | **341** | **119** | **460** | **Mitochondrial** |
| **R78** | **[CamS]** | **0** | **727** | **727** | **Nuclear** |
| **R90** | **[CamS]** | **684** | **226** | **910** | **Mitochondrial** |
| **R76** | **[CamR]*** | **0** | **707** | **707** | **Nuclear** |
| **R52** | **[CamS]** | **215** | **379** | **594** | **Mitochondrial** |

In the Table is provided the name and the phenotype of five independant revertant strains isolated from the original mutant cs 1370 (*mat a ade 2 trp1* $C^{R}_{321}$). * means that the resistance to chloramphenicol of R76 was very weak as compared to the one observed for the wild-type and R73 .
The revertant strains were crossed with the strain 1370-9 ( *Mat alpha ade2 leu2*) and in each case, the phenotype ( cold-sensitive [cr] or cold- resistant [cr] ) of individual diploids was checked. Since all diploids are homozygous for the recessive mutations cs 1370, an heterogeneous population of diploids indicates the presence of a mitochondrial suppressor.

This was indeed the case:

(i) the *nsu100* mutation, isolated as a nuclear suppressor of the *cs 909* mutation, allowed us to isolate new nuclear and mitochondrial suppressors. Among the mitochondrial suppressors , one class was different from the *cs 909* mutation, thus revealing a second interaction between the protein and the rRNA. A new nuclear gene was also identified (Daignan-Fornier et al,1988). These data allowed us to enlarge the number of interacting partners for the function which is modified in the *cs 909* mutation ( ribosome assembly probably ).
(ii) we were also interested to check if the same observation could be made with *cs 1370* and looked for cold-resistants revertants. Several independant clones with this phenotype were selected and analyzed (Table I). Both nuclear and mitochondrial suppressors with different phenotypes can be identified , indicating again the great diversity of mutations which are able to interact.

**II- selection of wild-type genes which can compensate the mitochondrial deficiency when they are over-expressed.**

This method is now used in a systematic way but is very often a by-product of cloning with a multicopy library, a necessity when one deals with recessive suppressors. Some of genes thus identified turn out to be additional multicopy suppressors and not the gene which was originally looked for. Such results, if they may appear disappointing at first are extremely informative. Complementary to the genetic approach, which reveals rather direct functional interactions, multicopy suppression reveals more indirect phenomena, and give informations on alternative pathway. This method has already been used with mitochondrial splicing mutants. For example, the NAM1 gene which was cloned (Ben Asher et al, 1989) turned out not to be the NAM1 gene defined by genetic analysis (Groudinski et al, 1981 ) but a multicopy suppressor . This new gene uncovered a relation between splicing and mitochondrial protein synthesis. Recently two new yeast genes -*MRS3* and *MRS4* ( Wiesenberger et al,1991)- which belong to the carrier family and suppress a splicing deficiency, have been identified by this method. This result shows the concentration of ions and/or polyamines within the mitochondria to be essential for a correct splicing event in vivo.

In addition to a search for second-site mutations as described above, we therefore decided to apply the multicopy suppression selection for several of our mitochondrial mutations.

In the case of the *cs 909* ribosomal mutation, we thus selected two new yeast genes *MBR1* and *MBR2*. These genes are related to the mitochondrial translation process since they can compensate the absence of growth on glycerol due to several mutations which have in common to affect the mitochondrial translation. They are not, however intrinsic components of the translation apparatus. Destruction of the chromosomal copy of these genes does not lead to an absence of growth on glycerol ( although a reduced growth on glycerol is observed for *MBR1* especially at low temperature) nor do these disrupted strains accumulate rho$^-$ cells as expected for such class of genes. These genes rather seem to act someplace in the pathway which drives a gene product from translation to assembly into a macromolecular complex. In good agreement with this hypothesis is the fact that only alleles which do not block completely gene expression can be suppressed.

In such a model, MBR1 which presents features reminiscent of a regulatory protein could regulate a transport, import or assembly pathway of nuclear genes necessary for mitochondrial biogenesis ; because this gene seems to be more important at low temperature and because translational mutations are suppressed by it, one could think for example of a control upon ribosome assembly.

It is interesting to note that *MBR1* also suppresses the absence of growth on glycerol due to a deletion of the *HAP2* or *HAP3* genes. These genes encode elements of a transcriptional activator complex which regulates the expression of several genes necessary for the mitochondrial function. The relation between these two set of genes is not explained at the moment.

Multicopy suppression experiments were also done with the *ts 932* mutation. We have seen previously that in addition to mitochondrial suppressors (second site intragenic suppressors), a nuclear suppressor has been isolated which might be a good candidate to encode the endonuclease responsible for this specific maturation step.

However, we were also able to uncover several genes which can compensate the absence of growth on glycerol of the *ts 932* mutation, when they are overexpressed on a multicopy plasmid.

Experimentally, we used a derivative of *ts 932* ( 932/12, *Mat a leu2 ade2* ) as a recipient strain for transformation with a YEP13 yeast genomic library constructed in the laboratory. Five different plasmids were found to suppress the phenotype due to *ts 932* (correlation analysis between the glycerol positive phenotype and the presence of the plasmid as well as restriction maps were performed to verify this point). Since these plasmids were obtained from a small number of transformants not representative of the total yeast genome, the number of different nuclear genes able to suppress this AsptRNA mutation is probably very large.

To rapidly analyze these yeast sequence we used insertional mutagenesis. This was done with a mini-Mu transposon derivative previously constructed in the laboratory (Daignan-Fornier and Bolotin-Fukuhara, 1989). Insertions which abolished the suppression effect were considered to be within the sequence necessary for expression of the gene, and the junction between the transposon and the yeast sequence was established. This method yields immediately about two hundred nucleotides within the coding sequence of the gene of interest which can be immediately compared to data bank.

Two kind of hypotheses could explain the multicopy suppression of the tRNA mutation: in the first one, an enzyme present in higher amount than normal, matures (in a specific or non specific way) the precursor tRNA, while in the second, the suppression takes advantage of the very low amount of mature tRNA which might be present but not detectable experimentally.

The following situations are relevant to the first class of hypothesis:

(i) The growth defect can probably be corrected by overproduction of the endonuclease gene itself. It is quite possible indeed that the inability to correctly mature the Asp tRNA is due to a lower affinity of the endonuclease for its substrate in the mutant strain. In such a case it is conceivable that overproduction of the enzyme could compensate the low affinity.

(ii) Genes encoding enzymes which interact with tRNA precursors and could contribute to their maturation should also be isolated. Although probably not as efficient as the specific endonuclease, they might recognize the tertiary structure of the precursor . If the amount of the gene product produced is much larger than the amount required for they normal function, they might produce enough tRNA for the cell to grow.

The second class could be explained by gene products which overcome the reduced translation rate taking place into the mitochondria in the mutant strain. There is only one Asp tRNA gene in the mitochondrial genome, and a mutation in

this gene cannot be directly compensated.
Those can be either:
(iii) genes modulators of the translational ambiguity. Translational errors allowing the insertion of Glu in place of Asp in the peptide chain might produce enzymes still functional, albeit poorly. Because of the genetic code (Asp= GAU, GAC ,Glu= GAA, GAG ), this is plausible.
(iv) genes which act at a limiting step during the transport , import or regulation of the gene products previously hypothesized. In that case, these steps being not limiting any more , the amount of gene product able to suppress the *ts932* will be larger than normal.
Finally, one cannot exclude partial mistargeting of nuclear genes which normally play an equivalent role but in the nuclear compartment .

Three yeast inserts have been submitted to this analysis. One was a new yeast sequence, while the two others were immediately identified with the mitochondrial Aspartyl-tRNA synthetase (Gampel and Tzagoloff,1989) and the mitochondrial elongation factor (Nagata et al,1983 ).
Several of the possible models we described previously could fit with these data: it is not highly probable that Ef-Tu acts directly as the elongation factor, since it is usually a very abundant factor which should not be limiting ( but this is not really known in the case of mitochondria). A function in the control of translational ambiguity has already been described for this factor ( references in Abdulkarim et al, 1991). It is also possible that both the tRNA synthetase and the elongation factor could help mature the precursor tRNA since they both recognize the tertiary structure of the tRNA.This has been shown to happen in vitro ( Beresten et al, 1992).

We have started this investigation to identify yeast genes which are involved in mitochondrial translation . Using both classical genetics and multicopy suppressors, we have been able to reveal new sets of interacting genes . These gene products participate to steps as different as the peptide translocation, the assembly of ribosomes or the maturation of precursors tRNAs. In addition, new yeast genes which may not act directly on mitochondrial translation but on related processes have been identified.
The final characterization of all these genes coupled with biochemical studies will hopefully lead us to a clear understanding of their mode of action. It will also allow us to study translation in an integrated way.

## Aknowledgments

We thank R. Zelikson for discussion and critical reading of the manuscripts. T.R. aknowledges the receipt of a fellowship from the foundation Cenci-Bolognetti. This work was supported by grants to URA D 1354 from the Centre National de la Recherche Scientifique , the Ministère de l'Education Nationale and the Ministère de la Recherche et de la Technologie ( Réseaux européens de laboratoires to M. B-F)

## References:

Abdulkarim F, Tuohy, TMF, Buckingham RH, Hughes D (1991) Missense substitutions lethal to essential functions of EF-Tu. Biochimie 73: 1457-1464

Ben Asher E, Groudinsky O, Dujardin G, Altamura N, Kermorgant M, Slonimski PP (1989) Novel class of nuclear genes involved in both mRNA splicing and protein synthesis in *Saccharomyces cerevisiae* mitochondria. Mol. Gen. Genet. 215: 517- 528

Berensten SM, Jahn M, Söll D ( 1992) Aminoacyl tRNA synthetase-induced cleavage of tRNA. Nucl. Acids Res. 20: 1523-1530

Bolotin-Fukuhara M, Faye G, Fukuhara H (1977) Temperature-sensitive respiratory-deficient mitochondrial mutations: isolation and genetic mapping. Mol. Gen. Genet. 152 : 295- 306

Bolotin-Fukuhara M. (1979) Mitochondrial and nuclear mutations that affect the biogenesis of the mitochondrial ribosomes of yeast. I. Genetics. Mol. Gen. Genet. 177 : 39-46

Bolotin-Fukuhara M, Sor F, Fukuhara H (1983) Mitochondrial ribosomal RNA mutations and their nuclear suppressors. In :"Mitochondria 83 ". Schweyen et al, eds. Walter de Gruyter, Berlin. pp. 455- 467

Contamine V, Bolotin-Fukuhara M (1984). A mitochondrial ribosomal RNA mutation and its nuclear suppressors. Mol. Gen. Genet. 193 : 280-285

Daignan-Fornier B, Su I, Contamine V, Bolotin-Fukuhara M (1988) A genetic approach of molecular interactions in yeast mitochondrial ribosomes. In "Genetics of Translation: New Approaches " M.Tuite, M.Picard and M. Bolotin-Fukuhara, eds. NATO/ASU Workshop . Springer-Verlag, Berlin, pp. 159-169

Daignan-Fornier B, Bolotin-Fukuhara M (1988) Mutational study of the rRNA in yeast mitochondria : functional importance of T 1696 in the large rRNA gene.

Nucleic Acids Research 16 : 9299- 9306
Daignan-Fornier B, Bolotin-Fukuhara M (1989) Functional exploration of the yeast (-*S. cerevisiae* ) genome : use of a mini-Mu transposon to analyse randomly cloned DNA sequences. Yeast , 5 : 259- 269.
Dujon B (1980) Sequence of the intron and flanking exons of the mitochondrial 21SrRNA gene of yeast strains having different alleles. Cell 20: 185-197
ElElj-Fridhi N, Pallier C, Guetari M, Bolotin-Fukuhara M (1991) Mutational studies of the major tRNA region of the S.cerevisiae genome. Current Genetics19: 301-308
Faye G, Bolotin-Fukuhara M, Fukuhara H (1976) Mitochondrial mutations that affect mitochondrial transfer ribonucleic acid in *Saccharomyces cerevisiae.* In : " The Genetics and Biogenesis of Chloroplasts and Mitochondria ". Ed. T. Bücher et al. Elsevier/North Holland Biomedical Press. Amsterdam. The Netherlands. pp : 613- 618
Gampel A, Tzagoloff A (1989) Homology of aspartyl- and lysyl-tRNA synthetases. Proc. Natl. Acad. Sci.USA 86: 6023-6027
Gietz RD, Sugino A (1988) New yeast-E.coli shuttle vectors constructed with in vitro mutagenized yeast genes lacking six-base pair restriction sites. Gene 74: 527-534
Grivell LA (1989) Nucleo-mitochondrial interactions in mitochondrial biogenesis. Eur. J. Biochem. 182: 477-493
Groudinski O, Dujardin G, Slonimski PP (1981) Long range control circuits within mitochondria and between nucleus and mitochondria. II. Genetic and biochemical analyses of suppressors which selectively alleviate the mitochondrial intron mutations. Mol. Gen. Genet.184:493-503
Guetari M, Zelikson R, Bolotin-Fukuhara M (1987) Nucleo-mitochondrial interactions involved in the biogenesis of mitochondrial ribosomes : identification of a possible conditional mutation of the Var 1 gene and its nuclear suppressors. Mol. Gen. (Life Sci. Adv. ) 6 : 29- 34
Julou C, Bolotin-Fukuhara M (1982). Genetics of mitochondrial ribosomes of yeast: mitochondrial lethality of a double mutant carrying two mutations of the 21S rRNA gene. Mol. Gen. Genet. 188 : 256- 260
Julou C, Contamine V, Sor F, Bolotin-Fukuhara M (1984) Mitochondrial ribosomal RNA genes of yeast : their mutations and a common nuclear suppressor. Mol. Gen. Genet. 193 : 275- 279
Myers AM, Pape L, Tzagoloff A. (1985) Mitochondrial protein synthesis is required for maintenance of intact mitochondrial genomes in Saccaromyces cerevisiae. EMBO J. 4: 2087-2092
Nagata S, Tsunetsugu-Yokota Y, Naito A, Kaziro Y (1983) Molecular cloning and sequence determination of the nuclear gene coding for mitochondrial elongation factor Tu of S. cerevisiae. Proc. Natl. Acad. Sci. USA 80: 6192-6196
Pfanner N, Neupert N (1990) the mitochondrial protein import apparatus. Ann. Rev. Biochem. 59: 331-353
Sor F, Faye G (1979) Mitochondrial and nuclear mutations that affect the biogenesis of the mitochondrial ribosomes of yeast II. Biochemistry. Mol. Gen. Genet. 177: 47-56
Valens M, Rinaldi T, Daignan-Fornier B, Bolotin-Fukuhara M (1991) Identification of nuclear genes which participate to mitochondrial translation in S.cerevisiae. Biochimie 12:1525-1532

Wiesenberger G, Link TA, von Ahsen U, Waldherr M, Schweyen R (1991) MRS3 and MRS4, two suppressors of mtRNA splicing defects in yeast, are new members of the mitochondrial carrier family. J. Mol. Biol. 217: 23-37
Zennaro E, Francisci S, Ragnini A, Frontali L, Bolotin-Fukuhara M (1989) A point mutation in a mitochondrial tRNA gene abolishes its 3' end processing. Nucleic Acids Research 17 : 5751-5764

# EARLY EVENTS IN PROTEIN IMPORT INTO MITOCHONDRIA

K. Kassenbrock, M. Douglas, and D. Cyr
Department of Biochemistry and Biophysics
University of North Carolina School of Medicine
Chapel Hill, North Carolina 27599-7260

## Introduction:

Assembly of newly synthesized precursor proteins into mitochondria involves events occurring both in the cytosol and at the surface of the organelle. Co-incident with or shortly following protein translation, recognition events must take place to allow a precursor to be correctly targeted to its final destination (Pfanner and Neupert, 1990; Glick and Schatz 1991). Cellular recognition of specific structural features of mitochondrial proteins takes several forms. Structural information must be recognized that specifies the destination of the protein, such as that contained in amino terminal pre-sequences that are sufficient to direct even foreign proteins to mitochondrial locations. In addition to this "positive" targeting information, the cell must recognize "negative" targeting information, i.e., properties of a protein which might impair its ability to be correctly localized. To be more specific, a precursor must be prevented from aggregating or assuming a tightly folded conformation that cannot be translocated linearly through a small diameter membrane channel (Rassow et al., 1990). Recognition of "positive" targeting information is the job of the receptors and translocation apparatus on the surface of the mitochondria (Pfanner et al., 1991). Recognition of "negative" targeting information, i.e. the maintenance of a "transport-competent" conformation, is the work of cytoplasmic protein chaperones such as hsp70 (Deshaies et al., 1988). In neither case are the molecular details well understood. In this paper we will describe work addressing both systems.

## Cytoplasmic recognition events:

Molecular chaperones are required in different intracellular spaces for translocation and movement of nascent or partially assembled

NATO ASI Series, Vol. H 71
Protein Synthesis and Targeting in Yeast
Edited by A. J. P. Brown, M. F. Tuite, and J. E. G. McCarthy

proteins. The role that these chaperones play in protein targeting and assembly in different organelles is an area of active study (for review, see Gething and Sambrook, 1992). In the cytosol, hsp70 members are thought to be required mainly for the maintenance of "transport competence", that is, to maintain precursors in a loosely folded state and prevent the formation of irreversible three dimensional structures including aggregates. The structural features recognized by hsp70 proteins are not understood in detail but *in vitro* studies suggest that short peptide sequences containing aliphatic residues are preferentially bound ( Flynn et al., 1991 ).

In E. coli, the hsp70 protein known as dnaKp has been studied in some detail and has been shown both genetically and biochemically to function in concert with two other proteins, the products of the dnaJ and grpE genes (for reveiw, see Ang et al., 1991). To further understand the function of cytoplasmic hsp70 in mitochondrial protein import, we have purified the major yeast cytosolic hsp70, SSA1p, and the yeast dnaJ homologue, YDJ1p, and studied their interactions *in vitro*.

## YDJ1p regulation of precursor binding by hsp70:

Genetic and biochemical studies in yeast have identified proteins which interact with nascent mitochondrial precursors at Stage 1, off membrane (Pfanner and Neupert, 1987) and Stage 2 (outer membrane bound) of protein import. The components at stage 1 are hsp70, the SSA1 gene product (Deshaies, et al., 1988) and a regulator of hsp70, the product of MAS5/YDJ1 (Caplan and Douglas, 1991; Atencio and Yaffe, 1992). Loss of function in either of these components results in a reduction in the rate of import of mitochondrial precursors.

The YDJ1 and the SSA1 gene products were purified for use in different assays to determine if they interact with each other, if they bind to nascent precursors and what the role of nucleotides is in this interaction. To document the direct interaction of these two proteins, we first observed that the ATPase activity of SSA1p ($hsp70^{SSA1}$) is stimulated by YDJ1p (Figure 1). The low ATPase activity of $hsp70^{SSA1}$ is typical of values reported for other hsp70 homologues (Flynn, et al., 1989; Liberek, et al., 1991; Kassenbrock and Kelly, 1989). At saturating

ATP concentrations, YDJ1p with no intrinsic ATPase activity of its own stimulates the ATPase of $hsp70^{SSA1}$ eight to ten-fold. Maximum stimulation by YDJ1p occurs at YDJ1p:$hsp70^{SSA1}$ molar ratio of 0.5-1.0.

Since nucleotide hydrolysis by hsp70 homologues has been

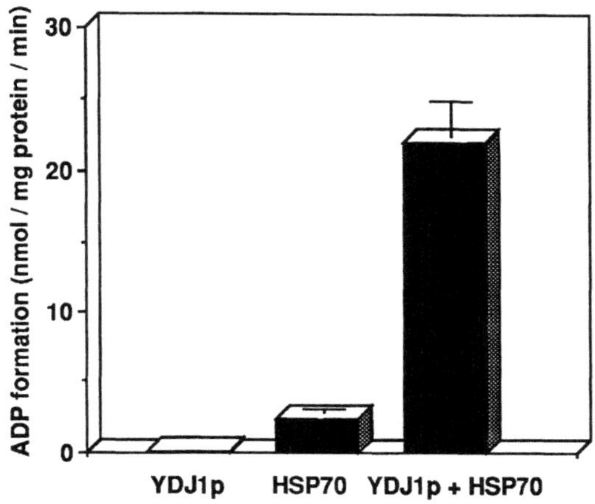


**Figure 1. YDJ1p stimulates ATPase activity of Hsp70.**
Hsp70 (SSA1p, 0.25 μM) and YDJ1p (0.16 μM) were incubated in reactions in 50 mM Hepes, pH 7.4, 50 mM NaCl, 1.0 mM $MgCl_2$, 50 μM ATP (α-32P) for 20 min at 30°C. ADP formation was quantitated using PEI-cellulose thin layer chromatography (see Cyr et al., 1992).

correlated with conformational changes to promote release of polypeptides (Kassenbrock and Kelly, 1989; Flynn, et al., 1989; Liberek et al., 1991), we determined the ability of the different purified components to bind to a synthetic peptide consisting of the first 51 residues of the $F_1$ATPase β-subunit precursor (Figure 2). This peptide has been shown to efficiently compete for protein import of a full length precursor (Cyr and Douglas, 1991). In other studies from this laboratory, the hsp70 protein can be demonstrated to bind to denatured α-carboxy methyl-lactalbumin (Cyr et al., 1992). In the absence of YDJ1p, $hsp70^{SSA1}$ associates with labeled $F_1$ß1-51 peptide to shift approximately 30% of the input peptide to the bound state as visualized on native gels (Figure 2, column 4). The majority of the peptide bound under these conditions is released only when ATP and YDJ1p are added together (Figure 2, column 6) but not when either is incubated independently in an otherwise complete reaction. In control studies, release of bound peptide requires ATP hydrolysis since substitution of a non-hydrolyzable analog fails to catalyze release. ATP- and YDJ1p-dependent release of bound peptide is the same as that observed for release of the unfolded control substrate. Maximal ATP-dependent release of peptide from SSA1p occurs at the same molar YDJ1p:SSA1p molar ratio observed for maximal stimulation of SSA1p ATPase (Cyr et al., 1992). These studies demonstrate that SSA1p can associate directly with nascent protein precursors as part of its role in protein import. The efficient release of SSA1p from precursors to promote proper "loose" folding or to initiate interaction with the outer membrane surface likely requires the action of YDJ1p which accelerates the ATPase activity of SSA1p. Mitochondrial import of precursors is less efficient when the YDJ1p is not farnesylated at its C - terminus (Caplan, Cyr and Douglas, submitted). The present observations do not rule out that YDJ1p may associate with nascent proteins independent of the SSA1p. The present observations do, however, demonstrate that the cycling of these chaperones on nascent mitochondrial precursors is an ATP-dependent activity which is a necesary to maintain their import compentency.

**Mitochondrial surface recognition events:**

A number of studies have identified putative receptors for mitochondrial precursor proteins in yeast. However to date, the known proteins on the surface of mitochondria which exhibit receptor activity *in vitro* (such as Mas70p) do not exhibit a strong phenotype when deleted. Such a lack of phenotype may mean either that the protein serves a non-essential function, or that it performs a function that can also be performed by another gene product, i.e., the protein is functionally redundant. In Neurospora crassa, the Mas70 homologue MOM72 has been shown *in vitro* to function in parallel with another receptor protein, MOM19 (Steger et al., 1990). Functional redundancy of surface receptors makes it difficult to precisely understand the pathway mitochondrial precursors take *in vivo*. Therefore, we have focussed our attention on the next step in the pathway, interaction with the translocation apparatus. Here the data are more clear; work in both yeast and N. crassa have identified homologous proteins that form part of the translocation channel of the mitochondrial outer membrane

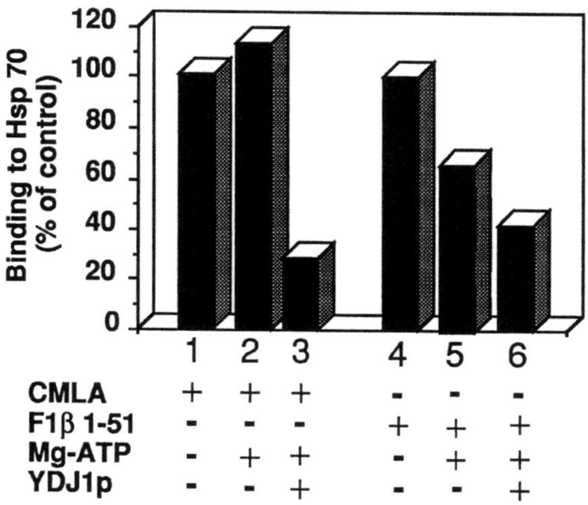


**Figure 2. YDJ1p stimulated peptide release from HSP70.** Reaction mixtures containing $^{125}$I-labeled peptides, Hsp70 and YDJ1p as indicated were incubated at 30°C in 50 mM Hepes, pH 7.4, 50 mM NaCl, 10 mM DTT, 0.1 mM EDTA, 0.4 μg/μl BSA. After 20 minutes reaction mixtures were analyzed by electrophoresis on 10-15% gradient native gels (Cyr et al., 1992).

(Vestweber et al, 1990; Baker et al, 1990). These proteins, termed import site protein 42 (ISP42) in yeast or MOM38 (Kiebler et al., 1990) in Neurospora, can be cross-linked to membrane-spanning mitochondrial translocation intermediates. Both proteins have been cloned, and the yeast protein has been shown to be essential for viability (Baker et al., 1990).

As an essential protein of the mitochondrial outer membrane translocation apparatus, ISP42 would appear to be an ideal starting place to search for additional components in the machinery. However, this protein is present in small amounts in mitochondria and biochemical characterization of the components associated with it is difficult (Vestweber and Schatz, 1988). To approach the problem of identifying additional components associated or interacting with ISP42, we have therefore turned to genetic methods in yeast.

**Suppressors of $ISP42^{ts}$:**

To genetically identify additional components of the mitochondrial translocation apparatus that interact with Isp42p, we have first prepared mutants in ISP42 which are conditional for function followed by selection and characterization of suppressors of these ISP42 conditional alleles. For these studies, we have constructed a strain of yeast in which the coding sequence of the chromosomal ISP42 gene has been completely deleted and is replaced by the HIS3 gene. The viability of this isp42::HIS3 deletion host is maintained by a temperature sensitive allele of ISP42 ($ISP42^{ts}$) on a CEN4TRP1 plasmid (Kassenbrock and Douglas, in preparation). This yeast mutant is able to grow at 23°C, but fails to grow at 35°C.

To select extragenic high copy suppressors, this mutant was transformed with a yeast genomic library on the yeast two micron URA3 vector, YEP24. $Ura^{+}$ transformants were selected and examined for their ability to restore growth at 35°C. In this manner ISP42 was recloned many times. However, about 10% of the transformants obtained in this genetic screen exhibit restoration of growth at 35° and contained no ISP42 DNA. The preliminary characterization of one of these suppressors in described here.

Starting from an initial series of about 10,000 transformants, one plasmid was recovered twice that could complement growth of the $ISP42^{ts}$ mutant at 35°C but not at 37°C, and contained a genomic insert of about 10kb. Using deletion analysis, the rescuing suppressor gene was mapped to a region of less than 700 base pairs.DNA sequence analysis revealed two overlapping reading frames of 61 and 58 residues respectively (Figure 3). Two experiments strongly indicate that the 61 amino acid open reading frame (ORF61) is the gene product expressed in yeast. First, the ORF61 and not the ORF58 when fused to E. Coli lacZ expressed β-galactosidase activity in yeast. Second, an expression plasmid containing the complete ORF61 but not capable of expressing the ORF58, was able to complement the $ISP42^{ts}$ mutant for growth at 35°.

Additional genetic studies were performed to determine if the ORF61 suppressor gene product was physically associated with the

Isp42p in the insertion site complex (ISC). Preliminary studies had suggested that the ORF61-lacZ gene product was localized to a

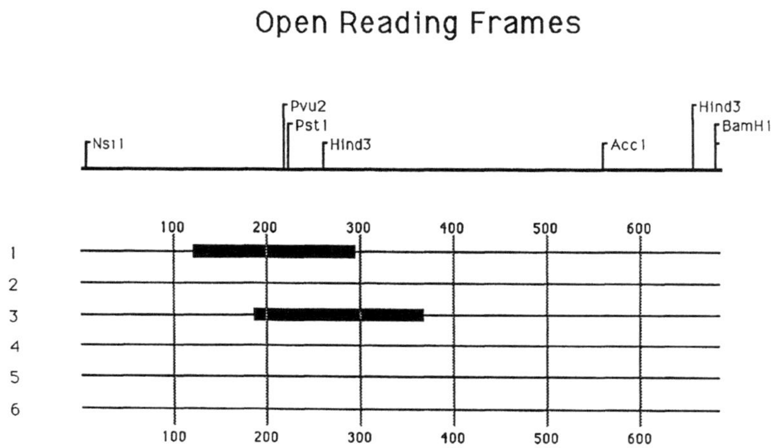


**Figure 3. Open reading frames encoding a high copy suppressor of ISP42$^{ts}$.**

particulate or membrane fraction in yeast consistent with its potential localization in the insertion site complex. In order to examine the physical association of Orf61p and Isp42p, genetic disruption of the ORF61 was performed. This disruption revealed that in the presence of a wild type ISP42 gene product, no phenotypic consequences of the ORF61 disruption were observed. Cells without the ORF61 gene product grew normally on all carbon sources and at all temperatures tested. Construction of the double mutant, however, which harbored a disruption of ORF61 and an ISP42$^{ts}$ failed to grow under any conditions. This observation of synthetic lethality is a strong indication that physical interactions exist between the ORF61 gene product and Isp42p. Studies to physically confirm the localization of the Orf61p are currently in progress.

The sequence of the ORF61 reveals the coding sequence for a methionine rich peptide which contains an even number of acidic and basic residues flanking a 20 amino acid potential membrane spanning domain between residues 32 and 52 (Figure 4). A scan of the available protein sequence data bases failed to reveal any significant motifs within ORF61 which would provide additional insight to its function. The predicted size of the Orf61p (about 6,400 daltons) is roughly similar to the molecular size of two outer membrane components (7 and 8Kda)

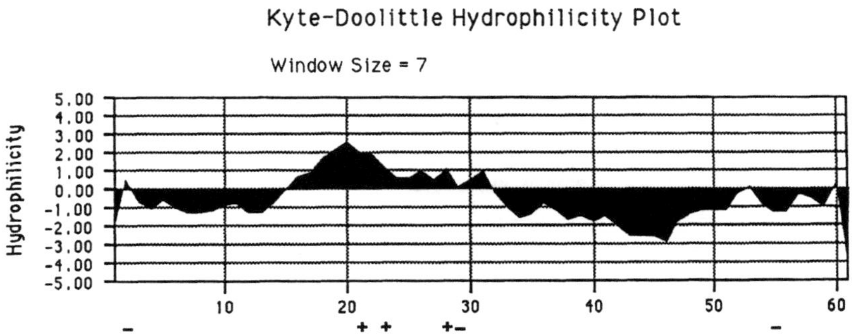


**Figure 4. Hydrophilicity plot and charge distribution of the Orf61p suppressor of ISP42$^{ts}$.**

which have been shown to co-purify with and cross-link to translocational intermediates within the insertion site complex of N. crassa (Sollner, et al., 1992). Determination of homology must await sequence analysis of the Neurospora proteins.

Further characterization of ORF61 as well as other suppressors that we are now characterizing should lead to additional insights into the molecular mechanisms of protein import and mitochondrial biogenesis.

**Acknowledgements:** These studies are supported by grants from NIH to K. K. (GM14266) and M. D. (GM36537) and American Heart (NC-90-F4) to D. C.

**References:**


Ang, D. Liberek, K. Skowyra, D., Zylicz, M., Georgopoulos, C. (1991) Biological role and regulation of the universally conserved heat shock proteins. J. Biol. Chem. 266:24233-24236.

Atencio, D., Yaffe, M. (1992) MAS5, a yeast homolog of dnaJ involved in mitochondrial protein import. Mol. Cell Biol. 12:283-291.

Baker, K., Schaniel, A., Vestweber, D., Schatz, G. (1990) A yeast mitochondrial outer membrane protein essential for protein import and cell viability. Nature 348:605-609.

Caplan, A., Douglas, M.G. (1991) Characterization of YDJ1: A Yeast Homologue of the Bacterial dnaJ Protein. J. Cell Biol. 114:609-621.

Cyr, D., Douglas, M. (1991) Early events in the transport of proteins into mitochondria: Import competition by a mitochondrial presequence. J. Biol. Chem. 266:21700-21708.

Cyr, D., Lu, X., Douglas M. (1992) Regulation of Hsp70 function by a eukaryotic dnaJ homologue. Submitted.

Deshaies, R., Koch, B., Werner-Washburne, M., Craig, E., Schekman, R. (1988) A subfamily of stress proteins facilitates translocation of secretary and mitochondrial precursor polypeptides. Nature 332:800-805.

Flynn, G Pohl, J. Flocco, M., Rothman, J. (1991) Peptide-binding specificity of the molecular chaperone BIP. Nature 353:726-30.

Flynn, G., Chappell, T., Rothman, J. (1989) Peptide binding and release by proteins implicated as catalysts of protein assembly. Science 245:385-389.

Gething, M-J., Sambrook, J. (1992) Protein folding in the cell. Nature 355:33-45.

Glick, B., Schatz, G. (1991) Import of proteins into mitochondria. Annual Rev. Genetics 25:21-44.

Kassenbrock, K., Kelly, R. (1989) Interaction of heavy chain binding protein (BIP/GRP78) with adenine nucleotides. EMBO J. 8:1461-1467.

Kiebler, M., Pfaller, R., Sollner, T., Griffiths, G., Horstmann, H., Pfanner, N., Neupert, W. (1990) Identification of a mitochondrial receptor complex required for recognition and membrane insertion of precursor proteins. Nature 348:610-616.

Liberek, K., Marszalek, J., Ang, D., Georgopoulos, C., Zylicz, M. (1991) E. coli dnaJ and grpE heat shock proteins jointly stimulate ATPase activity of dnaK. Proc. Natl. Acad. Sci. 88:2874-2878.

Lieberek, K., Skowyra, D., Zylicz, M., Johnson, C., Georgopoulos, C. (1991) The chaperone, the DnaK chaperone, the 70kDa heat shock protein equilivent, changes conformation upon ATP hydrolysis, thus triggering its dissociation from bound target protein. J. Biol. Chem. 266:14491-14996.

Pfanner, N., Neupert, W. (1990) The mitochondrial protein import apparatus. Ann. Rev. Biochem. 59:331-353.

Pfanner, N., Neupert, W. (1987) Distinct steps in the import of the ADP/ATP carrier into mitochondria. J. Biol. Chem. 262:7529-7536.

Pfanner, N., Sollner, T., Neupert, W. (1991) Mitochondrial import receptors for precursor proteins. Trends in Biochem. Sci. 16:63-67.

Rassow, J., Hartl, U., Guiard, B., Pfanner, N. and Neupert, W. (1990) Polypeptides traverse the mitochondrial envelope in an extended state. FEBS-Lett. 275:190-194.

Sollner, T., Rossow, J., Wiedmann, M., Schlossmann, J., Keil, P., Neupert, W., Pfanner, N. (1992) Mapping of the protein import machinery in the mitochondrial outer membrane by cross linking of translocation intermediates. Nature 355:84-87.

Steger, H., Sollner, T., Kiefler, M., Dietmeier, K., Pfaller, R., Trulzsch, K., Tropschug, M., Neupert, W., Pfanner, N. (1990) Import of ADP/ATP carrier into mitochondria: two receptors act in parallel. J-Cell-Biol. 111:2353-2363.

Vestweber, D., Brunner, J., Baker, A., Schatz, G. (1989) A 42K outer-membrane protein import site. Nature 341:205-209.

Vestweber, D., Schatz, G. (1988) A chimeric mitochondrial precursor protein with internal disulfide bridges blocks import of authentic precursors into mitochondria and allows quantitation of import sites. J. Cell Biol. 107:2037-2043.

# TARGETING PATHWAYS TO THE MITOCHONDRIAL INNER MEMBRANE

M.G. Cumsky, B.R. Miller, A.S. Gaikwad, and L.A. Jung
Department of Molecular Biology and Biochemistry
University of California, Irvine
Irvine, CA 92717

Residing within a functional mitochondrion are a diverse array of metabolic enzymes that catalyze reactions essential to the survival of nearly all eucaryotic cells. Only a few of the several hundred different mitochondrial proteins are products of the mitochondrial genome; the vast majority (>95%) are encoded on nuclear genes and imported into the organelle during or after translation in the cytosol. The correct delivery of these proteins to the mitochondrion, and their subsequent sorting to the proper mitochondrial compartment (outer membrane, inner membrane, intermembrane space, matrix), is therefore an essential problem within the field of mitochondrial biogenesis.

In recent years a large amount of effort has been devoted to the study of mitochondrial protein import (Hartl et al., 1989; Hartl and Neupert, 1990; Glick and Schatz, 1991). Results from a number of laboratories have made several aspects of the import process clear, and in addition, contributed significant insights into our overall understanding of protein translocation into and across all biological membranes.

We report here a summary of several aspects of our laboratory's work on yeast mitochondrial import. In particular we will be concerned with studies that have established the import pathway of a specific inner membrane protein, the precursor to subunit Va of the cytochrome *c* oxidase complex. We also summarize the results of cross-

NATO ASI Series, Vol. H 71
Protein Synthesis and Targeting in Yeast
Edited by A. J. P. Brown, M. F. Tuite, and J. E. G. McCarthy

linking experiments designed to identify proteins that specifically interact with mitochondrial targeting sequences.

**Mitochondrial Import Pathways**

A schematic illustration showing a current view of several established or proposed mitochondrial import pathways is presented in Fig. 1. Also represented on the figure are a number of proteins thought to function in the import process. Many mitochondrial precursors initially interact with proteinaceous components on the mitochondrial surface (Hartl and Neupert, 1990). While a majority of the evidence suggesting that these proteins are true receptors is largely indirect, 19 kD and 72 kD proteins (MOM19 and MOM72, respectively) have been identified as specific recognition elements in *Neurospora crassa* mitochondria (Hartl and Neupert, 1990; Kiebler et al., 1990). A protein of 70 kD (Mas70p), which is homologous to MOM72, has also been identified in yeast (Kiebler et al., 1990; Glick and Schatz, 1991, Fig. 1). To be transported into and across the mitochondrial membranes, precursor proteins must adopt an "import-competent conformation" which consists of a loose or unfolded state (Eilers et al., 1988). At present, it is thought that ATP, cytosolic hsp70 (Fig. 1), acidic phospholipids, and possibly other cytosolic factors help precursors achieve and/or maintain import competence.

Once associated with mitochondria, import-competent proteins appear to begin the translocation process by interacting with an integral protein of the outer membrane termed the general insertion protein or GIP. It has been suggested that the function of the GIP is contributed by a 38 kD outer membrane protein from *Neurospora* (MOM38) and a 42 kD outer membrane protein from yeast (ISP42) (Vestweber et al., 1989; Kiebler et al., 1990). From GIP, precursors are either inserted directly into the outer membrane or are transferred through inner-outer membrane contact sites to other compartments (Fig. 1). The latter step requires both the

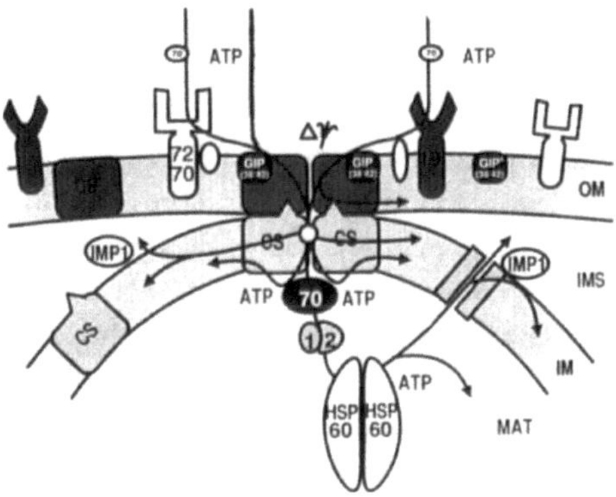


**Figure 1.** Mitochondrial import pathways. Arrows indicate known or proposed translocation routes of many mitochondrial precursor proteins. 70, hsp70; 1,2 heterodimer, matrix metalloprotease; IMP1, intermembrane space protease 1. Details concerning other mitochondrial proteins are given in the text.

potential across the inner membrane (Δψ) and the hydrolysis of ATP (Hartl and Neupert, 1990). It is important to note that while the existence of membrane contact sites for mitochondria, chloroplasts, and bacterial membranes has been clearly demonstrated, the composition and physical nature of these sites is poorly understood. However, recent results suggest that the contact sites are composed of discreet, separable channels in each membrane (Glick et al., 1991, Pfanner et al., 1992).

The contact sites are thought to be a switching point where translocation intermediates are diverted to several different pathways (Fig. 1). One pathway, referred to as "nonconservative" by Neupert and coworkers (Hartl and Neupert, 1990), involves direct partitioning of precursors into the inner membrane from contact sites (Fig. 1). A protein that follows a nonconservative pathway, the precursor to the ATP/ADP carrier (AAC), also requires the function of mitochondrial hsp70 (Fig. 1). However, as shown in the figure, we suggest that this may not be true in all cases (see below).

Precursor proteins may also be translocated completely across the inner membrane into the matrix. It has recently become clear that newly translocated polypeptides, and possibly proteins in the process of translocation, associate with several additional proteins located in the matrix or on the matrix face of the inner membrane. One such protein, already mentioned above, is a mitochondrial member of the hsp70 family (the SSC1 protein, Fig. 1). This protein is thought to bind to precursors as they enter the matrix and facilitate translocation (Baker and Schatz, 1991). Another matrix protein known to be involved in import is the protease responsible for cleavage of the presequence. This protease, often referred to as the matrix metalloprotease, is actually a heterodimer consisting of the MPP or MAS2 product and the PEP or the MAS1 product (Baker and Schatz, 1991).

The majority of the proteins studied thus far also interact with a matrix complex composed of 14 identical subunits of hsp60 (Hartl and Neupert, 1990). The hsp60 complex is a molecular chaperone that catalyzes the correct folding and assembly of matrix proteins and protein complexes. Furthermore, since some (but possibly not all, see Glick et al., 1992) proteins destined for both the inner membrane and intermembrane space may be routed through the matrix ("conservative sorting", Fig. 1), the hsp60 complex may also be involved in stabilizing the transport-competent conformation of these reexported proteins (Hartl and Neupert, 1990). It is important to note that the nature of the inner membrane channel through which conservatively-sorted proteins are reexported has not yet been defined.

**The subunit Va import pathway**

Our laboratory has been involved in studies of mitochondrial protein import in *Saccharomyces cerevisiae*. We have focused considerable attention on the import of an inner membrane protein, the precursor to subunit Va of the cytochrome *c* oxidase complex (pVa). Early results, obtained in vivo, demonstrated that subunit Va is efficiently targeted to mitochondria with minimal presequences (Glaser et al., 1988), including one of only 5 amino acids (Glaser et al., 1990). In addition, mutant presequences that fail to efficiently target their cognate protein (human ornithine transcarbamylase) direct efficient import of subunit Va (Glaser et al., 1990). When taken together, our combined results lead to two important (albeit related) conclusions regarding the import of pVa. First, they suggest that the mature protein contains targeting information that contributes to the extraordinary efficiency with which the protein is imported. Second, the results clearly demonstrate that the mature protein contains the information required for inner membrane localization of subunit Va.

When studied using the in vitro mitochondrial import assay, import of pVa is dependent upon both a membrane potential ($\Delta\psi$) and the hydrolysis of ATP (Miller and Cumsky, 1991). However, the levels of ATP necessary for the import of pVa are significantly lower than those required for the import of a different mitochondrial precursor protein, the ß subunit of the $F_1$-ATPase. The rate of import of pVa is also unaffected by temperature over the range $0^oC$ to $30^oC$, and is not facilitated by prior denaturation of the protein (Miller and Cumsky, 1991).

The above results, in conjunction with those described earlier (demonstrating that pVa is efficiently targeted to mitochondria with minimal presequences), strongly suggest that the subunit Va precursor normally exists in a loosely-folded conformation. We therefore speculated that import of pVa would not be dependent upon the function of cytosolic hsp70. Several attempts were made to demonstrate this point directly, using a strain in which hsp70 is expressed transiently (Deshaies et al., 1988). Because pVa is not stable outside the mitochondrion in vivo, the results of those experiments were inconclusive (B.R. Miller, unpublished observations). However, it has recently been shown that cell-free translation extracts prepared from wheat germ contain levels of hsp70 that are insufficient to support the import of certain mitochondrial precursors (Murakami et al., 1988). These extracts must therefore be supplemented with hsp70 if precursor proteins translated in them are to be used in mitochondrial import experiments (Murakami et al., 1988). When we translated pVa in an unsupplemented wheat germ system, we found that it was imported into mitochondria with normal efficiency (B.R. Miller and M.G. Cumsky, submitted). This result, although indirect, is in line with our earlier prediction that hsp70 function would not be required for the import of pVa, and provides further evidence that pVa exists in a loosely-folded conformation.

The import of pVa is also unusual in that it is independent of the function of protease-sensitive surface receptors. When mitochondria are pretreated with high concentrations of trypsin or proteinase K (1 mg per ml and 200 μg per ml, respectively) prior to import, not only is pVa imported, but strikingly, the rate of import, at both $0^{o}C$ and $30^{o}C$, is identical to that observed with the untreated organelles (Miller and Cumsky, 1991). In addition, pVa is properly sorted to the inner membrane after import at low temperature or into protease-treated mitochondria.

The results of the experiments described in the preceding paragraphs, and other results which indicate that pVa is imported through translocation contact sites (Miller and Cumsky, 1991) and is processed by the matrix metalloprotease (B.R. Miller, unpublished results) define, in part, the subunit Va import pathway. This pathway utilizes several well-established translocation steps, but, in its entirety, is distinct from either the receptor-independent pathway used by apocytochrome *c* (Hartl and Neupert, 1990), or the more general pathway used by a majority of mitochondrial precursor proteins (Hartl et al., 1989). Key questions, only recently resolved, concern the nature of the information in the mature protein required to deliver subunit Va to the inner membrane, and the type of pathway used by the protein to get there (conservative or nonconservative).

The signal necessary to localize subunit Va to the inner membrane was found to be a 22 amino acid membrane-spanning domain located within the C-terminal portion of the protein (Glaser et al., 1990). Deletion of this region, or changing a single hydrophobic amino acid to arginine, causes the protein to be redirected to the mitochondrial matrix (Glaser et al., 1990; L.A. Jung, B.R. Miller, and M.G. Cumsky, in preparation). When a short C-terminal portion of subunit Va (containing the localization domain) was appended onto a precursor protein that was normally imported to the matrix (a cytochrome $c_1$-dihydrofolate reductase fusion), the protein was instead found in the inner membrane (L.A. Jung, B.R. Miller, and M.G. Cumsky, in preparation). In contrast, the same fusion protein containing an internal deletion of the localization domain was found in the matrix (L.A. Jung, B.R. Miller, and M.G. Cumsky, in preparation). Taken together, these results show that this region of the Va protein is both necessary and sufficient for inner membrane localization.

Another important question concerning the import of pVa is whether the protein follows a conservative or nonconservative route from the contact sites to the inner membrane. Clearly, the localization sequence itself provides no answer, since there is precedent for such hydrophobic sequences functioning as both reexport (conservative sorting) or stop-transfer (nonconservative sorting) signals (Hartl and Neupert, 1990, Glick et al., 1992). To determine how pVa was sorted, we asked whether mitochondrial hsp60 and hsp70 were necessary to achieve efficient import and localization of the protein (the function of both chaperones should be required if pVa follows a conservative pathway). Using yeast strains that rendered the function of either hsp70 or hsp60 temperature-sensitive (the *ssc1* and *mif4* mutants, respectively; see Hartl and Neupert, 1990, Baker and Schatz, 1991), we found that pVa was imported normally without the function of either protein (B.R. Miller and M.G. Cumsky, submitted). In contrast, and as predicted, the ß subunit of the $F_1$-ATPase was found to require the function of both proteins (B.R. Miller and M.G. Cumsky, submitted). We suggest, therefore, that subunit Va moves from contact sites directly to the inner membrane. We also suggest that the function of the subunit Va localization domain is to stop the transfer of the precursor across the inner membrane. We note, however, that our results do not rule out the possibility that pVa normally utilizes hsp60 and hsp70 during import, but that it can still be efficiently taken up and localized without them.

**Cross-linking of a synthetic presequence to mitochondria**

Earlier studies from our lab demonstrated that a synthetic mitochondrial presequence peptide, corresponding to the N-terminal 22 residues of the *COX4* presequence (pL4(1-22)), blocks protein import into isolated mitochondria (Glaser and Cumsky, 1990a). The import block is competitive, reversible (it can be overcome by the addition of increased amounts of precursor), and does not result from the peptide dissipating the membrane potential (Glaser and Cumsky, 1990a). pL4(1-22) appears to be a general inhibitor of mitochondrial import in that it blocks uptake of several (and perhaps all) mitochondrial precursor proteins.

The results of additional studies demonstrated that pL4(1-22) can block import at a step subsequent to the initial recognition or binding of proteins to mitochondria (Glaser and Cumsky, 1990a). The peptide is taken up by mitochondria where it accumulates in the inner membrane or possibly within the contact sites of both membranes (Glaser and Cumsky, 1990b). Ultimately, it is cleared from the membrane fraction (we suspect that it is transported to the matrix where it is degraded), and, as the intramitochondrial concentration of peptide drops, the mitochondria regain the ability to import and process precursor proteins (Glaser and Cumsky, 1990b).

Taking advantage of the fact that pL4(1-22) accumulates within the mitochondrial membranes in high concentrations, a radiolabelled derivative of the peptide was synthesized and chemically cross-linked to intact mitochondria using the membrane-permeable reagent EGS. The pattern of cross-linked bands was precisely reproducible over many experiments. It showed a major cross-linked species of approximately 45 kD, strongly cross-linked bands at approximately 20 kD and 30 kD, and a group of cross-linked bands in the 60 kD to 70 kD range (A.S. Gaikwad and M.G. Cumsky, submitted). Cross-linking to each of these proteins was competable with cold peptide, suggesting that the interaction between the peptide and the proteins was specific (A.S. Gaikwad and M.G. Cumsky, submitted). Fractionation of cross-linked mitochondria demonstrated that each of the cross-linked proteins was located within the mitochondrial membranes. More recent results further indicated that the majority were specifically located in the outer membrane.

Two known yeast mitochondrial outer membrane proteins have molecular weights in the range of the predominant species cross-linked in our experiments. One, ISP42, is an essential outer membrane protein previously identified as being involved in protein import (Vestweber et al., 1989). The other, OMP45, is a non-essential protein whose function is presently unclear. We therefore carried out immunoprecipitation studies with both anti-ISP42 and anti-OMP45 antibodies. The results of these experiments clearly established that the predominant cross-linked protein was ISP42, as it was immunoprecipitable with the anti-ISP42 antibodies and not with the anti-OMP45 antibodies (A.S. Gaikwad and M.G. Cumsky, submitted). These results are important because they demonstrate, for the first time, a specific interaction between a mitochondrial targeting sequence and a distinct mitochondrial membrane protein.

**Acknowledgements**

Work from the authors' laboratory was supported by grants from the National Institutes of Heath and the American Cancer Society. MGC is an Established Investigator of the American Heart Association.

**References**

Baker KP, Schatz G (1991) Mitochondrial proteins essential for viability mediate protein import into yeast mitochondria. Nature 349:205-208

Deshaises RJ, Koch BD, Werner-Washburne M, Craig EA, Schekman R (1988) A subfamily of stress proteins facilitates translocation of secretory and mitochondrial precursor polypeptides. Nature 332:800-805

Eilers M, Hwang S, Schatz G (1988) Unfolding and refolding of a purified precursor protein during import into isolated mitochondria. EMBO J 7:1139-1145

Glaser SM, Cumsky MG (1989a) A synthetic presequence reversibly inhibits protein import into yeast mitochondria. J Biol Chem 265:8808-8816

Glaser SM, Cumsky MG (1989b) Localization of a synthetic presequence that blocks protein import into mitochondria. J Biol Chem 265:8817-8822

Glaser SM, Miller BR, Cumsky MG (1990) Removal of a hydrophobic domain within the mature portion of a mitochondrial inner membrane protein causes its mislocalization to the matrix. Mol Cell Biol 10:1873-1881

Glaser SM, Trueblood CE, Dircks L, Poyton RO, Cumsky MG (1988) Functional analysis of mitochondrial protein import in yeast. J Cell Biochem 36:275-287

Glick B, Schatz G (1991) Import of proteins into mitochondria. Ann Rev Genet 25:21-44

Glick B, Wachter C, Schatz G (1991) Protein import into mitochondria:two systems acting in tandem? TICB 1:99-103

Glick B, Brandt A, Cunningham K, Müller S, Hallberg RL, Schatz G (1992) Cytochromes $c_1$ and $b_2$ are sorted to the intermembrane space of yeast mitochondria by a stop-transfer mechanism. Cell 69:809-822

Hartl F-U, Neupert W (1990) Protein sorting to mitochondria: evolutionary conservations of folding and assembly. Science 247:930-938

Hartl F-U, Pfanner N, Nicholson DW, Neupert W (1989) Mitochondrial protein import. Biochim Biophys Acta 988:1-45

Kiebler M, Pfaller R, Söllner T, Griffiths G, Horstmann H, Pfanner N, Neupert W (1990) Identification of a mitochondrial receptor complex required for recognition and membrane insertion of precursor proteins. Nature 348:610-616

Murakami H, Pain D, Blobel G (1988) 70-kD heat shock-related protein is one of at least two distinct cytosolic factors stimulating protein import into mitochondria. J Cell Biol 107:2051-2057

Pfanner N, Rassow J, van der Klei IJ, Neupert W (1992) A dynamic model of the mitochondrial import machinery. Cell 68:999-1002

Vestweber D, Brunner J, Baker A, Schatz G (1989) A 42K outer-membrane protein is a component of the yeast mitochondrial import site. Nature 341:205-209

# Intra-mitochondrial sorting of precursor proteins

Rosemary A. Stuart, Hans Koll & Walter Neupert

Institut für Physiologische Chemie der Universität München, Goethestraße 33, 8000 München 2, Germany

Phone: +49 89 5996 312
Fax: +49 89 5996 270

## Introduction

With the exception of a limited amount of protein synthesis in mitochondria and chloroplasts, the majority of eukaryotic protein synthesis occurs in the cytoplasm. Consequently those proteins destined for the various cellular subcompartments other than the cytosol must be correctly targeted to their functional location. Many precursor proteins are thus translocated into or across biological membranes in order to reach their functional destination. This includes protein translocation into the endoplasmic reticulum, mitochondria, chloroplasts and peroxisomes.

The vast majority of mitochondrial proteins are encoded for by the cell nucleus and are synthesized in the cell cytosol as precursor proteins. These precursors are imported into mitochondria in a post-translational manner along a number of distinct, but sometimes overlapping pathways. Mitochondria are divided into four subcompartments outer membrane (OM), intermembrane space (IMS), inner membrane (IM) and matrix (M), each equipped with a specific set of proteins. Hence precursor proteins must contain the necessary information to target them not only to mitochondria but also for their correct submitochondrial sorting. Our knowledge of mitochondrial protein import has increased over the past years due to a number of detailed *in vitro* studies using mainly *Saccharomyces cerevisiae* and *Neurospora crassa* as model systems (for reviews see Hartl & Neupert, 1990; Baker & Schatz, 1991) .

In most organisms, lack of mitochondrial function is lethal to the cell. *S. cerevisiae,* however, can survive with defective respiratory chain and oxidative phosphorylation because it can use fermentable carbon sources for energy production. This ability to survive with defective mitochondria has rendered *S. cerevisiae* an excellent organism for the study of mitochondrial biogenesis. Furthermore, the genetic manipulation possibilities are greater in yeast than in other organisms such as *N. crassa*, and this has resulted in the identification of a number of mutants which have proved to be powerful in the analysis of mitochondrial protein import (for reviews see Grivell, 1989, Tzagoloff & Dieckmann, 1990, Bolotin-Fukuhara & Grivell, 1992).

NATO ASI Series, Vol. H 71
Protein Synthesis and Targeting in Yeast
Edited by A. J. P. Brown, M. F. Tuite, and J. E. G. McCarthy

## The mitochondrial import machinery

Mitochondrial precursor proteins contain targeting sequences which in the majority of cases are positively-charged, amino-terminal cleavable presequences. Not all precursors contain such presequences however, this includes all of the outer membrane proteins and some of those of the IMS. The first stage of mitochondrial protein import involves the interaction of precursors with surface receptor proteins on the outer surface of mitochondria. Recent studies using *N. crassa* have lead to the identification of the first mitochondrial import receptors, namely, MOM19 and MOM72 (Mitochondrial **O**uter **M**embrane proteins of 19kD and 72kD molecular weights, respectively (Söllner et al., 1989, 1990). MOM19 appears to act as the receptor for the majority of precursors tested to date, whilst MOM72 seems to be primarily involved in the binding of the inner membrane protein the ADP/ATP translocator. The yeast homolog of MOM72, termed MAS70, has also been identified and appears to have a broader specificity (Hines et al., 1990; Steger et al., 1990). Coimmuneprecipitation studies have indicated that MOM19 and MOM72 are present in a multi-subunit complex that catalyses both the recognition and the insertion of precursors into the outer membrane (Söllner et al., 1992). This outer membrane complex known as GIP, the **G**eneral **I**nsertion **P**ore (Pfaller et al., 1988). One component of this GIP complex, MOM38 is believed to be part of the OM translocation site (Kiebler et al., 1990). ISP42, the yeast equivalent of MOM38 has also been identified and has been shown to be essential for growth (Baker et al., 1990). MOM7, MOM8 and MOM30 are other components of the *N. crassa* GIP complex which have been identified, however, their functions to date are unknown.

Translocation of proteins across the inner membrane occurs at areas where the inner membrane is adjacent to the outer membrane. It has been estimated that *in vivo* most of the outer membrane is in close proximity to the inner membrane (Pfanner et al., 1992). Translocation across the inner membrane requires the presence of a membrane potential. Although the insertion of precursors into the outer membrane does not require a membrane potential per se, many proteins destined to cross the inner membrane cannot stably insert into the outer membrane in the absence of $\Delta\Psi$.

Passage of proteins across the inner membrane is also thought to occur through a proteinaceous pore, the components of which are not as yet identified. In addition to a $\Delta\Psi$, translocation through the inner membrane requires the cooperation of Hsp70 in the mitochondrial matrix (see below). The matrix-targeting presequences are proteolytically cleaved off, a reaction catalyzed by the matrix localized processing peptidase (MPP) in association with the processing enhancing protein (PEP) processing peptidase in the matrix. Folding of proteins in the matrix is mediated by Hsp60 in an ATP-dependent manner (see below).

*Role of mitochondrial Hsp70 in protein translocation and folding*

Translocation of proteins across the outer and inner membrane requires that precursor proteins have a loosely folded conformation. Cytosolic Hsp70 proteins have been implicated in playing a role in the maintenance of precursor in an unfolded conformation competent for translocation. As the precursor on transit through the inner membrane becomes exposed to the matrix, it forms an association with the mitochondrial Hsp70 (mt-Hsp70), also termed Ssc1p and is encoded for by the SSC1 gene. Interaction with mt-Hsp70 apparently is directly required for complete translocation into the matrix. In a temperature-sensitive yeast mutant affecting the gene coding for Ssc1p, the transfer of precursor proteins into mitochondria was shown to be defective (Kang et al., 1990). Precursor polypeptides were arrested during translocation spanning outer and inner membranes at contact sites. If however, the precursor proteins were artifically denatured with urea beforehand, import into mitochondria appreared to be mt-hsp70 independent. However, the precursor proteins imported under these conditions remained in a highly protease-sensitive, incompletely folded conformation. Therefore, it was concluded that the mitochondrial Hsp70 apparently has a dual role in translocation and folding of imported protein. Folding requires the transfer of the newly translocated polypeptides from mt-Hsp70 to Hsp60 (see below), a step that is blocked in the *ssc1* mutant. The energy resulting from binding the extended amino-terminus of the precursor protein to mt-Hsp70 is thought to be utilized to successively unfold parts of the precursor still outside the mitochondria. Release of mt-Hsp70 requires ATP hydrolysis in the matrix. Multiple molecules of mt-Hsp70 could bind to the traversing chain, thereby assisting it through the membrane.

*Role of Hsp60 in the folding and assembly of proteins imported into the matrix*

Evidence for the function of Hsp60 in protein assembly came from analysis of the temperature sensitive lethal yeast *mif4*, which is defective in the MIF4 gene coding for Hsp60. Proteins such as the ß subunit of the $F_1$-ATPase or the trimeric enzyme ornithine transcarbamylase are imported normally by the mutant mitochondria and are proteolytically processed, but cannot assemble into oligomeric complexes (Cheng et al., 1989). The unassembled subunits have a tendency to aggregate in the matrix compartment. The Hsp60 14-mer was also demonstrated to be required for not only its own assembly but also for the assembly of certain proteins which are targeted to the IMS, such as cytochrome $b_2$, which are imported via the matrix. Thus it is suggested Hsp60 plays a general role in protein assembly processes *in vivo*. A biochemical analysis of the interaction between imported proteins and mitochondrial Hsp60 revealed that the basic role of this chaperonin probably lies in mediating the folding of monomeric polypeptide chains. Using protein import into mitochondria as an experimental system to study protein folding, it could be demonstrated that the soluble, monomeric protein dihydrofolate reductase (DHFR) does not fold spontaneously when imported into mitochondria. Folding was demonstrated to occur in an ATP-dependent reaction mediated hy Hsp60 (Ostermann et al., 1989).

## Sorting of proteins to the mitochondrial IMS

Precursor proteins targeted to the mitochondrial matrix are imported along the general pathway described above. Other precursors destined for the other mitochondrial subcompartments initiate import along this general pathway, however, become deviated off at different points, depending on their final destination. For example porin, an outer membrane protein like the matrix-targeted proteins, uses the MOM19 receptor for initial mitochondrial recognition and binding and becomes inserted into the outer membrane in a GIP-dependent manner. Porin then becomes directly assembled into its characteristic trimeric structure in the outer membrane. Hence import and assembly of outer membrane proteins does not require the action the downstream components of the import machinery in the inner membrane and matrix.

The import of precursor proteins into the IMS represents the most complex situation for submitochondrial sorting. At least three distinct mechanisms of sorting to the IMS are known. The simplest pathway is taken by cytochrome *c*, where the precursor is imported directly across the outer membrane by a mechanism which is unique to this precursor (Stuart & Neupert, 1990). A second mechanism which has recently been shown to exist is the pathway taken by cytochrome *c* heme lyase (CCHL), which like porin uses the receptor/GIP complex (Lill et al., 1992). The third and probably most complex sorting pathway involves a multistep import pathway to the IMS (Hartl & Neupert, 1990).

*Import of cytochrome c*

The cytochrome *c* pathway differs markedly from the general route taken by the majority of other imported proteins, as it is characterized by the lack of involvement of surface receptors, GIP and energy requirements such as ATP hydrolysis and a membrane potential. Cytochrome *c* is synthesized as a precursor protein termed apocytochrome *c* and does not contain a cleavable targeting presequence (Stuart at al., 1987). No protease-sensitive components exist on the mitochondrial surface to mediate the binding of apocytochrome *c*. Instead, it is believed that due to the ability of apocytochrome *c* to spontaneously insert into lipid bilayers, it can become partially inserted into the outer membrane without the requirement of receptors etc. Upon insertion, apocytochrome *c* is recognized and binds in a complex with CCHL (Nicholson et al., 1988). Hence it is believed that CCHL acts as a specific binding site *in lieu* of a surface receptor. CCHL is the enzyme responsible for catalyzing the covalent attachment of heme to apocytochrome *c* and displays a requirement for NADH and flavins. Translocation across the outer membrane is believed to be driven by the refolding of the cytochrome *c* polypeptide as a result of the covalent heme addition, in a manner which does not require ATP hydrolysis (Stuart et al., 1990).

*Import of cytochrome c heme lyase*

Cytochrome *c* heme lyase is peripherally associated with the inner membrane, exposed to the IMS. Like apocytochrome *c* it is synthesized as a precursor lacking an amino-terminal cleavable presequence and consequently does not undergo cleavage upon import into mitochondria. Preliminary results suggest that an internal signal present in the central region of CCHL serves to target the precursor of CCHL to mitochondria (G. Kispal, B. Segui-Real, R. Lill & W. Neupert, in preparation). CCHL however, requires the function of the receptor/GIP complex for insertion into the outer membrane. Like the majority of precursors, CCHL appears to use the MOM19 as a receptor. In contrast to proteins targeted to the matrix, CCHL does not require the presence of an electrochemical membrane potential nor ATP hydrolysis to drive its import to the IMS. CCHL is imported directly across the outer membrane and does not traverse the inner membrane during its import (Lill et al., 1992). As CCHL does not appear to use the energy sources used by other precursor for import the question thus arises as to energetically drives the import events. One possible explanation is that the translocation of CCHL across the outer membrane is powered by the folding of the polypeptide chain upon emergence in the IMS.

*Sorting of other proteins to the IMS*

The third and probably most complex sorting pathway to the IMS exists for proteins such as cytochrome $b_2$, cytochrome $c_1$ and the Rieske Fe-S protein of the cytochrome $bc_1$ complex. Two models of the sorting events have been proposed, namely the "conservative sorting" mechanism and the "stop transfer" mechanism. The "conservative sorting mechanism" takes into account the evolutionary origin of the mitochondrion (Hartl et al., 1986, 1987; Hartl and Neupert, 1990). As is generally accepted, mitochondria have evolved from prokaryotic ancestors that were introduced into an ancestral eukaryotic host cell by an endosymbiotic event. The conservative sorting hypothesis proposes that the prokaryotic principles of membrane assembly and transport have been maintained during the evolution of the mitochondria. For example, *Rhodobacter* are probably close relatives of the endosymbiotic ancestor. Analysis of the synthesis and assembly pathway of the bacterial cytochrome $c_1$ reveals that this precursor is synthesized with a typical bacterial cleavable presequence and undergoes export from the cytoplasm across the plasma membrane to the periplasmic space. The conservative sorting hypothesis proposes that this export pathway has been conserved in mitochondria. Thus it is proposed that mitochondrial IMS proteins which follow this sorting pathway are initially targeted into the mitochondrial matrix and are subsequently exported into or across the inner membrane. Cytochromes $b_2$ and $c_1$ are initially synthesized as precursors with a bipartite amino-terminal targeting sequence which becomes processed in two steps upon import. The first domain is a positively-charged and serves as a "matrix targeting" domain and is cleaved by the matrix-localized processing peptidase, to generate an intermediate-size species. We believe that the second domain of the presequence contains the necessary information to direct the export of the protein from the matrix to the intermembrane space. This second domain is very similar to the leader or signal sequences of bacterial and eukaryotic secretory proteins, namely it contains several positively-charged residues at the amino-terminus which are followed by a hydrophobic segment of approximately 20 residues. These positive amino acids are of particular importance, as deletion of one or more of them has recently been shown to prevent the export of cytochrome $b_2$ from the matrix to the intermembrane space (E. Schwarz, T. Seytter & W. Neupert, in preparation).

Thus it is suggested that proteins such as cytochrome $b_2$, cytochrome $c_1$ and the Rieske Fe-S protein reach the intermembrane space by export from the matrix, which resembles the pathway of protein secretion into the bacterial periplasm. Under certain import conditions where export is retarded, however, such as low temperature, these proteins can be accumulated in the matrix as intermediate-size species (Hartl et al., 1987; Stuart et al., 1990). Recent evidence suggest that the export of these proteins from the matrix can occur in concert with completion of the import step (Koll et al., 1992). Using specially designed cytochrome $b_2$-DHFR fusion proteins, it could be shown that the sites of the protein export machinery can be in very close proximity to the sites of mitochondrial protein import, namely the translocation contact sites. This was demonstrated by the accumulation of Hsp60-bound intermediates of the cytochrome $b_2$-DHFR fusion proteins which were spanning three membranes, namely the outer and inner membranes at contact sites (the site of import) and again the inner membrane at the site of export. Using different protein fusion constructs, where the length of the cytochrome $b_2$ part was varied, it could be calculated that the minimal distance between the sites of import and those of export maybe as little as 30nm.

A second model for the mechanism of sorting of these proteins to the IMS, termed the"stop-transfer hypothesis" has been proposed by Schatz and co-workers (van Loon & Schatz, 1987). According to the "stop transfer" hypothesis, the second domain of the presequence serves to arrest the protein in the inner membrane upon import of the precursor into mitochondria. Following cleavage of the matrix targeting presequence, the intermediate species would be sorted to the IMS by a lateral movement in the inner membrane. Hence unlike the conservative sorting hypothesis, the stop-transfer hypothesis would predict that whilst on transit to the IMS, the mature part of the protein would never become exposed to the mitochondrial matrix. However, both sorting models would predict that the intermediate-size species would finally accumulate in the intermembrane space, membrane bound, prior to its final processing event to generate the mature-size species. Evidence in favour of the "stop transfer" model was recently published by Glick et al., where they failed to observe both intermediates accumulated in the matrix and a dependence of matrix ATP for correct sorting to the IMS (Glick et al., 1992). As explained above, however, if initiation of export of precursors can occur prior to completion of import, intermediates in the matrix would not be observed. Futhermore, only under certain experimental conditions where the kinetics of import are greater than those of export, (i.e. low temperature, or import of urea denatured precursors) can one accumulate such sorting intermediates in the matrix. As these are kinetic intermediates, they are labile and easily chased to their final location, exposed to the IMS whilst associated with the inner membrane (via the remaining uncleaved part of the presequence in the case of cytochrome $b_2$), a location which is also predicted for the intermediate-size species along the stop-transfer sorting pathway.

ATP hydrolysis in the matrix is required for the function of Hsp70 and Hsp60. In addition Glick et al., presented evidence which suggested that a fusion protein p$b_2$(167)-DHFR, consisting of 167 amino acids of cytochrome $b_2$ fused to mouse dihydrofolate reductase (DHFR), did not appear to require matrix ATP hydrolysis, and hence concluded that it was not being sorted via the matrix. This finding is in contrast to previously published results where authentic cytochrome $b_2$, under similar experimental conditions was shown to require matrix ATP, (Hwang & Schatz, 1989). The discrepancy here suggests that the apparent matrix ATP requirement reflects the IMS-targeted precursor protein being studied rather than the sorting pathway being taken. This strongly argues that the lack of an observable matrix ATP requirement cannot be used as an criterion in favour or against one sorting pathway or the other. Furthermore, following a study of the matrix ATP requirements for the IMS sorting of a series of cytochrome $b_2$-DHFR of increasing length, it would appear that longer precursor proteins display a greater requirement for matrix ATP than the smaller ones. This difference is most likely related to the level of involvement of Hsp70 that is required for the matrix import step. This seems to be reduced for the smaller precursors, as it appears that the export process, which can occur in concert with import, is sufficient to power the completion of import (A. Gruhler, W. Voos, unpublished results).

Evidence that Hsp60 could be involved in protein sorting to the IMS came from analyzing the Hsp60-deficient yeast mutant *mif4*. The mutant accumulated incompletely processed forms of cytochrome $b_2$ and of the Rieske Fe/S protein, which is believed to follow a similar route to the IMS. Recent evidence suggested a role for Hsp60 in maintaining precursor proteins imported into the matrix in an open conformation competent for export across the inner membrane (Koll et al., 1992). In contrast, Glick et al., fail to observe a requirement for Hsp60 for the sorting of cytochrome $b_2$ to the IMS. The cause for the discrepancy is, however, unclear and further experiments are clearly needed to resolve this situation. One possibiltiy is that maybe the requirement of Hsp60 is not absolute and depending on the import conditions used its function can be circumvented. One could imagine this to be the case under conditions where export is initiated prior to the completion of the import event, especially in the case of smaller precursor proteins.

Following export across the inner membrane, the proteins undergo a second proteolytic cleavage to generate the mature-size species and the proteins then assemble into their functional oligomeric complexes. An intermembrane space protease IMP1, which is involved with the cleavage of the second domain (the "sorting sequence") of the presequences of some of the IMS proteins has been identified and cloned. IMP1 is a membrane bound enzyme which is thought to be part of a multi-subunit complex and it was shown to be structurally related to the bacterial leader peptidase (Wolfe et al., 1983; Schneider et al., 1991; Behrens et al., 1991). The similarity of IMP1 to the bacterial peptidase is further support for the conservative sorting hypothesis which emphasizes the resemblance between the export of proteins from the matrix to the main pathway of protein secretion into the bacterial periplasm.

Initially the conservative sorting hypothesis was described for a subset of IMS proteins, but now following a recent study it has been extended to include a subset of inner membrane proteins. Proteins of the inner membrane which are synthesized as precursor proteins with cleavable amino-terminal extensions, such as the subunit 9 of the Fo-ATPase, have been shown to be initially imported into the mitochondrial matrix and following an interaction with Hsp60 become retranslocated into the inner membrane (Mahlke et al., 1990).

## Summary and future perspectives

The mitochondrial matrix may be envisaged as a dynamic milieu, as it serves as a temporary depot for many proteins that are on transit to the various submitochondrial locations. In addition to those proteins with their functional location in the matrix, some nuclear-encoded proteins destined for the inner membrane and intermembrane space, are initially imported into the matrix prior to subsequent sorting. Furthermore, proteins encoded by the mitochondrial genome are also synthesized in the matrix and from there are sorted to the IM and IMS. The mitochondrial matrix, hence, is the scene for many events that have important consequences for the correct sorting and assembly of precursor proteins encoded by both genomes.

A consequence of the "conservative sorting" pathway of proteins to the IMS, is that the inner mitochondrial membrane must contain a separate translocation machinery for the export of proteins into and across the inner membrane. An open question of course is whether the mitochondrially-encoded precursor proteins use the same export machinery as those nuclear proteins which undergo export from the matrix. The identification and characterisation of such a translocation machinery is thus of utmost importance and will the main focus of our attention in the next few years.

## References


Baker, K. & Schatz, G, (1991). *Nature*, **349**, 205-208.

Baker, K.P., Schaniel, A., Vestweber, D. & Schatz, G. (1990). *Nature*, **348**, 605-609.

Bolotin-Fukahara, M. & Grivell, L.A. (1992). *Antonie van Leewenhoek*, **62**, 131-153.

Cheng, M.Y., Hartl., F.-U., Martin, J., Pollock, R.A., Kalousek, F., Neupert, W., Hallberg, E.M., Hallberg, R.L. & Horwich, A.L. (1989). *Nature*, **337**, 620-625.

Glick, B., Brandt, A., Cunningham, K., Müller, S., Hallberg, R. L. & Schatz, G. (1992). *Cell*, **69**, 809-822.

Grivell, L.A. (1989). *Eur. J. Biochem.*, **182**, 477-493.

Hartl., F.-U. & Neupert, W. (1990). *Science*, **247**, 930-938.

Hwang, S.T. & Schatz, G. (1989). *Proc. Natl. Acad. Sci.*, **86**, 8432-8436.

Hines, V., Brandt, A., Griffiths, G., Horstmann, H., Brütsch, H. & Schatz, G. (1990). *EMBO J.*, **9**, 3191-3200.

Kang, P.J., Ostermann, J., Shilling, J., Neupert, W., Craig, E. A. & Pfanner, N. (1990). *Nature*, **348**, 137-142.

Kiebler, M., Pfaller, R., Söllner, T., Griffiths, G., Horstmann, H., Pfanner, N., & Neupert, W. (1990). *Nature*, **348**, 610-617.

Lill, R., Stuart, R.A., Drygas, M., Nargang, F.E. & Neupert, W. (1992). *EMBO J.*, **11**, 449-456.

Lill, R., Hergersberg, C., Schneider, H., Söllner, T., Stuart, R.A., & Neupert, W. (1992). in "Membrane Biogensis and Protein Targeting", eds., W. Neupert, & R. Lill, Elsevier, Amsterdam.

Mahlke, K., Pfanner,N., Martin, J., Horwich, A., Hartl, F.-U. & Neupert, W. (1990). *Eur. J. Biochem.*, **192**, 551-555.

Nicholson, D.W., Hergersberg, C. & Neupert, W. (1988). *J. Biol. Chem.*, **263**, 19034-19042.

Ostermann, J., Horwich, A.L., Neupert, W. & Hartl., F.-U. (1989). *Nature*, **341**, 125-130.

Pfaller, R., Steger, H., Rassow, J., Pfanner, N. & Neupert, W. (1988). *J. Cell. Biol.*, **107**, 2483-2490.

Pfanner, N., Rassow, J., van der Klei, I., & Neupert, W., (1992). *Cell*, **68**, 999-1002.

Söllner, T., Griffiths, G., Pfaller, R., Pfanner, N. & Neupert, W. (1989). *Cell*, **59**, 1061-1070.

Söllner, T., Pfaller, R., Griffiths, G., Pfanner, N. & Neupert, W. (1990). *Cell*, **62**, 107-115.

Söllner, T., Rassow, J., Wiedmann, M., Schlossmann, J., Keil, P., Neupert, W. & Pfanner, N. (1992). *Nature*, **355**,

Steger, H., Söllner, T., Kiebler, M., Dietmeier, K.A., Pfaller, R., Trülzsch, K.S., Tropschug, M., Neupert, W. & Pfanner, N. (1990). *J. Cell. Biol.*, **111**, 2353-2364.

Stuart, R.A. & Neupert, W. (1990). *Biochimie*, **72**, 115-121.

Stuart, R.A., Neupert, W. & Tropschug, M. (1987). *EMBO J.*, **6**, 2131-2137.

Stuart, R.A., Nicholson D.W. & Neupert, W. (1989). *Cell*, **60**, 31-43.

Stuart, R.A., Nicholson D.W., Wienheus, U. & Neupert, W. (1990). *J. Biol. Chem.*, **265**, 20210-20219.

Tzagoloff, A. & Dieckmann, C.L. (1990). *Microbiol. Rev.* **54**, 211-225.

van Loon, A.P.G.M. & Schatz, G. (1987). *EMBO J.*, **6**, 2441-2448.

# GENETIC CHARACTERIZATION OF THE INTERMEMBRANE SPACE SORTING DOMAINS OF YEAST CYTOCHROME $b_2$

Ellen M. Beasley
Department of Biochemistry
Biocenter
The University of Basel
Klingelbergstrasse 70
CH-4056 Basel
Switzerland

ABSTRACT. A genetic characterization to determine the domains of precytochrome $b_2$ involved in intermembrane space sorting has identified three domains necessary for efficient intermembrane space targeting. We are now using this information to try to identify the mitochondrial machinary involved in sorting to this compartment.

Mitochondria, the organelles of oxidative phosphorylation in eukaryotic cells, contain several hundred different proteins. Although the mitochondrion has its own genetic system that codes for some mitochondrial proteins, the majority of mitochondrial proteins are encoded in the nucleus and synthesized in the cytoplasm. These proteins must contain information that targets them to the mitochondria and interacts with the mitochondrial protein import machinary.

Mitochondria contain four compartments: the outer membrane, intermembrane space, inner membrane and matrix. Proper protein localization requires both targeting to the mitochondria and sorting to direct the protein to the correct submitochondrial compartment (reviewed in Glick & Schatz, 1991; Hartl & Neupert, 1990). Many nucleus-encoded mitochondrial proteins are synthesized as larger precursors with amino-terminal presequences. For many such precursors, it has been demonstrated that the targeting information is contained in this presequence which is then proteolytically removed after import.

For several precursors that are targeted to the intermembrane space these mitochondrial targeting and sorting functions appear to occupy distinct portions of bipartite presequences (e.g. van Loon et al., 1987). The amino-terminal

NATO ASI Series, Vol. H 71
Protein Synthesis and Targeting in Yeast
Edited by A. J. P. Brown, M. F. Tuite, and J. E. G. McCarthy

portion of these presequences resembles a matrix-targeting presequence, and in some cases has been shown to be able to function to target a protein to the matrix (van Loon et al., 1987). The carboxy-terminal portion of the presequence is also necessary for sorting to the intermembrane space. The obvious conserved feature of this intermembrane space sorting portion is a hydrophobic stretch (von Heijne et al., 1989). While there was a significant amount of information about the types of sequences that can act a matrix-targeting sequences, very little was known about the essential features of intermembrane sorting sequences.

We undertook a genetic study in order to investigate the determinants of cytochrome $b_2$ sorting to the intermembrane space. This analysis has defined three cytochrome $b_2$ presequence domains that are involved in intermembrane space sorting. We are now using this information to design experiments to identify the mitochondrial machinary that decodes this information.

## MATERIALS AND METHODS.

Strains. The *Escherichia coli* strain HB101 was used for plasmid selection and propagation.

The *Saccharomyces cereviseae* strain JKR101 (α, ade2, his4, leu2, ura3) was obtained from Dr. T. D. Fox, BYL2 (α, ade2, his4, ura3, *coxIV::LEU2*; Bibus et al., 1988), BLC1.3D (a, his4, lys2, ura3, *coxIV::LEU2*; Bibus et al., 1988), and W303-1A (a, ade2, ura3, his3, leu2, can$^R$, *cyb2::TRP2*; Guiard, 1985) was the kind gift of Prof. B. Guiard. The strain EMB4 (α, his4, lys2, ura3, *cyb2::LEU2*) was constructed for this study by replacing the *TRP1* gene in plasmid *cyb2::TRP1* (a gift from Prof. Guiard; Guiard, 1985) with the *LEU2* gene fragment from yEP13 and transforming the yeast strain BSL1-7 (α, his4, leu2, lys2, ura3). The disruption of the *cyb2* gene was confirmed by Southern blot.

Vectors. All the *E. coli*- yeast shuttle vectors contain the *URA3* gene as a selectable and counter selectable marker and a yeast centromere sequence. pEB3 contains the coding region for the first 167 amino acids of cytochrome $b_2$ fused in-frame to pseudo-mature CoxIV under the control of the ADH1 promoter (derived from A105; van Loon & Schatz, 1987). pGR401 which expresses the cytochrome $b_2$ gene under the control of the ADH1 promoter, was the kind gift of Dr. G. Reid. Mutant varients of pGR401 were constructed by replacing the 550 bp EcoRI-BamHI fragment that codes for the first 167 amino acids of cytochrome $b_2$ with the corresponding fragment from the pEB3 mutants.

Selection of Missorting Mutants. BYL2/pEB3 was maintained on SD (synthetic glucose medium; 0.67% yeast nitrogen base, 2% glucose) supplemented with 0.5% casamino acids and 20 $\mu$g/ml adenine. Spontaneous, respiratory competent mutants were selected at room temperature on SEG (synthetic ethanol glycerol medium; 0.67% yeast nitrogen base, 3% glycerol, 3% ethanol) supplemented with casamino acids and adenine.

Selection of Suppressor Mutations. W303-1A transformed with mutant varients of pGR401 was selected for lactate positive growth on SLac (synthetic lactate medium; 0.67% yeast nitrogen base, 2% lactic acid, 50 mM potassium phosphate, pH 5.8) supplemented with 0.5% casamino acids and 20 $\mu$g/ml adenine.

Classification of Missorting Mutants. Mutants were classified as plasmid or host-linked by analyzing their phenotype after curing of the original plasmid, by 5-fluororotic acid selection (Boeke et al., 1984), and retransforming. When retransformation restored the selected phenotype, strains had host-linked mutations. When retransformation failed to regenerate the selected phenotype, plasmids from the original isolates were used to transform the original host yeast strain. Plasmid-linked mutationss confer the selected phenotype to the original yeast strain.

Sequencing. Dideoxysequencing of the double stranded plasmid was carried out using the Sequenase version 2 kit (U.S. Biochemical). Sequencing primers were synthesized that prime 5'→3' in the ADH1 promoter and at the 3'-end of the cytochrome $b_2$ presequence coding sequence and 3'→5' primers that prime at the 5'-end of mature cytochrome $b_2$ and the 5'-end of CoxIV.

Miscellaneous Methods. Standard methods were used for *E. coli* transformation, plasmid purification, DNA digestion and ligation (Sambrook et al., 1989). Standard yeast methods were used for plasmid recovery (Baker & Schatz, 1987), genetic crosses and tetrad dissection (Campbell & Duffus, 1988). Total yeast protein was prepared according to the protocol of Yaffe and Schatz (1984). Immunoblots were reacted with polyclonal rabbit antisera, detected with [$^{125}$I]-protein A labeling and autoradiography (Haid and Suissa, 1983). Autoradiograms were quantified by using a computerized densitometer (Molecular Dynamics).

RESULTS.

Transforming BYL2 with pEB3 fails to complement the respiratory deficiency of the *coxIV* deletion because the chimeric precytochrome $b_2$-CoxIV protein is directed to the intermembrane space. Respiratory competent mutants were selected; a total of 97 independently-isolated respiratory competent mutants were characterized. Ninety of these proved to be plasmid-linked and 7 were host-linked.

Preliminary characterization of the host-linked mutants indicated that they were all informational suppressors and promoter-up mutations rather than mutations affecting the intermembrane space sorting machinary. We did not, therefore, characterize them further.

All of the plasmid-linked mutants were sequenced through the region coding for the cytochrome $b_2$ sequence. Eighty-one had mutations resulting in a single amino acid change in the presequence. Five had deletions, beginning in the second part of the cytochrome $b_2$ presequence and extending into mature cytochrome $b_2$ or the amino-terminal portion of CoxIV. Three of the mutants had no changes in the entire cytochrome $b_2$ sequence present in the plasmid. The specific single amino acid mutations and the number of times they were isolated is summarized in Table 1. The distribution of these mutations on the cytochrome $b_2$ presequence is illustrated in Figure 1.

TABLE 1. Mutations disrupting the intermembrane space sorting function of the cytochrome $b_2$ presequence.

| Mutation | Independent Isolates | Mutation | Independent Isolates |
|---|---|---|---|
| S46L | 1 | A63D | 2 |
| R47I | 1 | A63P | 1 |
| R49C | 2 | A64D | 1 |
| R49L | 12 | A64P | 1 |
| A60E | 2 | T65N | 30 |
| A60P | 1 | T65P | 5 |
| I61N | 1 | S66N | 1 |
| L62H | 1 | V68E | 1 |
| L62P | 5 | Y70N | 1 |
| L62Q | 1 | L71P | 1 |
| L62R | 2 | E81A | 1 |

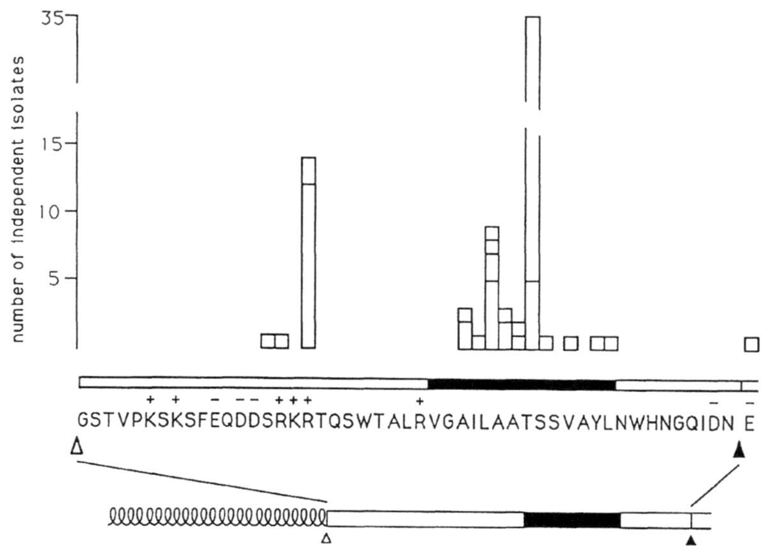


FIGURE 1. The distribution of intermembrane space sorting mutations is shown projected above the wild-type cytochrome $b_2$ presequence. Only the carboxy-terminal portion of the presequence is shown. The general structure of the presequence is illustrated at the botttom of the Figure. The amino-terminal portion of the presequence is postively charged and is predicted to form an amphipathic helix (shown as loops in the figure). The carboxy-terminal portion of the presequence has a hydrophobic stretch that is illustrated as a solid box and is underlined in the sequence. The presequence is processed sequentially by the matrix protease (open triangle) and inner membrane protease I (solid triangle).

The fact that most, 64%, of the mutant classes are represented by a single isolate indicates that we have not reached saturation. For this reason we can not comment on the absence of a particular mutant class.

The distribution of these mutations indicates that a minimum of three domains are involved in sorting to the intermembrane space: a cluster of basic residues, followed by a hydrophobic stretch and the residue at position 1 of mature cytochrome $b_2$. The mutations around the basic cluster do not necessarily result in a loss of charge but all of them involve the substitution of hydrophobic amino acids. Mutations in the hydrophobic domain all result in the introduction of less hydrophobic residues. However, the prevalence of proline and asparagine mutations was also suggestive of a structural constraint on the activity of this domain. In order to investigate this further, the missorting phenotype characterized.

The cytochrome $b_2$ presequence is processed in two steps and it was found that the processing state of the chimeric precursor accurately reflected its submitochondrial location (data not shown). The once-cleaved intermediate was found to be located in the matrix; while the twice-cleaved mature form was found to be in the intermembrane space. The relative degree of missorting was judged by western blot analysis, the degree of missorting was calculated as the amount of intermediate form relative to the total amount of CoxIV protein present. The missorting phenotype of the mutants is summarized in Table 2. The apparent degree of missorting for a particular mutation was reproducible within 10% between different, independent isolates and between independent preparations of the same isolate. While these missorting estimates do not take into account potential differences in the stability of the CoxIV protein in the matrix and the intermembrane space, it is likely that they reflect the relative strength of missorting.

The phenotype of mutations within the hydrophobic domain is not related to the extent of change in hydrophobicity. For example, when Leu62 is changed to Pro the missorting effect is much stronger than when it is changed to Arg; although the charge to Arg represents a much greater decrease in hydrophobicity.

TABLE 2. The extent of missorting of mutant precytochrome $b_2$-CoxIV is mutation specific. The percent missorting was calculated as the amount of CoxIV protein present as the once-processed, intermediate-form relative to the total amount (intermediate and mature).

| Mutation | % Missorting | Mutation | % Missorting |
|---|---|---|---|
| S46L | 82 | A63D | 81 |
| R47I | 71 | A63P | 98 |
| R49C | 55 | A64D | 99 |
| R49L | 63 | A64P | 96 |
| A60E | 36 | T65N | 50 |
| A60P | 16 | T65N | 85 |
| I61N | 57 | S66N | 45 |
| L62H | 64 | V68E | 27 |
| L62P | 96 | Y70N | 33 |
| L62Q | 98 | L71P | 28 |
| L62R | 30 | E81A | 85 |

That these missorting phenotypes do not simply reflect stablity differences between the different mutation was confirmed by a passenger protein exchange experiment. In order to rule out instability of CoxIV in the intermembrane space, we used an intermembrane space protein as an alternate passenger, reasoning that this protein would be stable in this compartment. We chose to assay the effect of the mutant presequences on the sorting of their native passenger, cytochrome $b_2$ which is required in the intermembrane space for growth on lactate medium. The restriction fragment containing the coding region for the first 167 amino acids of cytochrome $b_2$ was used to replace the corresponding fragment in the plasmid pGR401 creating a cytochrome $b_2$ expression vector with a mutation in the presequence coding region. These mutant plasmids were used to transform the cytochrome $b_2$ deletion strain and the ability of the transformants to grow on lactate medium was assayed. The results of this analysis are summarized in Table 3.

The results of the passenger exchange experiment and of western blot analysis of the transformants (data not shown) indicates that these mutations specifically affect sorting, rather than stability, of the cytochrome $b_2$ presequence. In fact, the apparent degree of missorting is greater with cytochrome $b_2$ as the passenger. Although the relative extent of missorting seems to be passenger protein independent, the absolute amount of missorting is affected by the passenger.

Table 3. Cytochrome $b_2$ presequence mutations result in the missorting of both CoxIV and cytochrome $b_2$ passenger proteins. Plasmids coding for precytochrome $b_2$-CoxIV or precytochrome $b_2$ were expressed in strains with *coxIV* or *cyb*$_2$ deletions. The growth phenotype of the transformants on SEG or SLac medium is indicated in the Table.

| YEAST | | | $\Delta$coxIV | $\Delta$cyt$b_2$ |
|---|---|---|---|---|
| CONSTRUCT | | | pCyt$b_2$-CoxIV | pre Cyt$b_2$ |
| PHENOTYPE | | | respiration | lactate growth |
| PLASMID | | none | – | – |
| PLASMID | MUTATION | WT | – | + |
| | | R49C | + | – |
| | | R49L | + | – |
| | | A63D | + | – |
| | | A63P | + | – |
| | | A64P | + | – |
| | | T65N | + | – |
| | | T65P | + | – |
| | | E81A | + | – |

The results of the passenger exchange experiment provide a tool for the further dissection of the sorting mechanism. We chose to take a genetic approach to identify both second-site suppressors within the cytochrome $b_2$ presequence that restore sorting and mutations in host functions that complement the missorting defects. Suppressor mutations were selected for their ability to grow on lactate medium. Thus far suppressors of the R49C and A63P mutations have been selected and characterized.

Sequencing the presequence region of several of the plasmid-linked suppressors of both the R49C and A63P mutations failed to identify intragenic suppressors of these mutations. The plasmid-linked mutants were either direct revertants to the wild-type presequence, 3 out of 7 R49C revertants and 9 out of 12 A63P revertants, or showed no change in the coding region for the presequence. While it is possible that these mutants have changes in the coding region for mature cytochrome $b_2$ that affect sorting, it is also possible that they represent *cis*-acting promoter-up mutations that complement the cytochrome $b_2$ deficiency. It is difficult to rule out this possibility because the intermediate size cytochrome $b_2$ that is localized to the matrix is apparently extremely unstable (data not shown), therefore an increase in the overall expression level can not be assessed by simply analyzing the steady state levels of cytochrome $b_2$ in the revertants.

The sorting of cytochrome $b_2$ to the intermembrane space must involve protein-protein interactions with the sorting machinary. In order to identify extragenic suppressors that complement the missorting defect by specifically affecting these protein-protein interactions we screened host-linked mutations to identify allele-specific suppressors. The phenotye of the host-linked mutants was characterized in the presence of the wild-type, R49C and A63P precytochrome $b_2$ coding plasmids.

All eight of the suppressors selected with the R49C plasmid analyzed were also able to suppress the A63P mutation. We are continuing to analyze suppressors of this mutation in order to identify allele-specific mutations.

Two of the 10 suppressors selected with the A63P mutations were unable to suppress the R49C mutation. Both of these mutants were also unable to grow on lactate medium when they were transformed with the wild-type cytochrome $b_2$ plasmid. These results rule out suppression via a *trans*-acting, promoter-up mutation and strongly suggest that the mutations specifically affect the sorting machinary that recognizes or interprets the sorting signal such that the specificity of this reaction is now changed to accommodate the A63P mutation.

Both of these mutations, 4 and 8, are recessive and segregate 2:2 in genetic crosses. One of the mutants, number 8, is temperature-sensitive respiratory deficient. The other mutant grows slowly under respiratory conditions.

The protein phenotype of these mutants is currently under investigation (data not shown). Preliminary data indicate that the once-processed, intermediate form of cytochrome $c_1$, which has a presequence similar in general structure to cytochrome $b_2$, can be detected in both of the mutants. Mutant 4 has additional changes in its mitochondrial protein profile that include: reduced levels of cytochrome c peroxidase and cytochrome oxidase subunit IV and a significant amount (approximately 20% of total) of detectable hsp60 precursor.

DISCUSSION.

The selection of mutations that allow a passenger protein directed to the intermembrane space by the cytochrome $b_2$ presequence to be mislocalized to the matrix has allowed us to define three domains in the cytochrome $b_2$ presequence that are essential for efficient intermembrane space sorting. The signal for intermembrane sorting is surprisingly complex and includes regions up and downstream from the hydrophobic domain that are not obviously conserved amoung different intermembrane space sorted presequences. The nature of the mutations in the hydrophobic domain indicates that this domain may have structural as well as hydrophobicity constraints. This suggests that the recognition of this domain involves protein-protein interactions.

We are now concentrating our efforts on the isolation and analysis of allele-specific, host-linked suppressors of the cytochrome $b_2$ missorting mutations. The cloning and characterization of these suppressors may allow us to identify the sorting machinary and to genetically define the sorting pathway.

ACKNOWLEDGEMENTS. I thank Jeff Schatz for his support and encouragement and Sabina Müller for exceptional technical assistance. This work was supported in part by an EMBO longterm postdoctoral fellowship.

REFERENCES.

Baker A, Schatz G (1987) Sequences from a prokaryotic genome or the mouse dihydrofolate reductase gene can restore the import of a truncated precursor protein into yeast mitochondria. Proc Natl Acad Sci USA 84:3117-3121.
Bibus CR, Lemire B, Suda K, Schatz G (1988) Mutations restoring import of a yeast mitochondrial protein with a nonfunctional presequence. J Biol Chem 263:13097-13102.
Boeke JD, Lacroute F, Fink GR (1984) A positive selection for mutants lacking orotidine-5'-phosphate decarboxylase activity in yeast: 5-fluoro-orotic acid resistance. Mol Gen Genet 197:345-346.
Campbell I, Duffus JH (eds.) (1988) Yeast: A Practical Approach. IRL Press, Oxford.
Guiard B (1985) Structure, expression and regulation of a nuclear gene encoding a mitochondrial protein: the yeast L(+)-lactate cytochrome c oxidoreductase (cytochrome $b_2$). EMBO J 4:3265-3272.
Haid A, Suissa M (1983) Immunochemical identification of membrane proteins after sodium dodecyl sulfate-polyacrylamide gel electrophoresis. Meth Enzymol 96:192-205.
Hartl F-U, Neupert W (1990) Protein sorting to mitochondria--evolutionary conservations of folding and assembly. Science 247:930-938.
Sambrook J, Fritsch EF, Maniatis T (1989) Molecular Cloning. Cold Spring Harbor Press, New York.
van Loon APGM, Schatz G (1987) Transport of proteins to the mitochondrial intermembrane space: the "sorting" domain of the cytochrome $c_1$ presequence is a stop-transfer sequence specific for the mitochondrial inner membrane. EMBO J 6:2441-2448.
van Loon AGPM, Brändli AW, Pesold-Hurt B, Blank D, Schatz G (1987) Transport of proteins to the mitochondrial intermembrane space: the "matrix-targeting" and the "sorting" domains in the cytochrome $c_1$ presequence. EMBO J 8:2433-2439.
von Heijne G, Steppuhn J, Herrmann RG (1989) Domain structure of mitochondrial and chloroplast targeting peptides. Eur J Biochem 180:535-545.

# MITOCHONDRIAL IMPORT OF CYTOCHROME *C*

Mark E. Dumont
Department of Biochemistry
University of Rochester School of Medicine and Dentistry
P.O. Box 607
Rochester, N.Y. 14642 U.S.A.

Cytochrome *c* is encoded in the nuclear genome, synthesized in the cytoplasm, and then imported into the mitochondrial intermembrane space. However, unlike most other proteins imported into this compartment, cytochrome *c* lacks cleavable targeting sequences and its import does not depend on maintenance of a potential across the inner mitochondrial membrane or on ATP hydrolysis. The presence of an excess of apocytochrome *c* does not inhibit import of other mitochondrial precursors and cytochrome *c* import does not involve proteolytically-sensitive receptors on the outer surface of mitochondria (for a review see Stuart and Neupert, 1990).

In spite of the differences between the mitochondrial import of cytochrome *c* and that of other proteins, several properties of cytochrome *c* make it an attractive system for studying protein translocation across membranes. Cytochrome *c* is a small, abundant protein that has been used extensively as a model for studies of protein folding and structure and has been the subject of an extensive genetic analysis in the yeast *Saccharomyces cerevisiae*. Heme is covalently attached to cytochrome *c* during or after import. By chemically removing the heme from the mature protein, holocytochrome *c*, it is possible to obtain large amounts of the pure precursor protein, apocytochrome *c*. The cellular machinery involved in cytochrome *c* import appears to be simpler than that required for import of other proteins, and can be manipulated using an apparent requirement for heme attachment.

## Attachment of Heme and Mitochondrial Import Pathway

The heme group of holocytochrome *c* is covalently attached via thioether linkages to two cysteine residues in the protein. Formation of the thioether bonds is catalyzed by the enzyme cytochrome *c* heme lyase (also referred to as holocytochrome *c* synthase; EC 4.4.1.17), encoded by the yeast gene *CYC3* (Dumont et al., 1987). A role for heme attachment in cytochrome *c* import was initially suggested by the demonstration that the presence of an excess of deuterohemin, a heme analog that can not be incorporated into cytochrome *c*, inhibits import of *in vitro*-translated apocytochrome *c* into isolated mitochondria (Hennig and Neupert, 1981).

NATO ASI Series, Vol. H 71
Protein Synthesis and Targeting in Yeast
Edited by A. J. P. Brown, M. F. Tuite, and J. E. G. McCarthy

We have investigated the relationship between heme attachment and cytochrome *c* import using yeast strains containing varying amounts of heme lyase, and using altered forms of apocytochrome *c* that vary in their abilities to have heme covalently attached. Cytochrome *c* translated *in vitro* can be imported with high efficiency into mitochondria isolated from normal yeast strains. However, no detectable cytochrome *c* is imported into mitochondria isolated from *cyc3*⁻ yeast strains, lacking heme lyase, even though such mutant strains import other mitochondria proteins with near-normal efficiencies (Dumont et al., 1988; also see Nargang et al., 1988). Altered forms of cytochrome *c* lacking one or the other of the cysteine residues that serve as the sites of covalent heme attachment are imported into normal mitochondria at efficiencies about 10% of those for normal cysteine-containing forms of cytochrome *c* (Dumont et al., 1988).

Heme attachment has also been found to affect accumulation of cytochrome *c* in yeast mitochondria *in vivo*. Yeast normally contain two isozymes of cytochrome *c*, iso-1-cytochrome *c*, encoded by the *CYC1* gene, normally comprising 95% of cellular cytochrome *c*, and iso-2-cytochrome *c*, encoded by *CYC7*, comprising the remaining 5%. The apo-form of iso-1-cytochrome *c* is unstable in cells, exhibiting a half-life of a few minutes. This makes it difficult to perform analyses of subcellular distributions of iso-1-cytochrome *c* in cells where import is defective. Fortunately, apo-iso-2-cytochrome *c* has a much longer half life (Dumont et al., 1990). Thus, import can be studied by immunoblotting of subcellular fractions in *cyc1*⁻ *CYC7-H3* strains that fail to express iso-1-cytochrome *c* but express elevated levels of iso-2-cytochrome *c* because of an alteration in the promoter for the *CYC7* gene.

Yeast strains with normal amounts of heme lyase contain virtually all of their immunologically detectable iso-2-cytochrome *c* inside mitochondria. However, strains with deletions of the *CYC3* gene, encoding heme lyase, contain most of their iso-2-cytochrome *c* in the cytoplasm, with none detectable in mitochondria. Cytoplasmic apocytochrome *c* is also found in yeast strains that contain normal levels of heme lyase but express an altered form of iso-2-cytochrome *c* that is incapable of having heme covalently attached because of the substitution of serine residues for the two cysteines that are the sites of heme attachment. However, in this case, low, but significant, levels of apocytochrome *c* are detected in a protease-protected compartment of mitochondria. This imported apocytochrome *c* can not be washed out of mitochondria, even after a 24 hour incubation. Surprisingly, accumulation of this mitochondrial apocytochrome *c* is dependent on the presence of heme lyase. Strains that lack heme lyase fail to accumulate even this low level of altered apocytochrome *c* whereas strains that express high levels of heme lyase accumulate altered apocytochrome *c* to levels that can reach 40% of the level of holocytochrome *c* in normal strains (Dumont et al., 1991).

The dependence of apocytochrome *c* accumulation in mitochondria on the presence of heme lyase suggests either that heme lyase plays a role in transport across the outer membrane, or that import occurs by a reversible mechanism and that apocytochrome *c* can be trapped in the intermembrane space by binding to heme lyase. In the former case, increased levels of heme lyase would lead to more efficient import into mitochondria, while in the latter case, the greater abundance of heme lyase would allowing trapping of stoichiometrically larger amounts of apocytochrome *c* (see Fig. 1). The trapping mechanism for the role of heme lyase in import was first proposed by Neupert and co-workers to explain import of normal apocytochrome *c*. In their proposal, a conformational change, triggered by attachment of heme inside the outer membrane, would lead to trapping of holocytochrome *c* in the intermembrane space (Korb and Neupert, 1978). More recently, evidence for reversible import of apocytochrome *c* into isolated mouse mitochondria has been presented (Hakvoort et al., 1990).

Two additional pieces of evidence support a role for heme lyase in trapping cytochrome *c* rather than in directly mediating transport: 1) The submitochondrial location of heme lyase. In order for heme lyase to play a direct role in transport across the outer membrane, it should be associated with this membrane in cells. Using anti-heme lyase antibodies, the enzyme was found instead to be preferentially associated with inner mitochondrial membranes. Heme lyase was also found in an intermediate-density fraction similar to one that has been reported to correspond to sites of contact between the inner and outer membrane (Pon et al., 1989), but there was no enrichment of heme lyase over the inner membrane marker cytochrome oxidase in this fraction (Dumont et al., 1991). 2) Physical association of apocytochrome *c* and heme lyase in mitochondria. When mitochondria containing apocytochrome *c* with serine residues substituted for cysteines are subjected to immunoprecipitation with anti-cytochrome *c* antibodies under non-denaturing conditions, heme lyase is specifically co-precipitated. The association between heme lyase and apocytochrome *c* can not be reconstituted by mixing the two separate components under the conditions used for immunoprecipitation. Furthermore, it is possible to cross-link heme lyase to altered apocytochrome *c*, to yield a complex that is immunoreactive with antibodies against both proteins and has an apparent molecular weight equal to their sum (K. Lapan, D. Nelson, C. Testa, and M.E.D., *in preparation*).

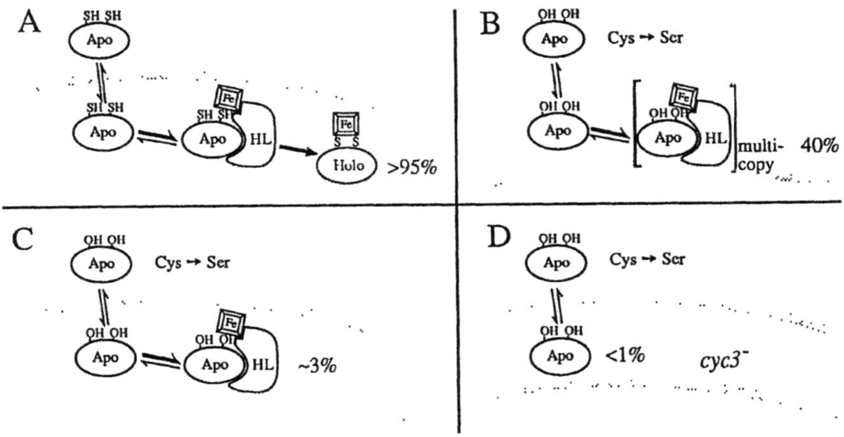


**Fig. 1.** Model for reversible mitochondrial import of normal cytochrome *c* and an altered form of the protein containing serine residues substituted for the cysteines that normally participate in thioether bonds to heme. The altered protein is encoded by the *CYC7-H3-67* allele. **A.** Import of normal apocytochrome *c* into normal mitochondria. The conformational change associated with covalent attachment of heme leads to trapping in the intermembrane space. **B.** Import of altered *CYC7-H3-67*-encoded apocytochrome *c* into mitochondria containing high levels of heme lyase expressed from a multicopy plasmid. Efficient accumulation in the intermembrane space is driven by binding to heme lyase. **C.** Import of the altered *CYC7-H3-67*-encoded apocytochrome *c* into mitochondria with normal levels of heme lyase. Accumulation is less efficient than in panel B because less heme lyase is available for binding. **D.** Absence of import of apocytochrome *c* into heme lyase-deficient mitochondria because there is no mechanism for trapping imported cytochrome *c*. Percentages in each panel refer to the fraction of immunologically detectable cytochrome *c* found in the mitochondria of each strain.

**Other cellular components involved in mitochondrial import of cytochrome c.**

Over the past thirty years, an extensive search for mutations that affect levels of cytochrome *c* in cells has been carried out by Fred Sherman and co-workers (See Rothstein and Sherman, 1980). In addition to the structural genes, *CYC1* and *CYC7*, encoding the two isozymes of cytochrome *c*, and the *CYC3* gene encoding heme lyase, an additional locus, *CYC2*, was identified. *Cyc2* mutations block cytochrome *c* biosynthesis at a post-translational step (Matner and Sherman, 1982). This gene has now been cloned, and its effects studied in yeast. It encodes a mitochondrial protein that affects the rate or efficiency of import of cytochrome *c* into yeast mitochondria, but is not strictly required for such import. *Cyc2* mutations act synergistically with of $\rho^-$ mutations, which are deletions of a substantial portion of the mitochondrial genome. However, the exact role of the *CYC2* gene product in import is not currently understood. (M.E.D., J.B. Schlichter, T.S. Cardillo, M.K. Hayes and F. Sherman, *submitted*).

It is possible that additional genetic loci affecting import of cytochrome *c* into mitochondria might have been missed in previous searches, since they were based on screens and selections that were specific for effects on cytochrome *c*. Thus additional cellular components may exist that affect import of cytochrome *c* as well as that of other mitochondrial proteins. In addition, previous genetic approaches might have failed to uncover genetic loci that, when altered, cause cell death.

A major unresolved question in understanding the import of cytochrome *c* into mitochondria is the mode of transport across the outer membrane. The possibility that apocytochrome *c* might partition directly into the lipid phase of the mitochondrial outer membrane is suggested by the absence of proteolytically-sensitive receptors on the outer surface of mitochondria (Nicholson et al., 1988). In addition, the fusion of cytochrome *c* sequences to the sequence of another mitochondrial protein, cytochrome $c_1$, allows import that bypasses the normal requirement of cytochrome $c_1$ for receptor binding (Stuart et al. 1990). Furthermore, purified apocytochrome *c* is capable of inserting into artificial lipid vesicles, even to the point of becoming exposed on the inner surface of vesicles (Dumont and Richards, 1984; see Jordi et al., 1992). However, the insensitivity of cytochrome *c* import to low temperatures (Nicholson et al., 1988; Hakvoort et al. 1990) appears to be inconsistent with direct interaction with the lipid phase.

**Targeting sequences of cytochrome c.**

Although the genes encoding cytochrome *c* have been the subject of extensive mutagenesis, resulting in hundreds of altered proteins, only three positions have been identified at which single amino acid substitutions are capable of causing a complete loss of cellular holocytochrome *c*. These positions are the two cysteine residues that are the sites of covalent heme attachment and His18 (mammalian numbering) that is immediately adjacent to the cysteines and serves as a ligand of the heme iron (Hampsey et al., 1988). While it is clear that mutations at these positions cause substantial defects in import *in vivo* and *in vitro*, apocytochrome *c* lacking both of the cysteine residues can be imported at low levels *in vivo* (Dumont et al., 1988; 1991). In addition, some protozoan cytochromes *c* normally contain only one of the two cysteine residues and two recent reports indicate the yeast can express low levels of holocytochrome *c* from genes encoding only one cysteine (Tanaka et al., 1990), or containing a substitution of Arg for His18 (Sorrell et al., 1989). Thus, while the two cysteines and His18 may be important for heme attachment and trapping of holocytochrome *c* in mitochondria, they may not be the sequences that are specifically involved in targeting to mitochondria.

Several sequence elements other than the site of heme attachment have been implicated in mitochondrial targeting of cytochrome *c*. Carboxyl terminal fragments of the protein, when present in excess, can block import into isolated mitochondria (Matsuura et al., 1981). Stuart et al., (1987) showed that a major alteration of the carboxyl terminal could block import. Nye and Scarpulla found sequences from both the amino terminal and a more carboxyl-terminal region (residues 63-80) that each appeared to be sufficient for mitochondrial targeting of fusion proteins containing cytochrome *c* sequences in yeast. However, the chloramphenicol acetyltransferase-containing fusion proteins used for these studies were not correctly targeted to the intermembrane space, presumably because of their tendency to fold prematurely (Nye and Scarpulla, 1990a,b). The existence of at least two separate targeting elements was also suggested by the results of *in vitro* assays of reversible import of mutated forms of *Drosophila* cytochrome c into mouse mitochondria (Sprinkle et al., 1990). The extreme carboxyl terminal of cytochrome *c* was not found to play a significant role in either of these studies. Thus, the exact nature of the mitochondrial targeting signals in cytochrome *c*, and the question of whether such signals take the form of particular sequences recognized by heme lyase or a tendency to adopt a structure that can interact with the membrane, remain to be resolved.

**Role of protein folding in mitochondrial import.**

Both genetic and biochemical experiments have demonstrated a role for heat shock proteins in maintaining mitochondrial precursors in a conformation that is competent for transport (see Pfanner and Neupert, 1990). One demonstration of the role of heat shock proteins in import was the inability of some mitochondrial precursors synthesized in wheat germ cell-free translation systems to be efficiently imported into isolated mitochondria. This was attributed to the low levels of heat shock proteins in wheat germ, as compared with the reticulocyte lysates that are usually used for such experiments. Addition of purified members of the Hsp70 heat shock family restored import from the wheat germ system (Murakami et al., 1988). In contrast to other mitochondrial precursors, apocytochrome *c* is efficiently imported into mitochondria following synthesis in the wheat germ system (Dumont et al., 1988). This may reflect the fact that apocytochrome *c* has the unusual property of behaving as an extended chain with little regular secondary structure, even in the absence of any denaturing treatment (Stellwagen et al., 1972; Fisher et al., 1973). This unfolded conformation is apparently maintained in the absence of any auxiliary heat shock proteins. We have recently found that non-covalent binding of heme to purified apocytochrome *c*, in the absence of heme lyase, is capable of triggering folding of the protein to a compact conformation different from that of mature holocytochrome *c*. Such a non-covalent complex may represent an intermediate in the folding of the newly-synthesized protein that is preferentially recognized by the mitochondrial import apparatus or heme lyase (Dumont and Corin, *submitted*).

**Degradation of cytoplasmic apocytochrome** *c*.

The stability of mitochondrial precursor proteins in the cytoplasm under conditions where import is blocked varies over a wide range (Reid and Schatz, 1982). Iso-1- and iso-2-cytochromes *c* provide particularly striking examples of this. Although their amino acid sequences are 80% identical, the half-life of apo-iso-1-cytochrome *c* in the cytoplasm is only a few minutes, while the half life of apo-iso-2 is on the order of hours. Substituting a central region from iso-2-cytochrome *c* into the corresponding region of iso-1, increases the stability of apo-1, though the apo-protein remains less stable than apo-iso-2 (Dumont et al., 1990). Alterations in the amino and carboxyl-terminal regions of apo-iso-1-cytochrome *c* also appear to be capable of increasing its stability (M.E.D. , Y. Kuang, T.S. Cardillo, and F. Sherman, *in preparation*). Thus, there either must be several different proteolytic systems with different specificities that are involved in degradation of cytoplasmic apocytochrome *c*, or the signal for recognition by a single protease must be a partially folded conformation can be disrupted by alterations in widely disparate regions of the sequence.

In the case of iso-1-cytochrome *c*, the predominant isozyme of cytochrome *c* in yeast cells, import and degradation appear to be in a closely balanced competition. Thus, the effects of mutations that slow import *in vitro* can be enhanced by the fact that, rather than accumulating in the cytoplasm, apocytochrome *c* will be rapidly degraded. Since the situation appears to be so different for apo-iso-2-cytochrome *c*, this provides an opportunity for the cell to regulate the two proteins differently. Observations of increased levels of iso-2-cytochrome *c*, compared with iso-1, following transitions from aerobic to anaerobic growth, or from glucose-repressed to derepressed conditions might be explained by rapid recruitment of cytoplasmic pools of apo-iso-2-cytochrome *c* under changing conditions (Dumont et al., 1990). Cytoplasmic degradation of apocytochrome *c* might also serve to prevent harmful accumulation of apo-iso-1-cytochrome *c* that, because of its membrane-active properties, might be incorrectly targeted to a greater extent than other mitochondrial precursors.

**Relation of cytochrome *c* import pathway to other pathways.**

While the pathway of cytochrome *c* import is different from that of other mitochondrial proteins, is becoming increasingly difficult to view mitochondrial import as occurring by one predominant pathway, with cytochrome *c* import as an isolated exception. The following proteins are transported by pathways that are either exceptions to the general scheme (see Glick and Schatz, 1991) or involve alternate sets of components: 1) The ADP/ATP translocator of the inner mitochondrial membrane appears to use outer membrane receptors that differ from those used by other mitochondrial proteins (Pfanner and Neupert, 1990). 2) The precursor to cytochrome oxidase subunit Va is imported by a pathway that appears to differ from that of most other inner membrane proteins, based on its insensitivity to low temperatures, its insensitivity to protease pre-treatment of mitochondria, and its unusual signal sequence requirements (Miller and Cumsky, 1991). 3) Cytochrome *c* heme lyase, the enzyme of the intermembrane space that catalyzes covalent attachment of heme to apocytochrome *c* is imported by a pathway that does not involve the inner membrane, since import does not require a potential across the inner membrane or hydrolysis of ATP (Lill et al., 1992). 4) Adenylate kinase, a protein of the intermembrane space, lacks a cleavable signal sequence and thus seems unlikely to be imported by a mechanism that involves initial targeting to the inner membrane (Magdolen et al., 1992). 5) Proteins of the outer mitochondrial membrane appear to be translocated via one or more pathways that are distinct from those followed by proteins destined for other compartments (see Glick and Schatz, 1991). Unique physical chemical properties of cytochrome *c* may have led to the evolution of a transport mechanism that has more in common with proteins destined for the outer mitochondrial membrane than it does with other proteins of the intermembrane space.

Acknowledgements: I wish to thank Fred Sherman for helpful discussions and assistance in preparing Fig. 1.

# LITERATURE REFERENCES


Dumont ME and Richards, FM (1984) Insertion of apocytochrome *c* into lipid vesicles. J. Biol. Chem. 259:4147-4156

Dumont ME, Ernst JF, Hampsey DM, and Sherman F (1987) Identification and sequence of the gene encoding cytochrome *c* heme lyase in the yeast *Saccharomyces cerevisiae*. EMBO J. 6:235-241

Dumont ME, Ernst JF, and Sherman F (1988) Coupling of heme attachment to import of cytochrome *c* into yeast mitochondria: Studies with heme lyase-deficient mitochondria and altered apocytochromes *c*. J. Biol. Chem. 263:15928-15937

Dumont ME, Mathews AJ, Nall BT, Baim SB, Eustice DC, and Sherman F (1990) Differential stability of two apo-isocytochromes *c* in the yeast *Saccharomyces cerevisiae*. J. Biol. Chem. 265:2733-2739

Dumont ME, Cardillo TS, Hayes MK, and Sherman F (1991) Role of cytochrome *c* heme lyase in mitochondrial import and accumulation of cytochrome *c* in *Saccharomyces cerevisiae*. Mol. Cell. Biol. 11:5487-5496

Fisher WR, Taniuchi H and Anfinson CB (1973) On the role of heme in the formation of the structure of cytochrome *c*. J. Biol. Chem. 248:3188-3195

Glick, BS, Brandt A, Cunningham K, Muller S, Hallberg RL and Schatz G (1992) Cytochromes $c_1$ and $b_2$ are sorted to the intermembrane space of yeast mitochondria by a stop-transfer mechanism. Cell 69:809-822

Glick B and Schatz G (1991) Import of proteins into mitochondria. Ann. Rev. Genet. 25:21-44

Hakvoort TBM, Sprinkle JR and Margoliash E (1990) Reversible import of apocytochrome *c* into mitochondria. Proc. Natl. Acad. Sci. USA 87:4996-5000

Hampsey DM, Das G and Sherman F (1988) Yeast iso-1-cytochrome *c*: Genetic analysis of structural requirements. FEBS Lett. 231:275-283

Hennig B and Neupert W (1981) Assembly of cytochrome *c*: Apocytochrome *c* is bound to specific sites on mitochondria before its conversion to holocytochrome *c*. Eur. J. Biochem. 121:203-212

Jordi W, Hergersberg C and de Kruijff B (1992) Bilayer-penetrating properties enable apocytochrome *c* to follow a special import pathway into mitochondria. Eur. J. Biochem. 204:841-846.

Korb H and Neupert W (1978) Biogenesis of cytochrome *c* in *Neurospora crassa*. Eur. J. Biochem. 91:609-620.

Lill R, Stuart RA, Drygas ME Nargang, FE and Neupert W (1992) Import of cytochrome *c* heme lyase into mitochondria: a novel import pathway into the intermembrane space. EMBO J. 11:449-456

Magdolen V, Schricker R, Strobel G, Germaier H and Bandlow W (1992) *In vivo* import of yeast adenylate kinase into mitochondria affected by site-directed mutagenesis. FEBS Lett 299:262-272

Matner RR and Sherman F (1982) Differential accumulation of two apo-iso-cytochromes *c* in processing mutants of yeast. J. Biol. Chem. 257:9811-9821

Matsuura S, Arpin M, Hannum C, Margoliash E, Sabatini DD and Morimoto T (1981) *In vitro* synthesis and posttranslational uptake of cytochrome *c* into isolated mitochondria, role of a specific addressing signal in the apocytochrome. Proc. Natl. Acad. Sci. USA 78:4368-4372

Miller, BR and Cumsky, MG (1991) An unusual mitochondrial import pathway for the precursor to yeast cytochrome *c* oxidase subunit Va. J. Cell Biol. 112:833-841

Murakami H, Pain D and Blobel G (1988) 70 kD heat shock related protein is one of at least two distinct cytosolic factors stimulating protein import into mitochondria. J. Cell Biol. 107:2051-2057

Nargang FE, Drygas ME, Kwong PL, Nicholson DW, and Neupert W (1988) A mutant of *Neurospora crassa* deficient in cytochrome *c* heme lyase activity can not import cytochrome *c* into mitochondria. J. Biol. Chem. 263:9388-9394

Nicholson DW, Hergersberg C, and Neupert W (1988) Role of cytochrome *c* heme lyase in the import of cytochrome *c* into Mitochondria. J. Biol. Chem. 263:19034-19042

Nye SH and Scarpulla RC (1990a) In vivo expression and mitochondrial targeting of yeast apoiso-1-cytochrome *c* fusion proteins. Mol. Cell. Biol. 10:5753-5762.

Nye SH and Scarpulla RC (1990b) Mitochondrial targeting of yeast apoiso-1-cytochrome *c* is mediated through functionally independent structural domains. Mol. Cell. Biol. 10:5763-5771.

Pfanner N and Neupert W (1990) The mitochondrial protein import apparatus. Ann. Rev. Biochem. 59:331-353

Pfanner N, Rasow J, Guiard B, Sollner T, Hartl F-U and Neupert W (1990) Energy requirements for unfolding and membrane translocation of precursor proteins during import into mitochondria. J. Biol. Chem. 265:16324-16329

Pon L, Moll T, Vestweber D, Marshallsay B Schatz G (1989) Protein import into mitochondria: ATP-dependent protein translocation activity in a submitochondrial fraction enriched in membrane contact sites and specific proteins. J. Cell Biol. 109:2603-2616

Reid GA and Schatz G (1982) Import of proteins into mitochondria: Yeast cells grown in the presence of CCCP accumulate massive amounts of some mitochondrial precursors polypeptides. J. Biol. Chem. 257:13056-13061

Rothstein RJ and Sherman F (1980) Genes affecting the expression of cytochrome *c* in yeast: Genetic mapping and genetic interactions. Genetics 94:871-889

Sorrell TN and Martin PK (1989) A novel, functional variant of cytochrome *c*: Replacement of the histidine ligand with arginine via site-directed mutagenesis. J. Am. Chem. Soc. 111:766-767

Sprinkle JR, Hakvoort TBM, Koshy TI, Miller DD and Margoliash E (1990) Amino acid sequence requirements for the association of apocytochrome *c* with mitochondria. Proc. Natl. Acad. Sci. USA 87:5729-5733

Stellwagen E, Rysavy R, and Babul G (1972) The conformation of horse heart apocytochrome *c*. J. Biol. Chem. 247:8074-8077

Stuart RA and Neupert W (1990) Apocytochrome *c* an exceptional mitochondrial precursor protein using an exceptional import pathway. Biochemie 72:115-121

Stuart RA, Neupert W and Tropschug M (1987) Deficiency in mRNA splicing in a cytochrome *c* mutant of *Neurospora crassa*: Importance of carboxy terminus for import of apocytochrome *c* into mitochondria. EMBO J. 6:2131-2137

Stuart RA, Nicholson DW and Neupert W (1990) Early steps in mitochondrial protein import: Receptor functions can be substituted by the membrane insertion activity of apocytochrome *c*. Cell 60:31-43

Tanaka Y, Kubota I, Amachi T, Yoshizumi H and Matsubara H (1990) Site-directed mutated human cytochrome *c* which retains heme *c* via only one thioether bond. J. Biochem. 108:7-8

# USING YEAST TO STUDY EXCHANGE OF MACROMOLECULES BETWEEN THE CYTOPLASM AND THE NUCLEUS

P. Silver

Department of Molecular Biology
Princeton University
Princeton, New Jersey 08544

## INTRODUCTION

Research over the past ten years has yielded information about the mechanism and some of the factors required for the specific uptake of certain proteins by the nucleus (see, for example, Garcia-Bustos et al., 1991; Silver, 1991; Stochaj and Silver, 1992 for recent review). Proteins move from their site of synthesis in the cytoplasm to the nuclear envelope. Specific targeting to the nucleus is accomplished in some cases by the presence of a short stretch of amino acids on the nuclear-destined proteins, termed an NLS for *n*uclear *l*ocalization *s*equence. The NLS may mediate the interaction of the transported protein with cytoplasmic carriers or "receptors." We know that proteins enter the nucleus via an aqueous channel formed by a complex of proteins in the nuclear envelope (termed the NPC for nuclear pore complex [Feldherr et al., 1984]). NLS-containing proteins may pass through the pore in association with the receptor or released from the receptor at the pore entry. The import process has been shown to require ATP in animal cells, *Xenopus* oocytes and yeast (Newmeyer and Forbes, 1988; Richardson et al., 1988; Breeuwer and Goldfarb, 1990; Garcia-Bustos et al., 1991).

Studies with the yeast *Saccharomyces cerevisiae* have played a critical role in these analyses. This article will review the characterization of some NLSs and components of the

NATO ASI Series, Vol. H 71
Protein Synthesis and Targeting in Yeast
Edited by A. J. P. Brown, M. F. Tuite, and J. E. G. McCarthy

nuclear import apparatus from yeast. Results of both biochemical and genetic experiments will be summarized.

## NUCLEAR LOCALIZATION SEQUENCES IN YEAST

NLSs are defined as short stretches of amino acids that are both necessary and sufficient for targeting proteins to the nucleus (for extensive review of NLSs, see Garcia-Bustos, et al., 1991). Since the first detailed description of NLSs in the early 1980s (i.e. Dingwall et al., 1982; Kalderon et al., 1984; Lanford and Butel, 1984), many reports of NLSs in yeast as well as higher cells have appeared. What follows is a brief description of what is known about the function of some NLSs in yeast.

### *GAL4*

Gal4 is a DNA binding protein required for induction of genes encoding enzymes required for galactose metabolism in yeast. The first 74 Gal4 amino acids were originally shown to be sufficient to target *E. coli* ß-galactosidase to the yeast nucleus (Silver et al., 1984). Moreover, a fusion protein missing the first 74 Gal4 amino acids but containing almost all of the remaining amino acids fused to ß-galactosidase remained in the cytoplasm. These results showed that the information contained in the Gal4 N-terminus was both sufficient and necessary to target Gal4 to the nucleus.

Interestingly, there was a difference in the level of expression of the various chimeric proteins that correlated with their location in the cell. Non-nuclear proteins were expressed at higher levels than their nuclear localized counterparts. On the basis of this expression difference, point mutations with GAL4, which reduced its nuclear localization and increased production of a normally nuclear GAL4-ß-galactosidase were isolated (Silver et al., 1988). All mutations corresponded to single amino acid changes throughout the first 74 GAL4 amino acids. Most, but not all, also affected the localization of the intact GAL4 protein.

Other results showed that it was possible to fuse only the first 29 amino acids of GAL4 to cytoplasmic invertase and produce a nuclear localized chimeric protein (Nelson and Silver, 1989). This would suggest that the GAL4 NLS is, in fact, contained within the first 29 GAL4 amino acids. However, the same 29 amino acids fused to ß-galactosidase did not yield a protein that was efficiently nuclear-localized. Taken together, these results indicate that the GAL4 NLS is more complicated than, for instance, the one from SV40 large T-antigen. This could be because the region identified as necessary for nuclear targeting also contains the GAL4 zinc finger DNA binding domain and, therefore, must fold in a certain manner.

### *NLSs from higher eukaryotes function in yeast*

The uptake of nuclear proteins is most likely a highly conserved process. One test of this is that NLSs from mammalian nuclear proteins also function in yeast. This was shown to be true for the NLS from SV40 large T-antigen (Nelson and Silver, 1989).

The NLS form SV40 large T-antigen is one of the best defined. A short stretch of amino acids with the sequence KKKRKV has been demonstrated to be both necessary and sufficient for nuclear protein targeting in animal cells (Kalderon et al., 1984). Also, a single amino acid change at the equivalent of lysine 128 in large T-antigen inactivates the NLS.

In yeast, a fusion protein containing the minimum SV40 NLS at the N-terminus of the cytoplasmic form of invertase is very efficiently localized within the nucleus (Nelson and Silver, 1989). Moreover, the single amino acid change in the T-antigen NLS inhibits the nuclear localization of the SV40 NLS-invertase fusion protein in yeast.

In separate experiment, fusions between nucleoplasmin and *E. coli* ß-galactosidase were efficiently localized to the nucleus in yeast (Sadler and Silver, unpublished results). Nucleoplasmin is an abundant nuclear protein of *Xenopus* laevis oocytes and its NLS has been shown to reside at the C-terminus (Dingwall et al., 1982). The nucleoplasmin NLS is more complex than that of SV40 large T-antigen and has been shown to be bipartite (Dingwall and Laskey, 1991).

Taken together, these results show that NLSs from higher organisms function correctly in the yeast *S. cerevisiae* and confirm the commonality of the mechanism of nuclear protein uptake.

## RECOGNITION OF PROTEINS FOR NUCLEAR IMPORT

As discussed in the preceding section, a variety of peptides can mediate the proper localization of proteins to the nucleus. Yeast proteins have been identified that bind NLS peptides of SV large T-antigen, *Xenopus* nucleoplasmin, and the yeast proteins Gal4 and histone H2B (Silver et al., 1989). One of these NLS-binding proteins of 70kd molecular mass has been purified 5,000-fold from yeast nuclei (Stochaj et al., 1991). Biochemical analysis suggests that at least some of this protein is associated with the nuclear surface. Nuclei preincubated with anti-NLS-binding protein antibodies no longer bind NLS-containing proteins *in vitro*.

Recent results indicate that a protein similar to the yeast NLS-binding protein exists in all species tested (Stochaj and Silver, 1992). Furthermore, all 70kd NBPs are phosphoproteins. Removal of the phosphate groups by treatment with alkaline phosphatase eliminates the NLS-binding activity. One model is that phosphorylated NBPs interact with nuclear-destined proteins in the cytoplasm. The complex then moves to the nucleus where the NBPs are dephosphorylated and, as a consequence, unload their substrates. Reversible modifications of NBPs may also serve as a way to globally regulate nuclear import during development and cell differentiation.

The conserved NBP is involved in nuclear protein import in higher cells. This was demonstrated by using semi-permeabilized *Drosophila melanogaster* tissue culture cells (Stochaj and Silver, 1992). In this system, proper import of NLS-containing proteins requires cytosol and ATP. In the absence of cytosol and ATP, NLS-containing proteins remain in the cytoplasm. Addition of cytosol and ATP results in their movement into the nucleus. Anti-NBP antibodies specifically inhibited this reaction. These results indicate that a phosphoprotein common to several eukaryotes acts as a receptor that recognizes NLSs before their uptake into the nucleus.

Other NBPs have been described in yeast. These include proteins of molecular mass 140kd, 95kd, and 59kd (Silver et al., 1989). These proteins bind NLSs from SV40 large T-antigen, yeast histone H2B, and *Xenopus* nucleoplasm. They have yet to be purified and their role in nuclear import has not been studied.

Lee et al., 1991 also described a 67kd yeast nuclear protein, termed Nsr1, that binds to NLSs from histone H2B and SV40 large T-antigen. The protein contains motifs typically associated with RNA binding proteins. Cells missing *NSR1* are viable and show normal distribution of nuclear proteins. This suggests that Nsr1 is either not directly involved in the process or is a member of a family of proteins responsible for recognizing NLSs. The latter seems likely given the potential number of NBPs now identified. It remains a question as to how a nuclear protein could be involved in nuclear import. Interestingly, others have described a nucleolar NBP in higher cells (Meier et al., 1990). Perhaps these NBPs actually shuttle between the cytoplasm and the nucleus and act as carriers for nuclear proteins. Consistent with this possibility is the observation that other nucleolar proteins have been shown to move in and out of the nucleus ().

## GENETICS OF NUCLEAR PROTEIN IMPORT

Isolation and characterization of mutants has proved invaluable for unraveling the process of protein export in yeast. Unfortunately, isolation of yeast mutants defective in nuclear protein import has lagged behind, in part, because their are no obvious chemical modifications that accompany uptake of proteins into the nucleus. Thus, unlike glycosylation in protein export, their are no obvious hallmarks to use to follow protein transit into the nucleus.

Our laboratory has developed a novel scheme for isolating yeast genes required for the proper localization of nuclear proteins (Sadler et al., 1989; Bossie et al., 1992). As such, we have devised a way to determine, based on growth characteristics, whether a protein is in the nucleus or not.

The selection for cells defective in nuclear protein localization is as follows (and outline in Figure 1). The screen relies on the missorting of a normally mitochondrial protein to the nucleus. Fusion of a NLS to cytochrome $c_1$ or $F_1$ß-ATPase results in preferential localization of these proteins to the nucleus, and thus, failure of mitochondrial function (i.e. cells missing the *ATP2* and *CYT1* genes cannot grow on glycerol). The hypothesis was then that an incomplete block in nuclear protein import would result in uptake of enough protein by the mitochondria to allow cells to grow on glycerol. However, a complete block in nuclear protein localization would result in cell death because, presumably, proper nuclear import is essential for cell growth. Thus, mutants were isolated that were Gly+ at 30$^{o}$C, but failed to grow at 36$^{o}$C. The resulting mutants were further analyzed for failure to correctly localize endogenous nuclear protein such as histones and nucleolar proteins. The mutants are termed *npl* for *n*uclear *p*rotein *l*ocalization defective. Several multiallelic complementation groups and single mutants have been identified by this scheme.

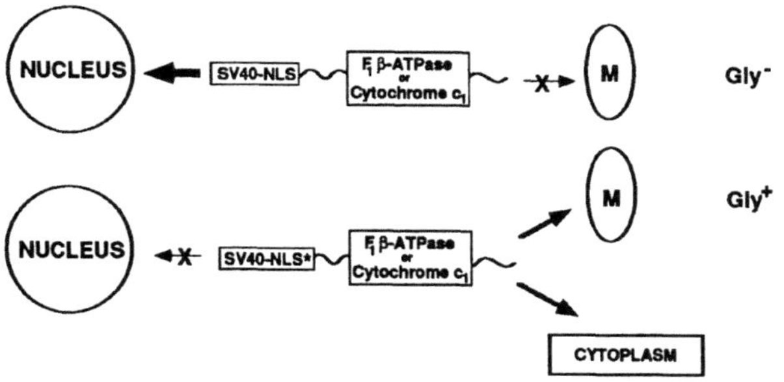


Figure 1. Scheme for the isolation of *npl* mutants

Several independently isolated mutant alleles of *NPL1* have been identified. *Npl1* mutants missort fusions containing either the GAL4 or SV40 NLS to cytochrome $c_1$ or $F_1$ß-ATPase. They also show varying degrees of mislocalization of endogenous nuclear proteins, although for most alleles the effect on endogenous proteins is not as strong as for fusion proteins. Some alleles also show unusual morphology of the nuclear envelope.

The translated amino acid sequence of *NPL1* predicts a membrane protein with a region of homology to DnaJ, an *E. coli* heat shock protein (Sadler et al., 1989). In *E. coli*, DnaJ interacts with and modulates the activities of DnaK, the HSP70 cognate. *NPL1* is also identical to *SEC63*, which is important for translocation of proteins across the ER (Rothblatt et al., 1989). However, our results show that all 6 *npl1* alleles are located in the highly charged cytoplasmic Npl1 C-terminus. These alleles are ts-lethal, but do not cause a strong secretion defect at any temperature. One model is that the acidic C-terminus may interact with basic NLSs or with another factor necessary for import (this interaction would be disrupted by the *npl1* mutations). Failure to correctly form this interaction results in the toxic accumulation of some factor(s) that can be relieved by mutations in *SNP1* (isolated as extragenic suppressors of *npl1* mutations (Nelson and Silver, 1992).

The sequence of *NPL3* predicts a protein with identity to a previously characterized class of RNA binding proteins (Bossie et al., 1992). All *npl3* mutants block import of several distinct nuclear proteins. We propose that this is due to jamming of the nuclear transport machinery by the mutant Npl3 protein. Preliminary results suggest that *npl3* mutants are also blocked in export of mRNA from the nucleus. The Npl3 protein is normally located in the nucleoplasm. However, some of the mutant Npl3 appears to accumulate at the nuclear periphery. We propose that it is the abnormal distribution of the mutant Npl3p that is antagonizing exchange of macromolecules between the cytoplasm and the nucleus. This could be occurring at the level of the nuclear pore complex or by sequestering some other factor important for protein and RNA exchange.

The sequence of *NPL6* predicts an acidic protein (Chiang and Silver, in preparation). The Npl6 protein is located within the nucleus and appears to contain a bipartite NLS. The original *npl6* mutants are temperature-sensitive for growth and for import of proteins into the nucleus. Cells completely missing *NPL6* (*Δnpl6*) grow poorly, are temperature-sensitive and have a defect in nuclear protein uptake. Thus, Npl6 is a nuclear protein essential for the efficient import of proteins into the nucleus.

Taken together, this genetic analysis has yielded new information about several factors that affect nuclear-cytoplasmic exchange. It remains to be shown precisely how these proteins are involved in the process. Several testable possibilities are predicted from the amino acid sequences of the various proteins. Future experiments with an *in vitro* nuclear import reaction developed with yeast components should help to address this more directly.

**CONCLUSION**

A brief summary of some results from studies of nuclear protein transport in yeast has been presented. These, in combination with other results accumulated over the past ten years, indicate that translocation of proteins across the nuclear envelope is highly conserved. Similar NLSs function in yeast and mammals (Nelson and Silver, 1989). Antibodies raised against mammalian nucleoporins recognize similar proteins in yeast (Davis and Fink, 1990). And a highly conserved NLS-binding protein is found in yeast, animal and plant cells (Stochaj and Silver, 1992). Future experiments will add to the already apparent conservation of the process of exchange of macromolecules between the nucleus and the cytoplasm.

## ACKNOWLEDGEMENTS

Work from the author's laboratory was supported by grants from the National Institutes of Health, National Science Foundation, and the American Heart Association.

## REFERENCES

Bossie, M., DeHoratius, C., Barcelo, G., and Silver, P. (1992). A mutant nuclear protein with similarity to RNA binding proteins interferes with nuclear import in yeast. Molecular Biology of the Cell, In press.

Breeuwer, M. and Goldfarb, D. (1990). Facilitated nuclear transport of histone H1 and other small nucleophilic proteins. Cell 60:999-1008.

Davis, L. and Fink, G. R. (1990). The *NUP1* gene encodes an essential component of the yeast nuclear pore complex. Cell 61:965-978.

Dingwall, C., Sharnick, S.V., and Laskey, R.A. (1982). A polypeptide domain that specifies migration of nucleoplasmin into the nucleus. Cell 30:449-458.

Dingwall, C. and Laskey, R.A. (1991). Nuclear targeting sequences - a consensus? Trends in Biochem. Sci. 16:478-481.

Feldherr, C., Kallenbach, E., and Schultz, N. (1984). Movement of a karyophillic protein through the nuclear pores of oocytes. J. Cell Biol. 99:2216-2222.

Garcia-Bustos, J., Heitman, J., and Hall, M. (1991). Nuclear protein localization. Biochim. Biophys. Acta. 1071:83-101.

Goldfarb, D.S., Gariepy, J., Schoolnik, G. and Kornberg, R. (1986). Synthetic peptides as nuclear localization signals. Nature 322:641-644.

Kalderon, D., Roberts, B.L., Richardson, W.D., and Smith, A.E. (1984). A short amino acid sequence able to specify nuclear location. Cell 39:499-509.

Lanford, R.E. and Butel, J.S. (1984). Construction and characterization of an SV40 mutant defective in nuclear transport of T antigen. Cell 37:801-813.

Lee, W.-C., Xue, Z., and Melese, T. (1991). The NSR1 gene encodes a protein that specifically binds nuclear localization sequences and has two RNA recognition motifs. J. Cell Biol. 113:1-12.

Meier, U.T. and Blobel, G. (1990). A nuclear localization signal binding protein in the nucleolus. J. Cell Biol. 111:2235-2245.

Nelson, M. and Silver, P. (1989). Context affects nuclear protein localization in *Saccharomyces cerevisiae*. Mol. Cell. Biol. 9:384-389.

Newmeyer, D.D. and Forbes, D. (1988). Nuclear import can be separated into distinct steps in vitro: nuclear pore binding and translocation. Cell 52:641-653.

Richardson, W.D., Mills, A.D., Dilworth, S.M., Laskey, R.A. and Dingwall, C. (1988). Nuclear protein migration involves two steps: rapid binding at the nuclear envelope followed by slower translocation through nuclear pores. Cell 52:655-664.

Rothblatt, J.A., Deshaies, R., Sanders, S., Dawn, G. and Schekman, R. (1989). Multiple genes are required for proper insertion of secretory proteins into the endoplasmic reticulum in yeast. J. Cell Biol. 109:2641-2652.

Sadler, I., Chiang, A., Kurihara, T., Rothblatt, J., Way, J. and Silver, P. (1989). A yeast gene important for assembly into the endoplasmic reticulum and the nucleus has homology to DnaJ, an E. coli heat shock protein. J. Cell Biol. 109:2665-2675.

Silver, P., Keegan, L., and Ptashne, M. (1984). Amino terminus of the yeast GAL4 gene product is sufficient for nuclear localization. Proc. Natl. Acad. Sci. 81:5951-5955.

Silver, P., Chiang, A. and Sadler, I. (1988). Mutations that alter both localization and production of a yeast nuclear protein. Genes Dev. 2:707-717.

Silver, P., Sadler, I. and Osborne, M. (1989). Yeast proteins that recognize nuclear localization sequences. J. Cell Biol. 109:983-989.

Silver, P. (1991). How proteins enter the nucleus. Cell 64:489-497.

Stochaj, U. and Silver, P. (1992). Nucleocytoplasmic traffic of proteins. Eur. J. of Cell Biol., In press.

Stochaj, U., Osborne, M., Kurihara, T., and Silver, P. (1991). A yeast protein that binds nuclear localization signals: Purification, localization, and antibody inhibition of activity. J. Cell Biol. 113:1243-1254.

Stochaj, U. and Silver, P. (1992). A conserved phosphoprotein that specifically binds nuclear localization sequences is involved in nuclear import. J. Cell Biol. 117:473-482.

# NUCLEOCYTOPLASMIC TRANSPORT IN RIBOSOME BIOGENESIS

Howard M. Fried
University of North Carolina
Department of Biochemistry & Biophysics
Chapel Hill, NC 27599-7260

The assembly of ribosomes in eukaryotic cells affords numerous opportunities to investigate mechanisms of nuclear-cytoplasmic transport. All stages of the assembly process involve both RNA and protein molecules moving into the nucleus, some being exported out, and still others being exported from the nucleus only to re-enter subsequently. What are the basic mechanisms that bring about the proper subcellular localization of ribosomal and non-ribosomal components involved in ribosome assembly? Are there aspects to the transport of ribosomal materials unique to these macromolecules and their multicomponent complexes or does nuclear-cytoplasmic transport of ribosomal components occur through the same pathways and mechanisms applicable to other cellular constituents? In addition, ribosome assembly takes place amid little if any pool of free ribosomal components; correspondingly, should a component either be produced in excess or prevented from assembly (usually through artificial conditions created by molecular biologists) that component is degraded very rapidly (Larson *et al.*, 1991); thus, one might also ask whether there are mechanisms that *coordinate* the nuclear-cytoplasmic traffic of ribosomal components and accessory assembly factors.

Most of ribosome assembly takes place within the nucleolus, a boundary-free subdomain of the nucleus, that owes much of its existence to the conjunction of nascent ribosomal RNA molecules, ribosomal proteins, and numerous non-ribosomal components in the act of pre-ribosomal RNA processing and protein-RNA binding. Since several nucleolar structures are evident cytologically prior to consolidation of the nucleolus, the "pre-nucleolus" may also bring with it its own structures that assist in organizing and promoting the orderly progression of ensuing assembly events. Traditionally then, ribosome biogenesis is discussed with reference to the major morphologically discernible structures within the nucleolus. Here I will adopt a less typical perspective, namely ribosome assembly in terms of the principal nuclear-cytoplasmic transport events that contribute to the overall process. The following is a

NATO ASI Series, Vol. H 71
Protein Synthesis and Targeting in Yeast
Edited by A. J. P. Brown, M. F. Tuite, and J. E. G. McCarthy

summary of what little is known about the mechanisms of these transport events, deduced mostly from studies in yeast and amphibian oocytes, and highlights of major unanswered questions. [See, among others, excellent reviews by Schwarzacher & Wachtler (1991), Fischer *et al.* (1991), and Jordan (1991) for detailed coverage of nucleolar structure and function.]

## General Considerations

The nucleus is separated from the cytoplasm by a double-membrane envelope enclosing a lumenal space 40-70 nm across; the cytoplasmic-facing membrane is continuous with the endoplasmic reticulum. At points in the nuclear envelope where the two membranes fuse there occur *nuclear pore complexes* (NPC). The NPC is the site through which macromolecular traffic into and out of the nucleus takes place. The NPC, itself a highly complex organelle of 25-50 X $10^6$ Da, possesses an aqueous channel that spans its length from cytoplasm to nuclear interior. While the NPC channel is large enough to permit passive diffusion of small molecules (< ~60 kDa), evidence suggests that even small endogenous macromolecules utilize an active transport mechanism to traverse the NPC (Breeuwer and Goldfarb, 1990). An active mechanism implies that transport substrates possess specific information recognized by receptors and other components of the transport machinery. Receptors for nuclear transport have been identified as part of the NPC as well as within the cytoplasm (and probably the nucleoplasm), suggesting that transport is a multistage process involving substrate binding to diffusible components that deliver their cargo to the NPC. [See articles by Silver and Hurt in this volume for more detailed information]. Nearly all of the studies of transport in ribosome assembly have concentrated on identifying signals for transport contained in various substrates.

## Transport Signals and Events in Eukaryotic Ribosome Biogenesis

Figure 1 presents a rather busy overview of the movement of materials during ribosome assembly, a depiction which in reality is probably a gross oversimplification of the actual events.

## NUCLEOCYTOPLASMIC TRANSPORT IN RIBOSOME BIOGENESIS

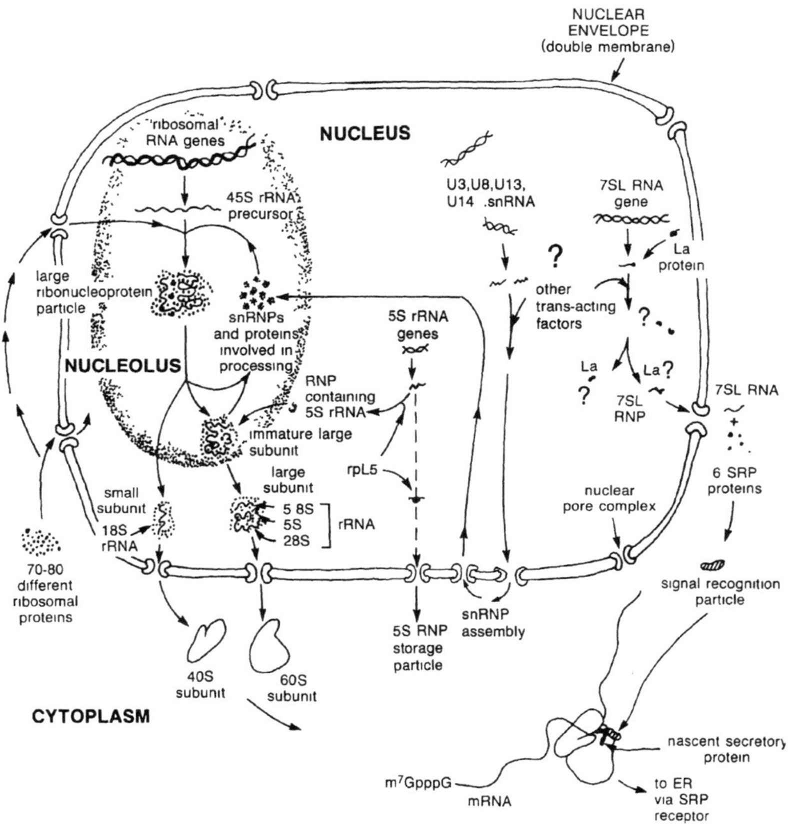


Figure 1. Nuclear-cytoplasmic transport steps in eukaryotic ribosome synthesis. Adapted in part from Alberts, B, Bray D, Lewis, J, Raff, M, Roberts, K, Watson, JD, *The Molecular Biology of the Cell*, Garland Publ., 1983 and Rawn, JD, *Biochemistry*, Patterson Publ., 1989.

(1) Ribosomal Protein Nuclear Import: Nuclear proteins contain so-called nuclear localization sequences (NLS), generally short (5-12) amino acid segments, often with a high proportion of positively charged residues, that appear to serve as ligands for transport receptors and that can act as independent domains to promote import. Thus, portions of yeast ribosomal protein L3, fused to a non-nuclear passenger protein, revealed an NLS consisting of the N-terminal 21 residues of L3 (Moreland *et al.*, 1985).

Neither further subdivision of this segment nor mutational analysis of fusions and native L3 were successful in isolating a smaller functional NLS (unpublished). Recently, a so-called *bipartite* NLS motif has been described (Dingwall & Laskey, 1991) in which functional residues in two basic domains are separated by a spacer region of relatively unimportant size. Single mutations in only one basic domain do not affect nuclear import but simultaneous substitutions in both domains abolishes import. The 21 residue segment of ribosomal protein L3 is strikingly homologous to the consensus bipartite NLS and we have found recently that amino acid substitutions in both basic segments are necessary to eliminate nuclear localization of a fusion protein containing this 21 residue L3 piece (unpublished result), supporting the idea that the rpL3 NLS is indeed bipartite.

In yeast ribosomal protein L29, two 7 residue NLS were identified, either of which targeted a fusion protein to the nucleus (Underwood & Fried, 1990). The two NLS appear to be of unequal potency within L29, since a single substitution in the N-terminally distal NLS, while abolishing fusion protein localization, was without effect on L29's ability to support ribosome assembly yet the same substitution in the proximal NLS severely reduced assembly. Interestingly, simultaneous mutation of both NLS resulted in a still greater decrease in ribosome assembly, indicating that both domains contribute cooperatively to promote nuclear import. Dingwall & Laskey have reconciled the occurrence of single and bipartite NLS motifs, both of which depend upon clusters of basic residues, by proposing that single motif NLS are merely exceptionally efficient variants of one of the two clusters in a bipartite motif. The two rpL29 NLS are separated by an acceptable number of residues as found in bipartite NLS; thus, L29 may contain a bipartite NLS in which the efficiency of the proximal domain is sufficient to support an adequate import rate on its own. Finally, Schaap *et al.* (1991) have described two NLS in yeast ribosomal protein L25. Residues 1 to 17 as well as 18 to 41 both promoted efficient import of a reporter protein. A more detailed dissection revealed that amino acid units 11-17 and 18-28 also showed import activity albeit less efficient than their longer segments. Although the 'spacing' between the basic clusters in these two segments is much smaller than in the consensus, it may be that L25 also possesses a bipartite NLS, notwithstanding the importance of adjacent residues.

The occurrence of more than one NLS in a protein is not uncommon nor is an effect on efficiency by adjacent residues. It remains to be seen whether occurrence of multiple or bipartite NLS is a general rule for ribosomal proteins. NLS contribute additively to the import rate of a protein and, perhaps, ribosomal proteins require very rapid import to sustain assembly without producing pools of unassembled polypeptides. Results for both L29 and L25 show that inactivation or deletion of one of their two NLS impairs growth, perhaps because of a reduced rate of nuclear entry of the proteins.

Finally, ribosomal proteins are rather small polypeptides (9-45 kDa) and potentially able to diffuse passively through the NPC. Being RNA binding proteins, one could argue that the NLS identified in ribosomal proteins do not actually promote active nuclear import at the level of the NPC but that the NLS are responsible simply for binding to rRNA following free diffusion of the protein into the nucleus. There are many reasons to discount this idea (see Fried, 1992), including results mentioned before indicating that some small proteins appear to be actively imported. Furthermore, Rutgers *et al.* (1991) showed that the rRNA binding site of L25 is confined to residues distal to residues 1-41 containing the proposed NLS; absence of residues 1-41 had no effect on the efficiency of (truncated) L25 to bind rRNA *in vitro*, suggesting that the NLS-containing segment does not contribute to nuclear localization by an RNA retention-type mechanism. Future studies might now be focussed on determining whether ribosomal proteins utilize the same pathway (*ie*, receptors) for nuclear import as other cellular proteins.

(2) Transport of 5S rRNA: In mammalian cells 5S rRNA associates transiently with La, a protein that may assist in termination of polymerase III transcription; La is displaced by ribosomal protein L5 and the resulting L5/5S RNP is incorporated into the assembling large ribosomal subunit (Steitz *et al.*, 1988). In amphibian oocytes, L5 also binds to 5S rRNA but the complex is exported out of the nucleus and stored in the cytoplasm; a small fraction of 5S rRNA binds to TFIIIA (the 5S gene transcription factor) instead of L5 and this complex is exported as well. Export is absolutely dependent upon binding of one or the other protein leading Guddat *et al.* (1990) to propose that L5 and TFIIIA represent a new functional class of proteins. If so, since they share no obvious structural or sequence homology, it will be of interest to determine just what are the signals in L5 and TFIIIA that target 5S rRNA out of the nucleus. The biogenesis of 5S rRNA in oocytes also raises the question of how the RNA presumably re-enters the nucleus. If it were to retain L5, then an RNP previously fated for export must become converted to one that is targeted into the nucleus. Alternatively L5 may dissociate from 5S rRNA leaving the RNA to interact with a different protein(s) required for re-entry. In either case, the cellular cues following fertilization that signal a conversion of the 5S RNP storage particle to become competent for nuclear re-entry should also be an interesting subject for investigation. Finally, even in somatic cells where it 5S rRNA appears to assemble directly into pre-ribosomal subunits in the nucleus, it can not be ruled out that a 5S RNP does not leave the nucleus for a very short stay in the cytoplasm (see Steitz *et al.* 1988).

(3) Transport of snRNAs Involved in Ribosome Biogenesis: The eukaryotic cell is host to a large number of small nuclear uridine-rich RNAs. As RNP particles, U1, U2, U4, U5, and U6 snRNPs form part of the spliceosome that excises introns from pre-mRNA; U7 snRNP is involved in histone mRNA 3' processing, while there are tantalizing indications that U11 and U12 may participate in mRNA polyadenylation. Except for U6, all of these RNA molecules are synthesized by RNA polymerase II, acquiring in the process the $m^7$GpppG cap characteristic of mRNA, whereupon the RNAs are exported from the nucleus. Efficient export of U snRNAs requires the $m^7$G cap. In the cytoplasm the U snRNAs bind a common set of proteins, the Sm antigens, as well as other proteins specific for particular RNAs. Binding of Sm proteins results in, and is required for, hypermethylation of the $m^7$G cap to a 2,2,7 trimethyl-guanosine (TMG) cap. The snRNPs are then re-imported into the nucleus where they assume their respective roles in RNA processing. In the case of U1 and U2, migration back into the nucleus requires *both* a TMG cap and bound Sm proteins; the two elements thus comprise an interesting bipartite NLS whose exact mode of action remains to be determined. In contrast U4 and U5 snRNPs have a less rigid requirement for cap trimethylation in nuclear import. [see Izaurralde & Mattaj, 1992 for a comprehensive review.]

As far as ribosome assembly is concerned, a number of nucleolar snRNAs appear to have roles in processing the 45S pre-rRNA transcript (Tollervey & Hurt, 1990). These include U3, U8, U13, and U14. All but U14 contain the TMG cap but none bind the Sm proteins; rather, all the RNAs (including U14) are found associated with a protein called fibrillarin, in addition to other proteins. Less is known about the nuclear import of these nucleolar RNAs. Studies with U3 (Baserga *et al.*, 1992) show that cap trimethylation and binding to fibrillarin can take place in the cytoplasm. Sequences near the 3' end of U3 are necessary for trimethylation but not fibrillarin binding. Surprisingly neither trimethylation nor association with fibrillarin is required for import, but deletion of the 3' sequences needed for trimethylation prevents transport. These observations suggest that a protein(s) whose binding to U3 requires sequences near the 3' end is necessary for cap modification and nuclear import, but protein binding alone is sufficient for import; fibrillarin binding is unnecessary for either trimethylation or import. Clearly, much remains to be clarified as to the actual pathway and transport signals involved in biosynthesis of U3 and the other nucleolar snRNPs. Furthermore, the mechanisms that specifically localize snRNPs to either the nucleolus or the nucleoplasm need to be determined.

(4) Ribosomal Subunit Nuclear Export: Somehow and at some point the nucleolus "knows" it has finished assembling a ribosomal subunit, which then must be exported to

the cytoplasm. How the completed subunits get to the nuclear envelope/nuclear pore complex is, of course, not known. Pre-ribosomal subunits could be released from the nucleolus and delivered to the nuclear pore complex (NPC) by a separable mechanism or conceivably final maturation of the pre-subunits may occur in more intimate association with the pore complex, a situation supported by cytological observations (see Fried, 1992). Nucleolar proteins have been suggested to play a role in transport. Nucleolin (also called C23) is found in the nucleolus in association with chromatin and with nascent pre-rRNA in pre-ribosomes. Although many functions including regulation of rDNA transcription, involvement in rRNA processing, and facilitation of pre-subunit assembly have been postulated for nucleolin, the protein was found unexpectedly to shuttle between the nucleus and cytoplasm (Borer *et al.*, 1989), leading to speculation that it is involved in pre-ribosomal subunit transport (or nuclear import of ribosomal proteins). Ribocharin (also known as B23) is a polypeptide specifically associated with precursors of the large subunit but not with mature cytoplasmic ribosomes (Hügle *et al.*, 1985). B23 likewise shuttles between the nucleus and cytoplasm (Borer, 1989) so it too may play a role in pre-ribosome or ribosomal protein transport.

To determine a property of ribosomal subunits that targets them for export, we injected radiolabeled subunits into *X. laevis* oocyte nuclei (Bataillé *et al.*, 1990). The RNPs entered the cytoplasm intact and at a physiologically significant rate. The rate of transport exhibited saturation kinetics and transport was inhibited either by low temperature or depletion of metabolic energy, all criteria indicative of a carrier-mediated transport process. Export of injected subunits appeared to be NPC-mediated, since transport was also inhibited by injection of wheat germ agglutinin (WGA), a lectin that binds to O-glycosylated proteins of the NPC and known to inhibit import of nuclear proteins. Most significantly, however, we found that prokaryotic ribosomal subunits were exported from the *X. laevis* oocyte nucleus in the same saturable, energy-dependent, WGA-inhibited manner as their eukaryotic counterparts. The finding that prokaryotic ribosomal subunits were transported in a facilitated fashion from a eukaryotic nucleus suggested to us that, at least for some types of RNPs, no particular set of proteins nor specific RNA sequences/structures is responsible for promoting transport; rather, ribosomal subunits might be recognized for export on the basis of a global or more general biochemical property. Just what that property is and how it is recognized remain to be elucidated. A possible clue may stem from work by Dworetzky & Feldherr (1988), who found that RNA-coated colloidal gold particles were transported to the cytoplasm in an apparently facilitated manner following nuclear injection. Whether proteins bound to the RNA-gold particles was unknown, but the complexes were unlikely to have duplicated the structures of naturally occurring RNPs. All ribosomal subunits, prokaryotic and eukaryotic, contain considerable amounts of solvent-exposed RNA; thus, transport of RNA-coated gold suggests that a globally accessible complement of rRNA may play a dominant role in promoting export of ribosomal RNPs.

(5) Nuclear Export of Signal Recognition Particle RNA (7SL RNA): While generally not considered in the context of ribosome assembly, we have recently become interested in biosynthesis of the signal recognition particle (SRP). SRP is a cytoplasmic RNP that associates with ribosomes translating secretory proteins so as to target those proteins to the endoplasmic reticulum. SRP consists of 6 polypeptides and an ~300 nucleotide RNA molecule known as 7SL (or SRP RNA). Due in large part to the elegant work of P. Walter and colleagues, we probably know more about the structure and function of SRP than just about any other RNP complex (Siegel & Walter, 1988 and elsewhere in this volume). However, little is known about SRP's birth *in vivo*. How does SRP assemble? Does nuclear exit of 7SL RNA depend upon specific structural features of the RNA molecule? Since 7SL RNA is an RNA polymerase III product, are there nuclear proteins that interact with 7SL RNA to promote its transport to the cytoplasm, just as there are for 5S rRNA, another Pol III transcript? If so, are such polypeptides novel or might they be any of the known components of cytoplasmic SRP?

The sequence of 7SL RNA is related to *Alu*, the most abundant family of middle repetitive sequences in the mammalian genome (Howard & Sakamoto, 1990). Rodent *Alu* elements are monomeric sequences of about 130 bp; the human *Alu* element is dimeric, formed by two 130 bp direct repeats with a 31 bp insertion in the second repeat. Human 7SL RNA represents the second human *Alu* repeat interrupted (at about position 100) by 140 bp of non-*Alu* sequence termed *S* (Ullu *et al.*, 1982). Some rodent *Alu* transcripts are found in the cytoplasm and a particular mouse *Alu* transcript was processed and exported to the cytoplasm when synthesized in *Xenopus* oocyte nuclei (Adeniyi-Jones & Zasloff, 1985). We wondered, therefor whether the *Alu* domain of 7SL RNA might contain information for its nuclear exit.

We have synthesized 7SL RNA *in vitro* and injected it into *Xenopus* oocyte nuclei. The RNA entered the cytoplasm and entry was blocked either at low temperature or in the presence of WGA, suggesting active transport. Preliminary results also indicate that 7SL RNA transport can be saturated, again reflecting a receptor mediated process. We then synthesized transcripts corresponding to only the *Alu* or *S* portions of 7SL RNA. Most of the *S* RNA was degraded when injected into nuclei but some appeared in the cytoplasm, even at low temperature suggesting that this molecule diffused freely out of the nucleus. The *Alu* domain also entered the cytoplasm but entry was blocked completely by low temperature, suggesting that the *Alu* fragment was transported in an active manner and that the *Alu* domain may be responsible for the facilitated transport of 7SL RNA. Transport of the *Alu* domain was rather less efficient than transport of

7SL RNA itself. To determine if both molecules utilized the same transport pathway (*ie*, same receptors) we co-injected an excess of one RNA along with a trace amount of the other. The two RNA species competed for transport, suggesting that indeed the *Alu* domain by itself was exported by the same mechanism as for intact 7SL RNA.

Assuming that a specific portion of 7SL RNA directs its transport, there probably exist *trans*-acting factors that bind to this region of the molecule to promote transport. Strub *et al.* (1991) mapped the binding site for the heterodimeric p9/p14 SRP proteins to the extreme end of 7SL RNA, *ie*, the *Alu* domain formed by the pairing between the 5' and 3' ends of 7SL RNA. Perhaps SRP9/14 migrates into the nucleus to bind newly transcribed 7SL RNA and escort it to the cytoplasm. Intriguingly, 7SL RNA synthesized with 4 extra nucleotides at its 5' end was exported slower that authentic 7SL RNA; the 5' end is contained in the SRP9/14 binding site and perhaps the additional nucleotides impaired loading of the proteins. Finally, like 5S rRNA, the 3' end of 7SL RNA also associates in the nucleus with La (Chambers *et al.*, 1983). The mouse *Alu* transcript was cleaved in the *Xenopus* nucleus to a 5' terminal 135 nucleotide product and a 3' terminal 75 nucleotide product, both of which entered the cytoplasm. The 75 nucleotide molecule was bound to La in both compartments, so perhaps La mediated its export. We can not detect any newly exported 7SL RNA associated with La, so either La is displaced prior to transport or 7SL RNA is exported in conjunction with La, the latter being displaced immediately by SRP9/14 upon entry into the cytoplasm.

## Summary

This brief overview highlights some of the the major active areas of research into mechanisms of nuclear-cytoplasmic transport in construction of the eukaryotic ribosome. Absent from the discussion is any consideration of the myriad of other yet to be identified protein and RNA factors that participate in the process. By analogy to pre-mRNA splicing, one may anticipate that numerous RNA helicases, RNAses, etc. will be found to contribute to the dynamic rearrangement of the emerging small and large subunits. The nucleolus contains unknown numbers of other proteins, some of which may constitute a framework upon which ordered assembly takes place and perhaps comprise highways for pre-ribosomal subunit delivery to the NPC. Obviously, all of the parts and tools for ribosome assembly must become properly localized. Unraveling the details by which this condition is achieved, and determining whether subcellular transport of ribosomal components can be coordinated, should keep cell and molecular biologists occupied for some time.

References

Adeniyi-Jones S, Zasloff M (1985) Transcription, processing and nuclear transport of a B1-Alu RNA species complementary to an intron of the murine alpha-fetoprotein gene. Nature (London) 317:81-84

Baserga, SJ, Gilmore-Herbert M, Yang EW (1992) Distinct molecular signals for nuclear import of the nucleolar snRNA U3 (in press)

Bataillé N, Helser T, Fried HM (1990) Cytoplasmic transport of ribosomal subunits microinjected into the Xenopus laevis oocyte nucleus: a generalized, facilitated process. J Cell Biol 111:1571-1582

Borer RA, Lehner CF, Eppenberger HM, Nigg EA (1989) Major nucleolar proteins shuttle between nucleus and cytoplasm. Cell 56:379-390

Breeuwer M, Goldfarb DS (1990) Facilitated nuclear transport of histone H1 and other nucleophilic proteins. Cell 60:999-1008

Chambers JC, Kurilla MG, Keene JD (1983) Association between the 7 S RNA and the lupus La protein varies among cell types. J Biol Chem 258:11438-11441

Dingwall C, Laskey RA (1991) Nuclear targeting sequences - a consensus? Trends Biochem Sci 16:478-481

Fischer D, Weisenberger D, Scheer U (1991) Assigning functions to nucleolar structures. Chromosoma 101:133-140

Fried H (1992) Transport of ribosomal proteins and rRNA, tRNA, and snRNA, in Nuclear Trafficking (Feldherr CA, ed), pp291-329. Academic Press, New York

Guddat U, Bakken AH, Pieler T (1990) Protein-mediated nuclear export of RNA: 5S rRNA containing small RNPs in Xenopus oocytes. Cell 60:619-628

Howard BH, and Sakamoto, K (1990) Alu interspersed repeats: selfish DNA or a functional gene family? New Biol 2:759-770

Hügle B, Scheer U, Franke WW (1985) Ribocharin: a nuclear $M_r$ 40,000 protein specific to precursor particles of the large ribosomal subunit. Cell 41:615-637

Izaurralde E, Mattaj IW (1992) Transport of RNA between nucleus and cytoplasm. Sem Cell Biol (in press)

Jordan EG (1991) Interpreting nucleolar structure: where are the transcribing genes? J Cell Sci 98:437-442

Larson DE, Zahradka P, Sells BH (1991) Control points in eukaryotic ribosome biogenesis. Biochem Cell Biol 69:5-22

Moreland RB, Nam HG, Hereford L, Fried HM (1985) Identification of a nuclear localization signal of a yeast ribosomal protein. Proc Nat Acad Sci USA 82: 6561-6565

Rutgers CA, Rientjes JMJ, van't Riet J, Raué HA (1991) rRNA binding domain of yeast ribosomal protein L25. Identification of its borders and a key leucine residue. J Mol Biol 218:375-385

Schaap PJ, van't Riet J, Woldringh CL, Raué HA (1991) Identification and functional analysis of the nuclear localization signals of ribosomal protein L25 from *Saccharomyces cerevisiae*. J Mol Biol 221:225-237

Schwarzacher HG, Wachtler F (1991) The functional significance of nucleolar structures. Ann Genet 34:151-160

Siegel V, Walter P (1988) Functional dissection of the signal recognition particle. Trends Biochem Sci 13:314-316

Steitz JA, Berg C, Hendrick JP, La Branche-Chabot H, Metspalu A, Rinke J, Yario T (1988) A 5S rRNA/L5 complex is a precursor to ribosome assembly in mammalian cells. J Cell Biol 106:545-556

Strub K, Moss J, Walter P (1991) Binding sites of the 9- and 14-kilodalton heterodimeric protein subunit of the signal recognition particle (SRP) are contained exclusively in the *Alu* domain of SRP RNA and contain a sequence motif that is conserved in evolution. Molec Cell Biol 11:3949-3959

Tollervey D, Hurt EC (1990) The role of small nucleolar ribonucleoproteins in ribosome synthesis. Molec Biol Rep 14:103-106

Ullu E, Murphy S, Melli M (1982) Human 7SL RNA consists of a 140 nucleotide middle-repetitive sequence inserted in an Alu sequence. Cell 29:195-202

Underwood M, Fried HM (1990) Characterization of nuclear localizing sequences derived from yeast ribosomal protein L29. EMBO J 9:91-99

# APPROACHES TOWARDS A GENETIC ANALYSIS OF THE NUCLEAR PORE COMPLEX IN YEAST

Christian Wimmer, Valérie Doye, Ulf Nehrbass, Nikolaus Schlaich and Eduard C. Hurt

EMBL
Postfach 1022.09
Meyerhofstrasse 1
6900 HEIDELBERG
F. R. GERMANY

## INTRODUCTION

The nuclear pore complex (NPC) mediates the nucleocytoplasmic transport through the nuclear envelope. It is a supramolecular, cylindrical structure with an overall diameter of ~ 130 nm, a thickness ~ 70 nm (Akey, 1989) and a molecular mass of 125 MDa (Reichelt et al. 1990). The pore consists of two concentric rings on the inner and outer face of the nuclear membrane, eight radial spokes and a central plug (Unwin and Milligan, 1982; Reichelt et al. 1990; Jarnik and Aebi, 1991). Recent EM studies revealed, besides a central channel, eight smaller peripheral channels with a diameter of 10 nm which could be used for diffusion of small molecules (Hinshaw et al. 1992). The NPC is associated with cytoskeletal structures: a basket-like network at the nucleoplasmic side and fibrillar elements directing into the cytoplasm (Akey, 1989; Reichelt et al. 1990).

The lectin wheat germ agglutinin is a potent inhibitor of nucleocytoplasmic transport in higher eukaryotes (Finlay et al. 1987; Newmeyer and Forbes, 1988). This is due to the fact that WGA binds

NATO ASI Series, Vol. H 71
Protein Synthesis and Targeting in Yeast
Edited by A. J. P. Brown, M. F. Tuite, and J. E. G. McCarthy

to a group of pore proteins called 'nucleoporins'. This family of proteins carrying O-linked N-acetyl glucosamine residues on serine/threonine contain, in the case of rat liver, 9 individual members with molecular weights ranging from 45 kD to 210 kD (Snow et al. 1987), but the actual number may be significantly higher. Up to now in yeast, two members of the nucleoporins termed NSP1 and NUP1 have been cloned and shown to be essential for cell growth (Hurt, 1988; Nehrbass et al. 1990; Davis and Fink, 1990). As revealed by sequence comparison, p62 is the mammalian homologue of NSP1 (Carmo-Fonseca et al. 1991). The evolutionary conservation of nucleoporins is striking and suggests similar transport mechanisms in yeast and man (Carmo-Fonseca and Hurt, 1991).

Proteins and snRNPs are imported into the nucleus, ribonucleoproteins, mRNA, snRNA, tRNA and pre-ribosomal particles are exported into the cytoplasm (Michaud and Goldfarb, 1991; Goldfarb and Michaud, 1991). Protein import consists of two distinct steps: binding of the karyophile to the nuclear periphery and ATP-dependent translocation through the pore channel (Newmeyer and Forbes, 1988; Moore and Blobel, 1992). The NPC is involved in all these processes and the central plug is considered to be the transporter (Richardson et al. 1988). Nuclear traffic differs from other intracellular transport mechanisms in two aspects:
(i) it is bi-directional; using gold particles covered either with mRNA or nuclear proteins, it has been shown by microinjection experiments that both directions of transport are performed by each pore (Dworetzky and Feldherr, 1988).
(ii) molecules are imported more than once into the nucleus, because the nuclear membrane disassembles and reassembles during mitosis and accordingly nuclear proteins are released in the mitotic cytoplasm during each round of the cell cycle (at least in higher eukaryotes).

Proteins are targeted into the nucleus by a nuclear localization sequence (NLS) (Dingwall and Laskey, 1991). The prototype of a nuclear targeting signal is the SV40 large T antigen NLS (Kalderon et al. 1984). Over the last ten years, many NLSs have been identified in nuclear proteins. Although not homologous, they are generally rich in positively-charged amino acids and can be as short as 7 residues. Recently, a bipartite pattern in NLS has been discovered in about 50% of the analyzed nuclear proteins; the NLS consists of two clusters of basic amino acids, separated by a 10 residue long spacer (Robbins et al. 1991).

To understand the function of the NPC for nucleocytoplasmic transport, its components and further elements of the import/export machinery need to be identified. The yeast *Saccharomyces cerevisiae* has been extensively used in the past as a model eukaryote to study the in vivo role of genes and their derived gene products and define their function in the context of the whole cell (Botstein and Fink, 1988). Here, we want to summarize several different genetic approaches to identify novel components of the nuclear pore complex in yeast.

## APPROACHES

A genetic system to analyze the nuclear pore complex is particularly desirable in the view that the pore is an extremely large biological structure exhibiting a molecular size of 125 MDa (Unwin and Milligan, 1982; Reichelt et al. 1990) which may not be easily amenable to biochemical or in vitro assembly approaches. We want to exploit the yeast *Saccharomyces cerevisiae* as an experimental organism easily amenable to genetic approaches to study the in vivo role of essential nuclear pore proteins for pore structure and pore function. In particular, (i) a structure/function relationship analysis of individual nuclear pore constituents is undertaken and (ii) different genetic approaches are exploited to identify the genetic network of interactions of a given nuclear pore protein with other (hopefully pore) constituents.

The yeast nuclear pore protein NSP1 belonging to a group of pore proteins termed 'nucleoporins' is a good start for such a genetic analysis since its gene is essential for cell growth (Hurt, 1988; Nehrbass et al. 1990).

*1) Generation of NSP1 mutants*

According to its amino acid sequence, the NSP1 protein was divided into different domains: a repetitive amino-terminal and middle domain, which are dispensable for cell growth, and an essential carboxy-terminal domain which is organized into hydrophobic heptad repeats typically found in e. g. intermediate filament proteins (Nehrbass et al. 1990). The carboxy-terminal domain mediates association with the nuclear pore complex and therefore could be involved in protein/protein interaction (Hurt, 1990). Mutants in the NSP1 carboxy-terminal domain were generated to exploit them for genetic analysis. These conditionally lethal mutants were made by site-specific (Nehrbass et al. 1990; Hurt et al. 1991) and random mutagenesis of the NSP1 gene. For random mutagenesis, the isolated gene encoding the NSP1 carboxy-terminal domain was chemically mutagenized with bisulfite and re-introduced into yeast lacking a functional NSP1 gene. Several thermosensitive mutant alleles could be recovered which upon sequence analysis showed single or multiple amino acid substitutions within the NSP1 carboxy-terminal domain (Ulf Nehrbass, unpublished data). These mutants grow at the permissive temperature (23°C) with variable generation times, but arrest cell growth at the restrictive condition (37°C). In the NSP1 mutant alleles either alpha-helix breaking residues such as prolines were generated or hydrophobic residues were changed into more hydrophilic amino acids which interfered with the heptad repeat pattern (see also above). For site-specific mutagenesis, amino acid substitutions were made within a predicted alpha-helical region of the carboxy-terminal domain which should inhibit the essential NSP1 function (Nehrbass et al. 1990). Another class of NSP1 mutant alleles was designed on the basis that mutations should not alter secondary structure organization or the hydrophobic heptad repeat pattern. Accordingly, such mutations in the carboxy-terminal domain should not affect protein folding, but impair protein/protein interaction. It has been shown earlier that changing clusters of charged amino acid residues into neutral alanines in the yeast actin gene yields many conditionally lethal actin mutants (ts, cs, dominant lethals) which are expected to be disturbed in protein assembly rather than protein folding (Wertman et al., 1991). Therefore, negatively charged amino acids within the NSP1 carboxy-terminal domain which tend to cluster in short predicted amphipathic alpha-helices (Hurt, 1990) and are predicted to be localized on the hydrophilic site of the helix were converted in neutral alanine residues which should not interfere with alpha-helical organization (Fig. 1).

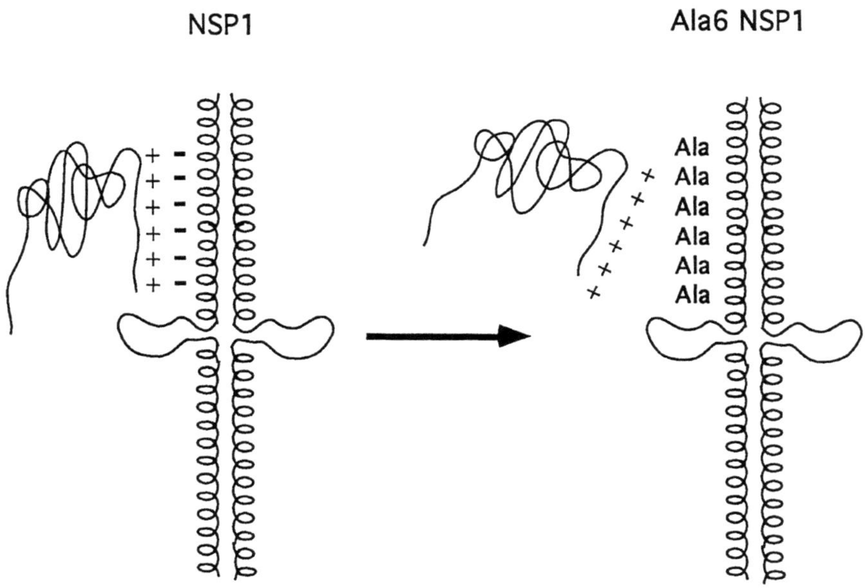


**Fig. 1 - Site-Specific Mutagenesis within the NSP1 Carboxy-Terminal Domain Changing Negatively-Charged Amino Acid Residues into Neutral Alanines**
The NSP1 carboxy-terminal domain is viewed in an assumed alpha-helical organization. In this model, it forms a dimer (whether it is a homo- or heterodimer is yet not known) along the hydrophobic axis of

the alpha-helix. In addition, negatively charged amino acids tend to cluster at the hydrophilic site of the alpha-helix and are shown to be involved in charge/charge protein interaction with another protein. By site-specific mutagenesis, six negatively-charged amino acids within the NSP1 carboxy-terminal domain were changed into neutral alanine residues.

One such mutant called Ala6 was temperature-sensitive for growth at elevated temperatures (V. Doye, unpublished results). This mutant is a good candidate for being blocked in protein assembly at the level of the nuclear pore and thus may be useful for suppressor analysis (see below). Besides a phenotypical characterization of these conditionally lethal NSP1 mutants including analysis of nucleocytoplasmic transport and nuclear pore assembly, genetic approaches will be applied to identify interacting components of NSP1.

*2) Suppressor approaches to identify interacting components of NSP1*

Conditionally lethal mutants have been extensively used in the yeast system to isolate suppressor genes including the genetic analysis of the yeast cytoskeleton (Huffaker et al. 1987). A mutationally impaired NSP1 function may be rescued by high copy number suppression, i. e. increasing the concentration of another protein which could be either functionally homologous to NSP1 when overproduced or may facilitate nucleoporin assembly if increased in its stoichiometric ratio. So far, no high copy number suppressor of ts nsp1 mutants was isolated. This could imply that overproduction of another protein in yeast cannot complement a NSP1 null or thermosensitive mutant. Alternatively, overproduction of such another component per se may be toxic for the yeast cell. This indeed has been found for the nucleoporins NUP1 (Davis and Fink, 1990) and NSP1 (Hurt, 1989). Other conditionally lethal alleles of NSP1 should be tested for high copy number suppression. In particular, cold-sensitive mutants may be useful for this approach (Huffaker et al. 1987). So far, cs mutant alleles of NSP1 could not be isolated.
Similarly, single copy extragenic suppressors of ts nsp1 alleles could not be found; intragenic reversion, however, easily occurred (Hurt et al.

1991). The ts nspl alleles chosen for suppressor analysis (it carries a proline mutation in the carboxy-terminal domain (Nehrbass et al. 1990)) may not be appropriate for extragenic suppression, because the mutant NSP1 protein could be denatured at the restrictive condition rather than being specifically inhibited in protein/protein interaction. Therefore, the ala6 ts nspl mutant (see above) which does not revert intragenically and may not be disturbed in its secondary structure was exploited for isolating EMS and UV-induced extragenic suppressors. Several suppressors were obtained which, if mated to the original ala6 mutant, are semi-dominant at the restrictive temperature. Genetic analysis suggests that two suppressors are encoded by single genes causing suppression of the ala6 mutant (V. Doye, unpublished data). Cloning of these suppressor genes is currently in progress. In conclusion, conditionally lethal mutants of NSP1 with low tendency of intragenic reversion and being not defective in protein folding may be exploited to identify the network of genetic interaction of NSP1 with other nuclear pore proteins.

*3) The approach of synthetic lethality to isolate nuclear pore components*

Alternatively, we considered the approach of synthetic lethality to identify components which are in genetic interaction with NSP1. A synthetic phenotype can be caused by the combination of mutant alleles of two different genes whereas one mutant allele alone gives viability to the cells (Huffaker et al. 1987). Synthetic lethality can provide genetic evidence for a physical interaction between gene products or overlapping function. For the approach of synthetic lethality the following scenario is discussed (Fig. 2). NSP1 forms a complex with another hypothetical nuclear pore protein (named NSP2) mediated by the essential carboxy-terminal domain. If both NSP1 and NSP2 would be mutated within their interaction domain, no complex formation can occur at any temperature and cells are not viable ('synthetic lethality'). However, only the single mutation in one of the two interacting components (either NSP1 or NSP2) still allows complex formation and cells thus can grow at the permissive temperature. Alternatively, NSP1 is in a subcomplex with several

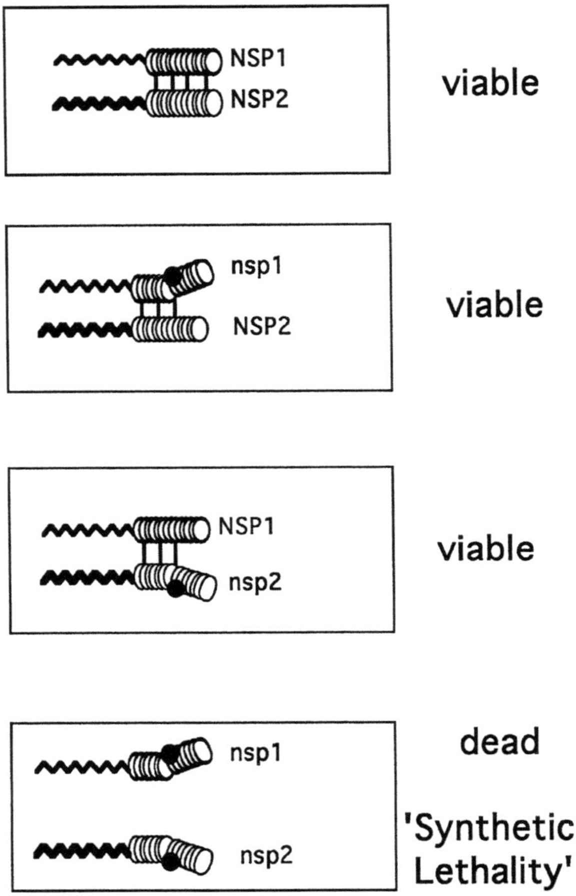


**Fig. 2- Synthetic Lethality to Identify Components in Genetic Interaction with NSP1**

NSP1 forms a complex with another pore protein tentatively assigned NSP2. NSP1 and NSP2 have a similar domain organization consisting of a C-terminal domain (alpha-helical drawing) and a repetitive amino- and central domain (wavy line). For heterodimer formation, the C-terminal domains are involved. Mutations in the carboxy-terminal domains are indicated by a black circle. If both NSP1 and NSP2 carry the mutation, heterodimer formation is blocked causing synthetic lethality.

other pore proteins performing an essential function at the level of the nuclear pore. Another nuclear pore subcomplex containing NSP2 performs a similar, but not identical function (overlapping function). Single mutations in one of the two subcomplexes is not harmful to the cell (viability), but the combination of two mutations causes synthetic lethality.

We identified synthetic lethals of ts nspl with the help of the ad2/ade3 red/white sectoring system in yeast. This approach has been initially used to determine plasmid stability in yeast and identify genes affecting plasmid stability (Koshland et al. 1985) and subsequently was adopted for synthetic lethal screens (Huffaker et al. 1987; Bender and Pringle, 1991; Costigan et al. 1992) or cloning yeast homologues of known higher eukaryotic genes (Kranz and Holm, 1990).
For this screening approach, an ade2/ade3 yeast strain was constructed which carries a disrupted NSP1 gene, but is complemented by the authentic wild type NSP1 allele inserted in a ADE3-containing plasmid; in addition, the strain carries a ts-allele of NSP1 on another plasmid lacking the ADE3 marker. This strain initially grows by forming red/white sectoring colonies at permissive temperature; it can afford to loose the ADE3 plasmid containing the wild type NSP1 gene since it can grow with the mutant nspl allele alone (functional at permissive temperature). This strain was mutagenized by UV light in order to hit chromosomal genes which give synthetic lethality with the mutant, but not wild type NSP1 gene. Such synthetic lethals of ts nspl can be directly identified on plate as red, non-sectoring colonies. Among 50 000 screened colonies, 14 red colonies were finally obtained which turned out to be synthetic lethals of ts nspl as shown by further genetic characterization.
Several genes giving rise to synthetic lethality have been cloned by transforming red strains with a yeast genomic plasmid library and screening for transformants which regain a red/white sectoring phenotype. One wild type allele of a gene giving rise to synthetic lethality has been sequenced; it encodes a novel protein with similarity to proteins belonging to the nucleoporin protein family (C. Wimmer, unpublished

data). Thus, the approach of synthetic lethality should allow us to identify novel pore proteins which interact with NSP1.

The role of the highly repetitive central domain of NSP1 consisting of 22 tandemly arranged peptide units with the central 'FSF' motif is not known. A molecular-genetic analysis of this domain is difficult to perform since it is not essential for growth and a truncated NSP1 protein lacking this domain is fully functional (Nehrbass et al. 1990). On the other hand, the repetitive 'FSF' motif is also found in other nuclear pore proteins (Davis and Fink, 1990) and is evolutionary conserved (Carmo-Fonseca et al. 1991). Therefore, the redundant 'FSF' motif in other nuclear pore proteins may have overlapping function.

We tested whether the NSP1 repetitive domain alone can interact with other proteins. A fusion protein consisting of 13 repetitive units of the NSP1 central domain fused to *E. coli* ß-galactosidase was expressed in yeast. By native immunoprecipitation using anti-ß-galactosidase antibodies, no evidence for a stable interaction with other yeast proteins was found. Therefore, a genetic approach was designed to identify genes which overlap in their function with that of the central repetitive domain of NSP1. We rationalized that in a genetic background in which only the NSP1 carboxy-terminal domain is expressed (lacking the central repetitive domain), a mutation in (an)other gene(s) exhibiting overlapping function may result in synthetic lethality. Accordingly, the ad2/ade3 red/white sectoring system was adopted to screen for mutants which require a full-length NSP1 protein for growth, but are synthetic lethal if NSP1 lacks the repetitive central domain. For the screening approach, a ade2/ade3 strain was constructed in which the chromosomal NSP1 gene was replaced by the gene encoding the carboxy-terminal domain and in addition carries an ADE3-containing plasmid with the full-length NSP1 allele. Under non-selective conditions, this strain forms red/white sectoring colonies even at higher temperatures. The strain was UV-mutagenized and about 30.000 colonies were plated. One red colony was found which is a good candidate for being synthetic lethal in combination with the carboxy-terminal domain, but not with the full-length NSP1 protein (N. Schlaich, unpublished data). It will be interesting to see whether the cloned gene(s) which complement synthetic lethality encode nuclear pore proteins with the 'FSF' motif.

## CONCLUSIONS

The NPC is one of the largest macromolecular assemblies currently under investigation. Its mass corresponds to approximately 50 ribosomes. The nuclear pore may fulfil many more functions than currently assumed. As shown, yeast genetics offer the potential to identify structural components of this organelle and arrange them in a functional context. Furthermore, the yeast system should also allow the cloning of the homologues from higher eukaryotic systems by functional complementation of yeast mutants. The combination of both genetic, biochemical and structural approaches in the various systems will then be a powerful way of unraveling the principles of nuclear pore structure and function.

## REFERENCES

Akey CW (1989) Interactions and structure of the nuclear pore complex revealed by cryo-electron microscopy. J Cell Biol 109:955-970

Bender A, Pringle JR (1991) Use of a screen for synthetic lethal and multicopy suppressee mutants to identify two new genes involved in morphogenesis in *Saccharomyces cerevisiae*. Mol Cell Biol 11:1295-1305

Botstein D, Fink GR (1988) Yeast: an experimental organism for modern biology. Science 240:1439-1443

Carmo-Fonseca M, Hurt EC (1991) Across the nuclear pores with the help of nucleoporins. Chromosoma 101:199-205

Carmo-Fonseca M, Kern H, Hurt EC (1991) Human nucleoporin p62 and the essential yeast nuclear pore protein NSP1 show sequence homology and a similar domain organization. Eur J Cell Biol 55:17-30

Costigan C, Gehrung S, Snyder M (1992) A synthetic lethal screen identifies SLK1, a novel protein kinase homolog implicated in yeast cell morphogenesis and cell growth. Mol Cell Biol 12:1162-1178

Davis LI, Fink GR (1990) The NUP1 Gene Encodes an Essential Component of the Yeast Nuclear Pore Complex. Cell 61:965-978

Dingwall C, Laskey RA (1991) Nuclear targeting sequences-A consensus? Trends Biochem Sci 16:478-481

Dworetzky SI, Feldherr CM (1988) Translocation of RNA-coated gold particles through the nuclear pores of oocytes. J Cell Biol 06:575-584

Finlay DR, Newmeyer DD, Price TM, Forbes DJ (1987) Inhibition of in-vitro nuclear transport by a lectin that binds to nuclear pores. J Cell Biol 104:189-200

Goldfarb DS, Michaud N (1991) Pathways for the nuclear transport of proteins and RNAs. Trends Cell Biol 1:20-24

Hinshaw JE, Carragher BO, Milligan RA (1992) Architecture and design of the nuclear pore complex. Cell 69:1133-1141

Huffaker TC, Hoyt MA, Botstein D (1987) Genetic Analysis of the Yeast Cytoskeleton. Annu Rev Genet 21:259-284

Hurt EC (1988) A novel nucleoskeletal-like protein located at the nuclear periphery is required for the life cycle of Saccharomyces cerevisiae. EMBO J 7:4323-4334

Hurt EC (1989) NSP1, a yeast protein located at the nuclear periphery is required for the cell-cycleperiphery, is required for the cell cycle of *Saccharomyces cerevisiae.* J Cell Sci 12:243-252

Hurt EC (1990) Targeting of a cytosolic protein to the nuclear periphery. J Cell Biol 111:2829-2837

Hurt EC, Mutvei A, Carmo-Fonseca M (1991) The nuclear envelope of the yeast Saccharomyces cerevisiae. Int Rev Cytol in press:

Jarnik M, Aebi U (1991) Toward a more complete 3-D structure of the nuclear pore complex. J Struct Biol 107:291-308

Kalderon D, Roberts BL, Richardson WP, Smith AE (1984) A short amino acid sequence able to specify nuclear location. Cell 39:499-509

Koshland D, Kent JC, Hartwell LH (1985) Genetic analysis of the mitotic transmission of minichromosomes. Cell 40:393-403

Kranz JE, Holm C (1990) Cloning by function: an alternative approach for identifying yeast homologs of genes from other organisms. Proc Natl Acad Sci USA 87:6629-6633

Michaud N, Goldfarb DS (1991) Multiple pathways in nuclear transport: The import of U2 snRNP occurs by a novel kinetic pathway. J Cell Biol 112:215-223

Moore MS, Blobel G (1992) The two steps of nuclear import, targeting to the nuclear envelope and translocation through the nuclear pore, require different cytosolic factors. Cell 69:939-950

Nehrbass U, Kern H, Mutvei A, Horstmann H, Marshallsay B, Hurt EC (1990) NSP1: A Yeast Nuclear Envelope Protein Localized at the Nuclear Pores Exerts its Essential Function by its Carboxy-terminal Domain. Cell 61:979-989

Newmeyer DD, Forbes DJ (1988) Nuclear import can be separated into distinct steps in vitro: nuclear pore binding and translocation. Cell 52:641-653

Reichelt R, Holzenburg EL, Buhle EL, Jarnik M, Engel A, Aebi U (1990) Correlation between structure and mass distribution of nuclear pore complex components. J Cell Biol 110:883-894

Richardson WD, Mills AD, Dilworth SM, Laskey RA, Dingwall C (1988) Nuclear protein migration involves two steps: rapid binding at the nuclear envelope followed by slower translocation through nuclear pores. Cell 52:655-664

Robbins J, Dilworth SM, Laskey RA, Dingwall C (1991) Two interdependent basic domains in nucleoplasmin nuclear targeting sequence: Identification of a class of bipartite nuclear targeting sequence. Cell 64:615-623

Snow CM, Senior A, Gerace L (1987) Monoclonal antibodies identify a group of nuclear pore complex glycoproteins. J Cell Biol 104:1143-1156

Unwin PN, Milligan RA (1982) A large particle associated with the perimeter of the nuclear pore complex. J Cell Biol 93:63-75

Wertman KF, Dubrin DG and Botstein D (1991) Mutational analysis of the actin gene. Abstract book of CSH Meeting 'Yeast cell biology', pp3

# THE SRP-DEPENDENT PROTEIN TARGETING PATHWAY IN SACCHAROMYCES CEREVISIAE

Peter Walter, Stephen C. Ogg & Byron C. Hann
Department of Biochemistry and Biophysics
University of California, Medical School
San Francisco, CA 94143-0448

In mammalian cells, the signal recognition particle (SRP) and the SRP receptor are required for the targeting of nascent secretory proteins to the endoplasmic reticulum (ER) membrane. Homologues of SRP and SRP receptor subunits have recently been identified in Saccharomyces cerevisiae and their functions have been characterized in vivo. Surprisingly, in S. cerevisiae, neither SRP nor SRP receptor are essential for cell growth, though cells deficient in these components grow poorly. Cells that lack both SRP and SRP receptor show identical growth rates to those in which either component alone is missing, indicating that SRP and SRP receptor function in the same pathway in vivo. Yeast cells depleted of SRP or SRP receptor show impaired translocation of soluble and membrane proteins across the ER membrane. Interestingly, the degree of the translocation defect varies for different proteins. These results indicate that proteins can bypass the SRP/SRP receptor mediated targeting pathway to the ER membrane, and that there is, in yeast, an alternate targeting pathway. Together with the recent identification of SRP and SRP receptor homologues in procaryotes and archeae, these results suggest that the structural components and mechanisms of the SRP-dependent protein targeting pathway are evolutionarily conserved.

Mammalian SRP acts to target ribosomes that synthesize secretory and membrane proteins to the ER membrane where they are translocated across the membrane into the ER lumen. In vitro assays have allowed the development of a detailed model describing the function of SRP in mammalian cells (see Walter and Lingappa, 1986, for review). In brief, signal sequences that are expressed as part of a nascent protein chain emerging from the ribosome bind to SRP in a binding site contained on the 54 kDa SRP subunit (SRP54) (Krieg et al., 1986; Kurzchalia et al., 1986). This interaction causes SRP to bind tightly to the ribosome, effecting an elongation arrest or pause in the translation of the nascent secretory protein.

NATO ASI Series, Vol. H 71
Protein Synthesis and Targeting in Yeast
Edited by A. J. P. Brown, M. F. Tuite, and J. E. G. McCarthy

Interaction of the engaged SRP with the SRP receptor, a heterodimeric membrane protein, that is anchored in the ER membrane (Gilmore et al., 1982a; Meyer et al., 1982; Tajima et al., 1986), releases the translational arrest and allows the ribosome that is synthesizing the protein chain to become bound to other components in the ER membrane. These components, collectively known as the "translocon", allow the transfer of the growing protein chain across the membrane, presumably through a transient aqueous pore in the membrane (Gilmore and Blobel, 1985; Simon and Blobel, 1991). SRP and SRP receptor act catalytically in this process; they are not part of the ribosome - membrane junction that mediates the transfer of the protein chain across the lipid bilayer. Rather, they function to bring the ribosome to the correct target membrane and are then released from the ribosome (Gilmore and Blobel, 1983).

Interestingly, SRP54 and the α-subunit of the SRP receptor, SRα, both contain a GTPase domain (G-domain) (Bernstein et al., 1989; Connolly and Gilmore, 1989; Römisch et al., 1989). The G-domains of SRP54 and SRα are homologous to one another and define a new family of GTPases. The β-subunit of the SRP receptor, SRβ, also contains a GTPase domain, which is structurally distinct from those of SRP54 and SRα (J. Miller & PW, unpublished). The precise mechanistic role of each of the three GTPase domains that interact during the SRP-dependent protein targeting reaction remains to be elucidated. Cycles of GTP hydrolysis by SRP54 and SRP receptor may control the assembly of the SRP-signal sequence-ribosome complex and the assembly of components of the translocon (Ogg et al., 1992a). Thus, conceptually similar to the initiation of protein synthesis, GTP hydrolysis may enhance the fidelity of and provide unidirectionality to the targeting reaction. Point mutations in the GTP binding sites of SRP54 and SRα block the translocation of proteins across the ER membrane (Bernstein and Walter, unpublished; Rapiejko and Gilmore, 1992), indicating that at least these two GTP binding sites are functionally important in the pathway.

Recently, SRP and SRα have been identified and characterized in S. cerevisiae (Hann et al., 1992; Hann and Walter, 1991; Ogg et al., 1992b; Stirling and Hewitt, 1992). The yeast SRP is a 16S cytoplasmic ribonucleoprotein that is important for the targeting of membrane and secretory proteins to the endoplasmic reticulum membrane. S. cerevisiae SRP is not essential for cell growth, though cells in which SRP is genetically disrupted grow poorly. The translocation of a number of soluble and membrane proteins into the lumen of the ER is impaired in SRP deficient cells. Surprisingly, different proteins show translocation defects of varying severity. After

SRP depletion, the membrane integration of dipeptidyl aminopeptidase B (DPAP-B) and the translocation of preKAR2 are severely inhibited, whereas the translocation of preprocarboxypeptidase Y (preproCPY) is not affected.

Similar to yeast SRP54, yeast SRα was also identified as a structural homologue of mammalian SRα (Ogg et al., 1992b). Its subcellular localization is consistent with the notion that yeast SRα, like its mammalian homologue, is a peripheral ER membrane protein. Several lines of evidence suggest that the yeast SRα also serves an analogous function. First and most important, yeast cells in which SRα has been depleted show defects in the translocation of proteins across the ER membrane. Second, cells lacking both SRα and SRP show a growth defect that is indistinguishable from cells in which either component alone was deleted. This suggests that yeast SRα and SRP function in the same pathway. Moreover, the severity to which different proteins are compromised in their translocation across the ER in SRα deficient cells resembles that of SRP deficient cells. After shut-off of SRα mRNA synthesis from a regulated promoter, the translocation of KAR2p and DPAP-B was severely impaired; the translocation of invertase and pro-α-factor showed intermediate defects; and the translocation of proCPY was not affected. There are some quantitative differences in the relative magnitudes of the translocation defects. The translocation of DPAP-B, for example, was impaired to over 90% in SRP-depleted cells, yet only impaired to about 50% in SRα-depleted cells. Most likely, however, these subtle differences arose because different strains were used in these studies.

Originally, SRα was purified from mammalian cells (Gilmore et al., 1982b) based on its activity to release the SRP mediated translation arrest of pre-secretory proteins (Walter and Blobel, 1981). These in vitro studies led to the notion that SRα is required for releasing SRP-induced translational arrest. In the absence of SRα, protein synthesis would not proceed beyond a short "arrested fragment" of the presecretory protein. To date, such a tight arrest has only been observed in vitro assays composed of heterologous components, e.g. mammalian SRP and a wheat germ translation system. In assays composed of exclusively mammalian components, the SRP-induced elongation arrest is transient thus causing a kinetic delay in protein elongation (Wolin and Walter, 1989). It is possible that the effects of SRP on presecretory protein synthesis helps to maintain proteins in a translocation competent state and/or provides a regulatory point to couple translation to the secretory needs of the cell.

Notably, presecretory protein synthesis is not detectably reduced as compared to overall protein synthesis when SRα is depleted from yeast cells. Thus, the lack of SRP receptor in vivo does not lead to a tight elongation arrest that is not released in the absence of SRα. In future studies, it will be interesting to measure directly the elongation rates of presecretory protein synthesis in SRα-depleted cells to test whether yeast SRP causes elongation pausing in vivo as observed in the mammalian in vitro system. Interestingly, in a different yeast, *Yarrowia lipolytica* , it was recently shown that mutations in SRP RNA can cause an inhibition of the translation of a secretory protein (He et al., 1992; Yaver et al., 1992).

In yeast, protein targeting to the ER can occur by redundant pathways. In the absence of SRP and SRP receptor, proteins must be targeted to and translocated across the ER membrane by alternative, SRP and SRP receptor independent targeting pathways. Although the translocation of some proteins is severely impaired upon disruption of the SRP/SRP receptor dependent targeting pathway, the mutant cells are viable. Thus, every protein must cross the ER membrane during its biogenesis and which is essential for cell viability can use an alternate targeting pathway sufficiently well to sustain cell growth.

Most of the proteins analyzed were impaired in their translocation across the ER membrane in SRP/SRP receptor depleted cells. We consider it likely that these proteins are co-translationally targeted by SRP and SRP receptor to the ER membrane in wild type cells (Fig. 1, pathway A), and that they become rerouted into alternative targeting pathways only in the mutant cells lacking SRP and/or SRP receptor function. Some preproteins, e.g. preproCPY, that can be targeted as efficiently in the absence of SRP as in its presence may have evolved signal sequences that do not interact efficiently with SRP. Thus, preproCPY may not use the SRP dependent targeting pathway even in wild type cells (Bird et al., 1987; Hann and Walter, 1991).

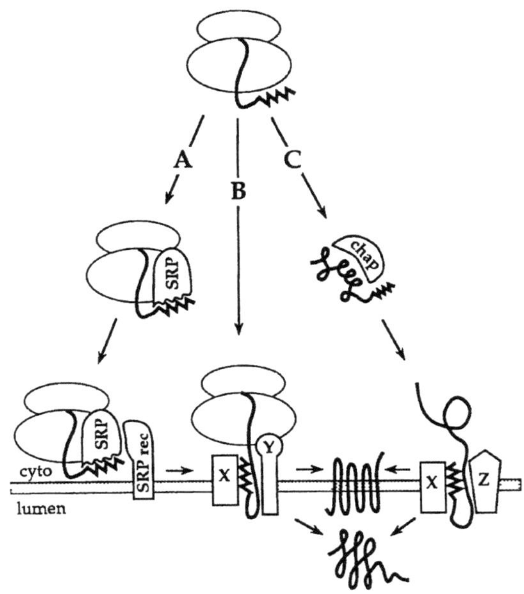


Figure 1. Model for Different Putative Targeting Pathways To the Yeast ER (reproduced from Hann and Walter, 1991).

Pathway A shows SRP-dependent co-translational targeting. The signal sequence is recognized on the ribosome by SRP and, after interaction with the SRP receptor, the ribosome becomes engaged with the translocon (components X and Y). The nascent protein becomes translocated into the ER lumen or integrated into the ER membrane. Pathway B shows SRP-independent co-translational targeting. SRP and SRP receptor are bypassed and the ribosome attaches to the ER membrane via an interaction indistinguishable from that achieved in pathway A. Attempts to identify such a pathway across mammalian ER were negative (Garcia and Walter, 1988). Pathway C shows SRP-independent post-translational targeting. For translocation to be post-translational, protein synthesis has to be terminated and the precursor has to be released from the ribosome into a soluble pool. Translocation competence may be maintained by interactions with chaperonins. The preproteins interact with the translocon (components X and Z) which is not necessarily identical to that in used in pathways A and B to effect its translocation or integration.

The molecular details of alternative targeting pathways in these mutant cells is presently unclear. There are two conceptually distinct, but not mutually exclusive possibilities. First, SRP/SRP receptor independent targeting could occur post-translationally: precursor proteins are released from ribosomes and maintained soluble and translocation competent by interactions with cytosolic chaperonins and other putative targeting factors (Fig. 1, pathway C). The folding characteristics of a particular preprotein may thus determine how efficiently it can be maintained in a translocation competent form. This could explain why translocation defects observed in SRP and SRP receptor deficient cells vary in magnitude for different proteins. Consistent with this hypothesis, in vitro studies have shown that some fully synthesized precursor proteins can be efficiently translocated across yeast microsomal membranes. Prepro-α-factor, for example, can be efficiently translocated post-translationally in vitro (Hansen et al., 1986; Rothblatt and Meyer, 1986; Waters and Blobel, 1986) and in vivo (Rothblatt et al., 1989) and shows only minor translocation defects in SRP and SRP receptor depleted cells in vivo.

According to a second hypothesis, SRP/SRP receptor independent targeting could occur co-translationally: ribosomes synthesizing precursor proteins engage with the ER membrane independent of SRP and SRP receptor, but prior to termination of protein synthesis (Fig. 1, pathway B). If co-translational targeting is obligate for a given precursor protein, then the kinetics of its elongation would affect the efficiency of its membrane translocation. If elongation is slow, for example, then a longer time window would be available for the nascent precursor protein to engage with the ER membrane before it is completed and released from the ribosome.

From pulse chase experiment we can conclude that the accumulated pool of one tested cytosolic preprotein (preKAR2p) in SRP-depleted as well as in SRα-depleted cells is not translocation competent (Hann and Walter, 1991; Ogg et al., 1992b). Although we cannot rule out that some of these preKAR2p molecules are translocated very slowly, such slow kinetics cannot account for the rapid appearance of translocated KAR2p at early time point. These data are consistent with either model outlined above. The observed translocation of preKAR2p may occur by a rapid post-translational mechanism, and a factor mediating this reaction may be limiting. Therefore, only a portion of the precursor proteins would be able to interact with this factor and be translocated. The other portion would be rapidly rendered translocation incompetent. Alternatively, it is possible that preKAR2p requires co-translational targeting and cannot be maintained translocation competent if released from the ribosome. In this case the translocated portion of the KAR2p molecules would be synthesized exclusively by ribosomes which are engaged with the ER membrane during KAR2p synthesis.

Unexpectedly, cells which have been grown in the absence of SRP or SRP receptor for a prolonged time "adapt" to the lack of these components. Although these cells still grow with the much increased doubling time characteristic of the SRP or SRP receptor depleted cells the accumulation of preproteins is reduced in some cells (strain W303) to almost undetectable levels. In another strain (TR1) a similar adaptation is observed which, however, is less pronounced. The molecular nature of the adaptation process is presently unknown. Genetic analyses indicate, however, that the adaptation is not due to a suppressor mutation, but rather, may reflect a change in cell physiology. According to the two models discussed above, translocation in adapted cells could be enhanced either because limiting components mediating alternate targeting pathways are induced and/or because the rate of protein elongation is reduced. The latter possibility would provide a plausible mechanism by which translocation can be improved, but cell growth remains impaired. More efficient protein translocation would be obtained at the expense of efficient protein synthesis. Further studies of the molecular nature of the adaptation process are required to distinguish between these possibilities and may provide valuable information as to the mechanism of SRP/SRP receptor independent protein targeting. From the available data, it is also conceivable that the efficiency of protein translocation only appears to be improved in adapted cells. Cells may have improved a "housekeeping" pathway which allows mislocalized precursor proteins to be degraded more rapidly. This would change the observed precursor to product ratio which presently is the only measure for translocation efficiency.

The results presented here confirm in vivo the role of SRP in the early steps in protein secretion in eukaryotic cells. These experiments, however, also unveil unanticipated complexities such as the existence of redundant targeting pathways and the unexpected differential sensitivity of individual translocation substrates to disruption of the SRP-dependent pathway. Preceding studies have identified putative SRP and SRP receptor homologues in evolutionarily distant species ranging from mammalian cells to yeast, archeae and eubacteria (Bernstein et al., 1989; Römisch et al., 1989; Lauffer et al., 1985; Hann et al., 1989; Ogg et al., 1992b; Byström et al., 1983; Gill et al., 1986; Ramirez and Matheson, 1991). In recent studies the SRP54 homologue was depleted from E. coli cells in vivo which led to severe translocation defects of different E. coli periplasmic and outer membrane proteins (Phillips and Silhavy, submitted). As discussed here for yeast, E. coli translocation substrates were impaired in their translocation efficiency to varying degrees. Complementary cell free studies demonstrate that the E. coli SRP54 homologue can functionally recognize signal sequences in vitro (Bernstein, Zopf, Freymann & Walter, submitted; Luirink, High, Wood, Giner, Tollervey & Dobberstein, submitted). Taken together, these results leave little doubt that the E. coli SRP homologue functions similar to the mammalian and yeast counterparts in mediating co-translational targeting of periplasmic and membrane proteins to the translocation apparatus. These results suggest that the structural components and the mechanism of the SRP-dependent protein targeting pathway are evolutionarily conserved among all living cells.

This work was supported by predoctoral fellowships from the Howard Hughes Medical Institute to S.C.O. and from the NIH Medical Scientist Training Program to B.C.H., and by grants from the NIH and the Alfred P. Sloan Foundation to P.W.

References

Bernstein, H. D., M. A. Poritz, K. Strub, P. J. Hoben, S. Brenner and P. Walter. 1989. Model for signal sequence recognition from amino-acid sequence of 54k subunit of signal recognition particle. *Nature.* 340:482-486.

Bird, P., M. J. Gething and J. Sambrook. 1987. Translocation in yeast and mammalian cells: not all signal sequences are functionally equivalent. *J. Cell Biol.* 105:2905-2914.

Byström, A. S., K. J. Hjalmarsson, P. M. Wikström and G. R. Björk. 1983. The nucleotide sequence of an Escherichia coli operon containing genes for the tRNA($m^1G$)methyltransferase, the ribosomal proteins S16 and L19 and a 21-K polypeptide. *EMBO J.* 2:899-905.

Connolly, T. and R. Gilmore. 1989. The signal recognition particle receptor mediates the GTP-dependent displacement of SRP from the signal sequence of the nascent polypeptide. *Cell.* 57:599-610.

Garcia, P. D. and P. Walter. 1988. Full-length prepro-alpha-factor can be translocated across the mammalian microsomal membrane only if translation has not terminated. *J. Cell Biol.* 106:1043-1048.

Gill, D. R., G. F. Hatfull and G. P. C. Salmond. 1986. A new cell division operon in *Escherichia coli. Mol. Gen. Genet.* 205:134-145.

Gilmore, R. and G. Blobel. 1983. Transient involvement of signal recognition particle and its receptor in the microsomal membrane prior to protein translocation. *Cell.* 35:677-685.

Gilmore, R. and G. Blobel. 1985. Translocation of secretory proteins across the microsomal membrane occurs through an environment accessible ot aqueous perturbants. *Cell.* 42:497-505.

Gilmore, R., G. Blobel and P. Walter. 1982a. Protein translocation across the endoplasmic reticulum. I. Detection in the microsomal membrane of a receptor for the signal recognition particle. *J. Cell Biol.* 95:463-469.

Gilmore, R., P. Walter and G. Blobel. 1982b. Protein translocation across the endoplasmic reticulum. II. Isolation and characterization of the signal recognition particle receptor. *J. Cell Biol.* 95:470-477.

Hann, B., C. Stirling J. and P. Walter. 1992. *SEC65* gene product is a subunit of the yeast signal recognition particle required for its integrity. *Nature.* 356:532-533.

Hann, B. C., M. A. Poritz and P. Walter. 1989. Saccharomyces cerevisiae and Schizosaccharomyces pombe Contain a Homologue to the 54-kD Subunit of The Signal Recognition Particle That in S. cerevisiae Is Essential for Growth. *J. Cell Biol.* 109:3223-3230.

Hann, B. C. and P. Walter. 1991. The Signal Recognition Particle in S. cerevisiae. *Cell.* 67:?

Hansen, W., P. D. Garcia and P. Walter. 1986. In vitro protein translocation across the yeast endoplasmic reticulum: ATP-dependent post-translational translocation of the prepro-a factor. *Cell.* 45:397-406.

He, F., J.-M. Beckerich and C. Gaillardin. 1992. A Mutant of 7SL RNA in *Yarrowia lipolytica* Affecting the Synthesis of a Secreted Protein. *The Journal Biological Chemistry.* 267:1932-1937.

Krieg, U. C., P. Walter and A. E. Johnson. 1986. Photocrosslinking of the signal sequence of nascent preprolactin to the 54-kilodalton polypeptide of the signal recognition particle. *Proc. Natl. Acad. Sci. USA.* 83:8604-8608.

Kurzchalia, T. V., M. Wiedmann, A. S. Girshovich, E. S. Bochkareva, H. Bielka and T. A. Rapoport. 1986. The signal sequence of nascent preprolactin interacts with the 54K polypeptide of the signal recognition particle. *Nature.* 320:634-636.

Lauffer, L., P. D. Garcia, R. N. Harkins, L. Coussens, A. Ullrich and P. Walter. 1985. Topology of the SRP receptor in the endoplasmic reticulum membrane. *Nature.* 318:334-338.

Meyer, D. I., E. Krause and B. Dobberstein. 1982. Secretory protein translocation across membranes- the role of the docking protein. *Nature.* 297:647-650.

Ogg, S., M. Poritz and P. Walter. 1992b. The signal recognition particle receptor is important for growth and protein secretion in Saccharomyces cerevisiae. *Molec. Biol. Cell.* submitted:

Ramirez, C. and A. T. Matheson. 1991. A gene in the archaebacterium *Sulfolobus solfataricus* that codes for a protein equivalent to the alpha subunits of the signal recognition particle receptor in eukaryotes. *Mol. Micro.* 5:1687-1693.

Rapiejko, P. J. and R. Gilmore. 1992. Protein translocation across the endoplasmic reticulum requires a functional GTP binding site in the $\alpha$-subunit of the signal recognition particle receptor. *J. Cell Biol.* 117:493-503.

Römisch, K., J. Webb, J. Herz, S. Prehn, R. Frank, M. Vingron and B. Dobberstein. 1989. Homology of the 54K Protein of signal-recognition particle, docking protein, and two E. Coli proteins with putative GTP-binding domains. *Nature.* 340:478-482.

Rothblatt, J. A., R. J. Deshaies, S. L. Sanders, G. Daum and R. Schekman. 1989. Multiple Genes are Required for Proper Insertion of Secretory Proteins Into the Endoplasmic Reticulum in Yeast. *J. Cell Biol.* 72:61-68.

Rothblatt, J. A. and D. I. Meyer. 1986. Secretion in yeast: Translocation and glycosylation of prepro-a factor in vitro can occur via ATP-dependent post-translational mechanism. *EMBO J.* 5:1031-1036.

Simon, S. M. and G. Blobel. 1991. A protein-conducting channel in the endoplasmic reticulum. *Cell.* 65:371-380.

Stirling, C. J. and E. W. Hewitt. 1992. The *Saccharomyces cerevisiae SEC65* gene encodes a component of the yeast signal recognition particle with homology to human SRP19. *Nature.* in press:

Tajima, S., L. Lauffer, V. L. Rath and P. Walter. 1986. The signal recognition particle is a complex that contains two distinct polypeptide chains. *J. Cell Biol.* 103:1167-1178.

Walter, P. and G. Blobel. 1981. Translocation of proteins across the endoplasmic reticulum. III. Signal recognition protein (SRP) causes signal sequence and site specific arrest of chain elongation that is released by microsomal membranes. *J. Cell Biol.* 91:557-561.

Walter, P. and V. R. Lingappa. 1986. Mechanism of protein translocation across the endoplasmic reticulum membrane. *Annu. Rev. Cell Biol.* 2:499-516.

Waters, G. and G. Blobel. 1986. Secretory protein translocation in a yeast cell free system can occur post-translationally and requires ATP hydrolysis. *J. Cell Biol.* 102:1543-1550.

Wolin, S. L. and P. Walter. 1989. Signal recognition particle mediates a transient elongation arrest of preprolactin in reticulocyte lysate. *J Cell Biol.* 109:2617-2622.

Yaver, D. S., S. Matoba and D. M. Ogrydziak. 1992. A Mutation in the Signal Recognition Particle 7S RNA of thge Yeast *Yarrowia lipolytica* Preferentially Affects Synthesis of the Alkaline Extracellular Protease: *In vivo* Evidence for Translational Arrest. *J. Cell Biol.* 116:605-616.

# SIMILARITIES BETWEEN *S. CEREVISIAE* SEC61p AND *E. COLI* SECY SUGGEST A COMMON ORIGIN FOR PROTEIN TRANSLOCASES OF THE EUKARYOTIC ER AND THE BACTERIAL PLASMA MEMBRANE.

Colin J. Stirling
Department of Biochemistry & Molecular Biology,
Medical School, University of Manchester,
Oxford Rd., Manchester.
England M13 9PT

## Introduction

The first stage in the eukaryotic secretory pathway is the translocation of polypeptides across the endoplasmic reticulum (ER) membrane. This process has been extensively studied in mammalian systems using an *in vitro* assay which faithfully reproduces the cotranslational translocation of specific precursor proteins into the ER lumen (Blobel, & Dobberstein, 1975). Such biochemical analyses have revealed that translocation requires both cytosolic and membrane-associated factors, including the cytosolic ribonucleoprotein complex, signal recognition particle (SRP), and its cognate membrane-bound receptor, SRP-receptor (or "Docking protein"; Walter & Blobel, 1980: Meyer *et al.*, 1982). Evidence suggests that SRP binds to the signal sequence of a nascent secretory protein as it emerges from the ribosome, and in doing so effects a reduction in the rate of polypeptide chain elongation ("elongation arrest", Walter and Blobel, 1981). The arrested complex is then targeted to the ER membrane *via* an interaction between SRP and the integral membrane protein SRP-receptor (Gilmore *et al.*, 1982; Meyer *et al.*, 1982). SRP-receptor then mediates the GTP-dependent displacement of SRP from the signal sequence/ribosome complex (Connolly & Gilmore, 1989), thus releasing elongation arrest, and enabling the co-translational translocation of the targeted precursor across the ER membrane. Whilst the SRP-dependent targeting cycle is relatively well characterised, the mechanism by which a targeted polypeptide actually penetrates the lipid bilayer remains obscure. There exists a wealth of data implicating integral membrane proteins in the translocation process, however, the identification and characterisation of these proteins has proven problematic. Several candidates for components of the mammalian ER translocase have been identified in a series of crosslinking studies in which a translocating polypeptide is first trapped in the membrane, and then cross-linked to those proteins in closest proximity. Such analyses have identified an ER membrane protein termed signal sequence receptor (SSR),

NATO ASI Series, Vol. H 71
Protein Synthesis and Targeting in Yeast
Edited by A. J. P. Brown, M. F. Tuite, and J. E. G. McCarthy

which interacts directly with the signal peptide of nascent pre-proteins (Weidmann *et al.*, 1987). Current evidence suggest that SSR exists as an oligomeric complex comprising at least two integral membrane glycoproteins, namely SSRα (34K), and SSRβ (22K) (Rapoport, 1990). Moreover, SSRα remains in close proximity to the mature portion of a translocating pre-protein, and may therefore represent a constituent of the translocon *per se* (Weidmann *et al.*, 1989; Prehn *et al.*, 1990; Rapoport, 1990). More recent cross-linking studies have identified a 37K ER protein (P37) which can be cross-linked to the signal anchor sequence of a type I membrane protein during its insertion into the membrane (High *et al.*, 1991). However, despite the identification of these putative translocon components, their actual contribution to the translocation process remains speculative. A more potent case can be made for the recently identified integral membrane glycoprotein TRAM (for translocating chain-associating membrane protein; Görlich *et al.*, 1992). TRAM represents the major species found crosslinked to short nascent chains, and may therefore be involved in the early stages of translocation. TRAM is an abundant ER-protein, present at a level at least equivalent to the number of membrane-bound ribosomes, raising the possibility that TRAM represents a constitutive component of the active translocase. A direct role for TRAM in translocation is supported by the observation that addition of the purified protein stimulates the translocation of preprolactin into reconstituted microsomes to levels approaching those of native ER microsomes (Görlich *et al.*, 1992).

**Protein Translocation in Yeast**

The experimental tractability of *Saccharomyces cerevisiae* provides an excellent model system for the genetic dissection of the eukaryotic translocation machinery. We have previously reported two genetic selections designed to identify conditional-lethal yeast mutants which are specifically defective in ER translocation (Deshaies & Schekman, 1987; Stirling *et al*, 1992). The isolation of such mutants might then identify essential components of the translocation apparatus, including those integral membrane proteins which have proven so elusive in other systems. These selections have led to the identification of four complementation groups of conditional-lethal translocation mutants namely, *sec61, sec62, sec63,* & *sec65* (Deshaies & Schekman, 1987; Stirling *et al.*, 1992; Stirling & Schekman, 1992). Mutants in all four classes accumulate untranslocated precursor forms of various secretory proteins including pre-pro-α-factor, invertase, & acid phosphatase etc. In addition, *sec61* and *sec65* mutants are severely defective in the insertion of integral membrane proteins into the ER membrane (Stirling *et al.*, 1992). Using similar selection techniques, other workers have identified further alleles of *sec63*

(*ptlI;* Toyn *et al.*, 1988), plus 3 apparently novel mutants defective in membrane protein assembly (*sec70, 71 and 72;* Green & Walter, 1992).

The *SEC61, 62, 63* and *65* genes have been cloned and sequenced, and their products characterised in some detail. Of these, *SEC65* encodes a 32K polypeptide representing the yeast homologue of the 19K subunit of human SRP (Stirling & Hewitt, 1992). Genetic and biochemical data now strongly suggests the presence of a mammalian-like SRP, and SRP-receptor, in *Saccharomyces cerevisiae* (Hann *et al.*, 1989; Amaya & Nakano, 1991; Hann and Walter, 1991; Hann *et al*, 1992; Stirling & Hewitt, 1992; Ogg *et al*, submitted). The remaining three *SEC* genes, *SEC61, 62, & 63,* are all essential, and all encode ER-specific integral membrane proteins (Deshaies & Schekman, 1989; 1990; Sadler *et al.*, 1989; Stirling *et al*, 1992). *SEC61* encodes a 53K polypeptide (SEC61p) whose deduced amino-acid sequence predicts an extremely hydrophobic molecule with up to 10 potential transmembrane domains (Stirling *et al*, 1992). The membrane topology of SEC61p is currently unknown, however those of SEC62p (32K), and SEC63p (73K), have been empirically determined, confirming the existence of two and three transmembrane domains respectively (Deshaies & Schekman, 1990; Feldheim *et al.*, 1992).

**A multi-subunit "SEC-complex" in the ER membrane**

A number of genetic observations led to predictions that the *SEC61, 62,* and *63* gene products might interact directly with one another (Deshaies & Schekman, 1989; Rothblatt, *et al.*, 1989). Biochemical evidence in support of such interactions has come from cross-linking studies which have identified a membrane-associated, multi-subunit complex comprising SEC61p, SEC62p, SEC63p, together with a 31.5K glycoprotein (gp31.5), and a 23K protein (p23). The same complex, minus SEC61p, can be immunoprecipitated from detergent solubilised membranes under non-denaturing conditions, suggesting that the association with SEC61p is labile (Deshaies *et al.*, 1991). Whilst the identities of gp31.5 and p23 are unknown, their observed relative molecular weights are reminiscent of those reported for the sub-units of the dog pancreas signal sequence receptor, SSRα (34K), and SSRβ (22K; Rapoport, 1990). Alternatively, gp31.5 may represent the yeast equivalent of mammalian TRAM. The molecular characterisation of gp31.5 and p23 is awaited with interest.

**SEC61p is in contact with a translocating chain**

The cross-linking approaches which have identified mammalian SSR, P37, and TRAM, have very recently been extended to *Saccharomyces cerevisiae*. Two groups have developed techniques with which to trap a translocating precursor (pre-pro-α-factor), within the yeast ER membrane *in vitro.* Musch *et al* (1992) employ a truncated pre-pro-α-factor mRNA lacking a stop codon at its 3' end. Translation of this mRNA results in the ribosome failing to dissociate at the 3' end, instead it remains attached to the message and, more importantly, the nascent polypeptide chain remains tethered to the ribosome *via* a peptidyl-tRNA. When such a complex is added to yeast microsomes, translocation is initiated, as demonstrated by the signal-processing, and glycosylation of the nascent chain. However, translocation cannot be completed due to the physical block imposed by the tethered ribosome. That this trapped precursor represents a genuine translocation intermediate has been demonstrated by stimulating the release of the ribosome with puromycin, whereupon translocation is rapidly completed. When a photoactivatable derivative of Lysine is incorporated into the nascent chain, the trapped intermediate can be cross-linked to any membrane proteins in close proximity to the translocating polypeptide. In an alternative approach, Sanders *et al* (1992), exploit the fact that pre-pro-α-factor is capable of efficient post-translational translocation into yeast microsomes *in vitro.* This permits the covalent coupling of an internally disulphide-bonded moiety (avidin) to the C-terminus of pre-pro-α-factor prior to initiation of translocation. The resultant pre-pro-α-factor-Avidin conjugate can initiate translocation, but fails to complete the process, presumably due to the disulphide-bonded avidin domain being unable to penetrate the translocase.

In either case, when the trapped translocation intermediate is cross-linked to its nearest neighbours, a single major crosslinked species is observed. Using specific antisera, this cross-linked species has been identified as SEC61p. The cross-linking of precursor to SEC61p is absolutely dependent upon the presence of ATP; a factor known to be required for translocation of pre-pro-α-factor in yeast (Hansen *et al.*, 1986; Waters & Blobel, 1986; Rothblatt & Meyer, 1986). In the absence of ATP the major cross-linked species observed is SEC62p (Musch *et al.*, 1992; Sanders *et al.*, 1992). These data suggest that the initial membrane attachment of a secretory pre-protein may involve a transient interaction with Sec62p, but that its subsequent ATP-dependent translocation involves a close association with SEC61p. Given the labile (perhaps transient) interaction detected between Sec61p and a Sec62/63-containing complex (see earlier; Deshaies *et al.*, 1991), then the role of the Sec62/63 complex may involve the direct presentation of a membrane associated precursor to the Sec61p-containing translocase. Such a catalytic role for Sec62p is consistent with the finding that it is present at approximately one tenth the level of Sec61p (Deshaies *et al.*, 1991). This convergence of genetic and biochemical data provides overwhelming evidence that these Sec proteins are directly involved in the translocation process, and strongly suggests that Sec61p represents a component of the translocase *per se*.

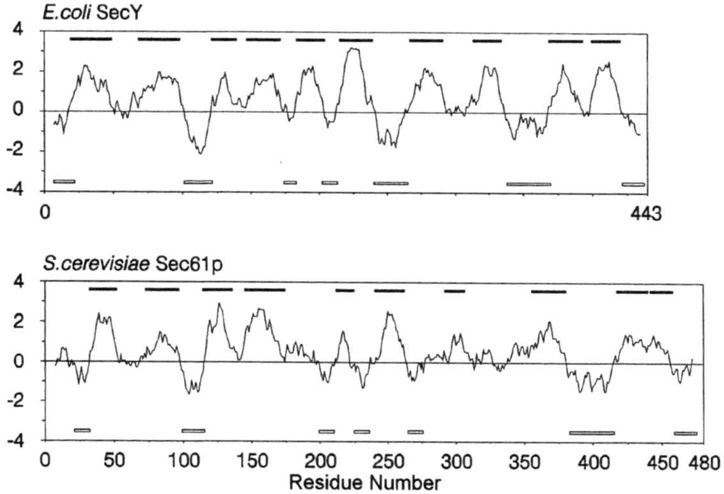


**Figure 1** A comparison between the mean hydropathy profiles of Sec61p and *E.coli* SecY suggests a similar pattern of transmembrane domains. Mean hydropathies are calculated over a window of 15 residues according to Kyte and Doolittle (1982). The 10 putative transmembrane domains are indicated by black bars, whilst significantly hydrophilic regions are highlighted with open bars.

### Similarities between Sec61p and *E.coli* SecY

In several ways Sec61p resembles *E.coli* SecY (PrlA), a component of the bacterial plasma membrane translocase (Cerretti *et al.*, 1983; Akiyama & Ito, 1987). Mutations in either *SEC61* or *secY* result in severe defects in the translocation of both secretory and integral membrane precursors. Both proteins are of similar size, and their hydrophobic profiles suggest a similar arrangement of transmembrane domains (fig. 1). *E.coli* SecY is a polytopic integral membrane protein comprising 10 transmembrane (TM) domains, with both the N- and C- termini oriented toward the cytoplasm (Akiyama and Ito, 1987; 1989). In Sec61p there are 10 clusters of hydrophobic residues, but several of these are significantly shorter than would be required to span a lipid bilayer in an α-helical conformation (most notably putative TM domains V and VII). Therefore, if Sec61p does span the bilayer 10 times, then some of its TM domains must do so in a non-α-helical conformation, and/or several intra-membranous domains must possess a low net hydrophobicity; this latter property might be predicted of a polypeptide designed to form a hydrophilic channel through a lipid bilayer. It should be noted that, like many similarly hydrophobic polypeptides, Sec61p and SecY exhibit aberrant gel mobilities in SDS-Page. In particular, the 53K Sec61p migrates with a relative molecular weight of 38K, a value strikingly similar to that of mammalian P37 (Stirling *et al.*, 1992; High *et al.*, 1991).

Like Sec61p, *E.coli* SecY associates with other integral membrane proteins required for pre-protein translocation. Brundage *et al* (1990; 1992), have demonstrated that SecY forms a stable, stoichiometric complex with SecE and, the as yet undefined, band 1 protein. This complex is required for the restoration of translocation activity in reconstituted microsomes. As previously stated, Sec61p can be found in association with a complex containing Sec62p, Sec63p, gp31.5, and p23. However, unlike the secY/E-band I complex, the association between Sec61p and the Sec62/63-containing complex is relatively labile, and non-stoichiometric. The possibility remains that further components of a more stable translocase complex remain to be defined in yeast.

**Sequence similarities between Sec61p and various bacterial SecY proteins**

The SecY protein appears to have been highly conserved amongst the eubacteria (including the mycoplasma *M.capricolum*), with further homologues having been identified in the genomes of photosynthetic plastids present in the algae *Cyanophora paradoxa,* and *Cryptomonas Φ* (see table 1). A more distantly related homologue has also been identified in the archaebacterium *Methanococcus vannielii,* which shares only 20-25% sequence identity with the various eubacterial proteins (Table 1). The 20% sequence identity between *E.coli* SecY and the *M.vannielii* protein is statistically insignificant; especially given the constraints imposed by the largely hydrophobic nature of these two proteins Indeed, equivalent percentage identities can be obtained when comparisons are made after shuffling either sequence. Nonetheless, despite their limited sequence similarity, these two proteins appear to be functionally equivalent since the expression of *M. vannielii* SecY effectively complements the temperature sensitive growth defect in an *E.coli* $secY^{ts}$ mutant (IQ292, Auer *et al.*, 1992). This apparent conservation of function is all the more remarkable given the very different lipid compositions of archaebacterial *versus* eubacterial plasma membranes Whilst the suppression of the *E.coli* growth defect strongly suggests that the *M.vannielii* gene encodes a genuine SecY equivalent, it should be noted that this evidence is by no means compelling. Firstly, it remains to be determined whether or not the suppression of the growth defect correlates with a suppression

| | Bsu | Lla | Eco | Mlu | CΦ | Mca | Cpa | Mva | Sec61 |
|---|---|---|---|---|---|---|---|---|---|
| **Bsu** | ----- | 69.0 | 65.5 | 67.3 | 67.2 | 65.6 | 61.0 | 55.5 | 52.2 |
| **Lla** | **48.7** | ----- | 63.9 | 64.0 | 64.9 | 66.4 | 59.3 | 53.1 | 49.4 |
| **Eco** | **42.3** | **39.1** | ----- | 65.2 | 64.6 | 61.2 | 58.2 | 51.5 | 48.5 |
| **Mlu** | **39.8** | **36.7** | **40.2** | ----- | 64.5 | 60.2 | 58.7 | 51.4 | 50.4 |
| **CΦ** | **37.2** | **38.3** | **38.8** | **38.4** | ----- | 63.2 | 63.1 | 52.4 | 49.6 |
| **Mca** | **39.4** | **39.0** | **34.0** | **29.8** | **35.3** | ----- | 57.4 | 54.6 | 50.2 |
| **Cpa** | **30.1** | **31.4** | **26.7** | **29.9** | **39.3** | **27.8** | ----- | 52.0 | 44.5 |
| **Mva** | **24.5** | **20.8** | **20.2** | **22.8** | **21.2** | **23.7** | **22.7** | ----- | 61.7 |
| **Sec61** | **24.9** | **22.7** | **20.4** | **21.4** | **20.0** | **25.7** | **19.2** | **32.8** | ----- |

**Table 1. Comparison of the degrees of similarity between various bacterial SecY proteins and yeast Sec61p.** Percentage sequence identities are given in bold in the lower left-hand portion of the table. Percentage sequence similarities, including chemically similar residues, are listed in the top right-hand portion of the table. All calculations are based upon the optimal alignment of a pair of proteins along their entire length; this tends to bias the results slightly downward. Code: Bsu, *Bacillus subtilis* (Nakamura *et al.*, 1990a; Suh *et al.*, 1990); Lla, *Lactobacillus lactis* (Koivula *et al.*, 1991); Eco, *Escherichia coli* (Cerretti *et al.*, 1983); Mlu, *Micrococcus luteus* (Ohama *et al.*, 1989); CΦ, plastid from *Cryptomonas Φ* (Douglas, 1992); Mca, *Mycoplasma capricolum* (Ohkubo *et al.*, 1987); Cpa, cyanelle plastid from *Cyanophora paradoxa* (Michalowski *et al.*, 1990); Mva, *Methanococcus vannielii* (Auer *et al.*, 1991); Sec61, *Saccharomyces cerevisiae* Sec61p (Stirling *et al.*, 1992).

of the mutant's secretory defect; for example the *M.vannielii* protein may render the mutant cells more able to withstand their secretory defect at the elevated temperature. Secondly, any suppression of the secretory defect might itself be indirect, perhaps due to some gross affect on membrane structure leading to a restoration of function due to some stabilising influence upon the mutant *E.coli* SecY protein. These caveats would be nullified should the *M.vannielii* gene be found to rescue an *E.coli secY* null mutant.

Further evidence for functional conservation amongst SecYs comes from the observation that overexpression of *B.subtilis* SecY suppresses the secretory defect in *E.coli secY24* cells. However, in this case the temperature sensitive growth defect of the mutant cells is not restored. This latter observation implies that the suppression of *secY24* is incomplete, but may be due to dominant detrimental affects resulting from overexpression of the heterologous protein (Nakamura *et al.*, 1990b).

In *E.coli* the *secY* gene is located at the 3' end of the *spc* operon which encodes a number of ribosomal protein sub-units. It is notable that all bacterial and plastid "*secY*" genes identified to date, are located in a similar position relative to the homologous ribosomal operon. Whilst the functional significance of this genetic linkage between *secY* and ribosomal genes is unclear, it strongly supports the view that the archaebacterial, eubacterial, and plastid *secY* genes are homologues derived from a common ancestor.

```
Sec61p     1 MSSNRVLDLFKPFESFLPEVIAPERKVPYNQKLIWTGVSLLIFLILGQIP 50
                      :||:  ::|||  | : | : :|: |||: |::::||| |
MvSecY     1 .......MKIKPILELIPEVKRPLKGVSFKEKIQWTGLVLILYFILGTID 43

Sec61p    51 LYGIVSSETSDPLYWLRAMLASNRGTLLELGVSPIITSSMIFQFLQGTQL 100
             :| : : |    : ::  : ||  |||: ||::||:| ::|:|:| | :|
MvSecY    44 IY.MGGAEMPAMFAFWQTVTASKMGTLITLGIGPIVTAGIIMQLLVGSEL 92

Sec61p   101 LQIRPESKQDRELFQIAQKVCAIILILGQALVVVMTGNYGAPSDLGLPIC 150
             : :      :| |||  ||: :|:| : :|:: |  | :|    :   :
MvSecY    93 ISLDLSKPMNRALFQGLQKLFGIFLCFLEAVMFVGAGAFGV...VNSTLA 139

Sec61p    51 LLLIFQLMFASLIVMLLDELLSKGYGLGSGISLFTATNIAEQIFWRAFAP 200
             |:|::|| :: ::|: |||::|: ||:||||:|| |  :|: ||  ||
MvSecY   140 LILVLQLALGAILVIYLDEIVSR.YGIGSGIGLFIAAGVAQTIFVGAF.. 186

Sec61p   201 TTVNSGRGKEFEGAVIAFFHLLAVRKDKKRALVEAFYRTNLPNMFQVLMT 250
                         ||  :: ::         |:  :     :  :: :| |
MvSecY   187 ............GAEGYLWKFFS.......AMSVGSLGIAFEYILPILST 217

Sec61p   251 VAIFLFVLYLQGFRYELPIRSTKVRGQIGIYPIKLFYTSNTPIMLQSALT 300
             : :|| |:|::::| |:|:   :|:| :| ||||::| || |::|  ||
MvSecY   218 LFVFLVVVYVESIRVEIPLAHGRVKGAVGKYPIKFIYVSNLPVILAAALF 267

Sec61p   301 SNIFLISQILFQKYPTNPLIRLIGVWGIRPGTQGPQMALSGLAYYIQPLM 350
              ||      |  :    |::             :   |:||:|||:
MvSecY   268 ANIQLWGMFL.DRMGY.PILGQY..........SNGTAVSGIAYYFSTPY 305

Sec61p   351 SLSEALLDPIKTIVYITFVLGSCAVFSKTWIEISGTSPRDIAKQFKDQGM 400
             ::|: : ||:  | |  :::  | :|:  |:| ||  :: :|| : : :|
MvSecY   306 GISNIISDPLHAIFYTLMMVIFCILFGLFWVETSGLDAKSMAKKLGNLDM 355

Sec61p   401 VING..KRETSIYRELKKIIPTAAAFGGATIGALSVGSDLLGTLGSGASI 448
              | |  | : ||   ||: |     :|:| :| |  : |: | ||:| ::
MvSecY   356 AIKGFRKSQKSIEQRLKRYIKPITVMGSAFVGFLAAAADFTGALGGGTGV 405

Sec61p   449 LMATTTIYGYYEAAAKEGGFTKNLVPGFSDLM.... 480
             |:    :|  ||    |     :| |:   ::
MvSecY   406 LLTVSIVYRLYEQLVQEQ..LSELHPAVAKFVGKR  439
```

**Figure 2.** Optimal sequence alignment between *S. cerevisiae* Sec61p and *Methanococcus vannielii* SecY (MvSecY). Amino acid identities are highlighted with a vertical line (|), and chemically similar residues denoted by double dots (:). Gaps inserted to maximise the alignment are represented by dots within the linear sequence. Stretches of hydrophobic residues corresponding to putative transmembrane domains are indicated by bold underlined text. The two sequences are 32.8% identical, with 62% similarity overall when conservative substitutions are included.

Sec61p shares little sequence identity with *E.coli* SecY, but is strikingly similar to that of *M. vannielii* with 33% identity, extending to 62% similarity when conservative substitutions are included (see table 1 and figure 2). Obviously, when aligning two sequences with numerous hydrophobic stretches, the possibility exists that spurious alignments may arise due to the large numbers of hydrophobic residues present. The overall level of sequence identity, plus the fact that hydrophilic regions of Sec61p are also conserved in *M.vannielii* SecY (see fig 2), suggests that these two sequences are genuinely related.

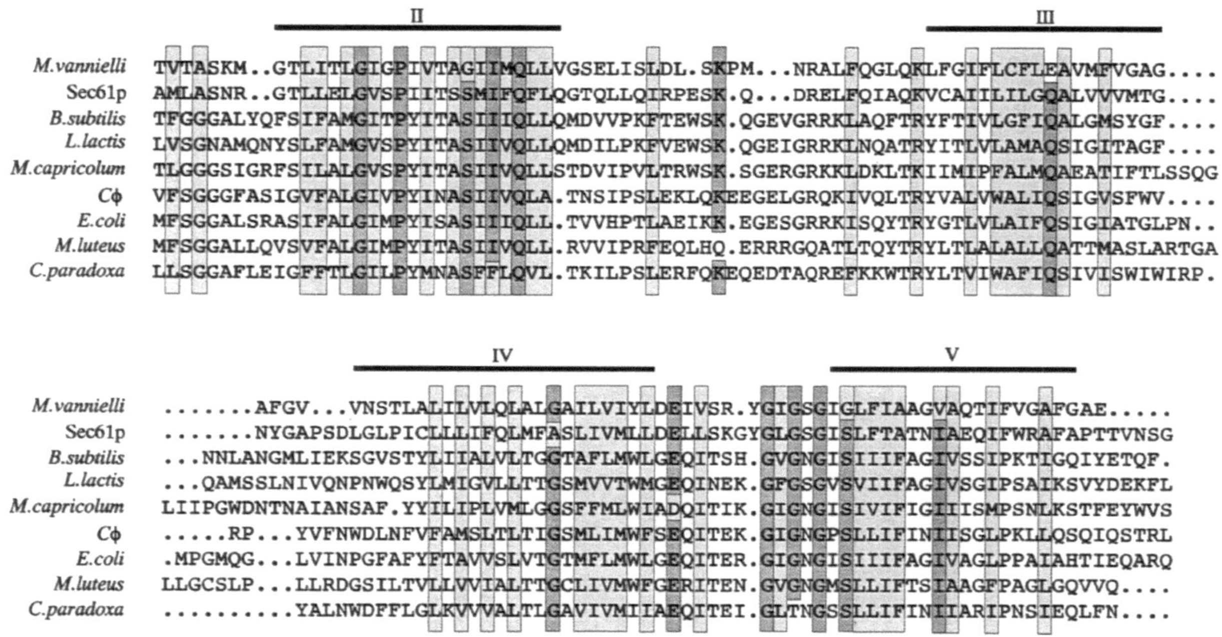


**Figure 3** Multiple sequence alignment encompassing the most highly conserved regions between *S.cerevisiae* Sec61p and eight SecYs from various sources (see text). Residues conserved between at least eight of the nine sequences are highlighted with shaded boxes (▩). Chemically similar residues are highlighted using lighter shading (□). The position of putative transmembrane domains II-V are indicated using black bars above the sequence. The region shown corresponds to residues 68-206 of Sec61p. For references see legend to table 1.

When all nine available sequences are compared the extent of their divergence is apparent. Nonetheless, they do share certain structural similarities; all are similar in size, and all possess a similar pattern of 10 (putative) transmembrane domains. Only a handful of residues appear to have been absolutely conserved amongst all nine representative sequences, with the majority of these lying within the region encompassing TM domains II to V (fig. 3). The relatively high degree of conservation observed within transmembrane domain II suggests a critical role for this region in the structure and function of SecY/Sec61p. The other most notable region within the alignment corresponds to a cluster of glycines within the very short region between TM domains IV and V (fig. 3). The significance of these particular residues is difficult to assess since they may be constrained by the need for a flexible turn close to the surface of the membrane. It might therefore be the spacing between transmembrane domains IV and V which is essential for function, and not necessarily the individual glycine residues themselves.

**Conclusions**

The pattern of relatedness amongst the various SecY/Sec61p homologues is consistent with them having arisen from an ancient progenote, pre-dating the divergence of eukaryotes, archaebacteria, and eubacteria. Moreover, the high degree of similarity between algal plastid SecYs and those of eubacteria, accords with the view that plastids are descended from eubacterial endosymbionts. Recently a cDNA encoding a mammalian homologue of Sec61p has been identified which shares 56% sequence identity with the yeast protein, confirming that Sec61p has been highly conserved amongst eukaryotes (Tom Rapoport, personal communication). These findings suggest that the mechanism of protein translocation across a number of different membranes may be fundamentally similar, involving at least one common component derived from an ancient translocase. Nonetheless, the clear sequence divergence between eukaryotic Sec61p, and eubacterial SecY, suggests that these proteins have become adapted to meet specific requirements within their different systems. These differences may reflect the emergence of novel interacting components required to regulate, or refine, the activity of the translocase.

In eukaryotes, a number of membranes are recognised as being translocation competent, including the ER, the peroxisomal membrane, the inner and thylakoid membranes of chloroplasts (plastids), and the inner mitochondrial membrane. It remains to be seen whether or not Sec61/SecY-like proteins are universally present in these various translocation systems, or whether alternative translocases are involved.

**References**

Akiyama Y and Ito K (1989). Export of Escherichia coli alkaline phosphatase attached to an integral membrane protein, SecY. J. Biol. Chem. 264; 437-442

Akiyama Y et al. (1987). Topology analysis of the SecY protein, an integral membrane protein involved in protein export in Escherichia coli. EMBO J. 6; 3465-3470

Amaya Y and Nakano A (1991). SRH1 protein, the yeast homologue of the 54KDa subunit of signal recognition particle, is involved in ER translocation of secretory proteins. FEBS Lett. 283; 325-328

Amaya Y, Nakano A, Ito K and Mori M (1990). Isolation of a yeast gene, SRH1, that encodes a homologue of the 54K subunit of mammalian signal recognition particle. J Biochem (Tokyo) 107 (3); 457-463

Auer J, Spicker G and Boeck A (1991). Presence of a gene in the archaebacterium Methanococcus vannielii homologous to secY of eubacteria. Biochimie 73; 683-68

Blobel G and Dobberstein B (1975). Transfer of proteins across membranes. I. Presence of proteolytically processed and unprocessed nascent immunoglobulin light chains on membrane-bound ribosomes of murine myeloma. J. Cell Biol. 67, 835-851.

Brundage L, Fimmel CJ, Mizushima S and Wickner W (1992). SecY, SecE, and Band 1 form the membrane-embedded domain of Escherichia coli preprotein translocase. J Biol. Chem. 267; 4166-4170

Brundage L et al. (1990). The purified E. coli integral membrane protein secY/E is sufficient for reconstitution of SecA-dependent precursor protein translocation. Cell 62; 649-657

Cerretti DP et al. (1983). The spc ribosomal protein operon of Escherichia coli; sequence and co-transcription of the ribosomal protein genes and a protein export gene. Nuc. Acids Res. 9; 2599-2615

Connolly T and Gilmore R (1989). The signal recognition particle receptor mediates the GTP-dependent displacement of SRP from the signal sequence of the nascent polypeptide. Cell 57 (4); 599-610

Deshaies RJ, Sanders SL, Feldheim DA and Schekman R (1991). Assembly of yeast Sec proteins involved in translocation into the endoplasmic reticulum into a membrane-bound multisubunit complex. Nature 349; 806-808

Deshaies RJ and Schekman R (1987). A yeast mutant defective at an early stage in import of secretory protein precursors into the endoplasmic reticulum. J Cell Biol 105 (2); 633-645

Deshaies RJ and Schekman R (1989). SEC62 encodes a putative membrane protein required for protein translocation into the yeast endoplasmic reticulum. J Cell Biol 109; 2653-2664

Deshaies RJ and Schekman R (1990). Structural and functional dissection of Sec62p, a membrane-bound component of the yeast endoplasmic reticulum protein import machinery. Mol Cell Biol 10 (11); 6024-6035

Douglas SE (1992) A secY homologue is found in the plastid genome of *Cryptomonas Φ*. FEBS Letts 298; 93-96

Feldheim D, Rothblatt J and Schekman R (1992). Topology and functional domains of Sec63p, an E.R. membrane protein required for secretory protein translocation. Mol. Cell. Biol. 12; 3288

Gilmore R, Walter P and Blobel G (1982). J. Cell Biol. 95,470-477, Protein translocation across the endoplasmic reticulum. II Isolation and characterisation of the signal recogntion particle.

Gorlich D, Hartmann E, Prehn S and Rapoport TA (1992). A protein of the endoplasmic reticulum involved early in polypeptide translocation. Nature 357; 4747-4752

Green N, Fang H and Walter P (1992). Mutants in three novel complementation groups inhibit membrane protein insertion into and soluble protein translocation across the endoplasmic reticulum membrane of Saccharomyces cerevisiae. J. Cell Biol. 116; 597-604

Hann BC, Poritz MA and Walter P (1989). Saccharomyces cerevisiae and Schizosaccharomyces pombe contain a homologue to the 54-kD subunit of the signal recognition particle that in S. cerevisiae is essential for growth. J Cell Biol 109; 3223-3230

Hann BC, Stirling CJ and Walter P (1992). SEC65 gene product is a subunit of the yeast signal recognition particle required for its integrity. Nature 356; 532-533

Hann BC and Walter P (1991). The signal recognition particle in S. cerevisiae. Cell 67; 131-144

Hansen W, Garcia PD and Walter P (1986). In vitro protein translocation across the yeast endoplasmic reticulum: ATP-dependent post-translational translocation of the prepro-alpha-factor. Cell 45; 397-406

High S et al. (1991). The identification of proteins in the proximity of signal-anchor sequences during their targeting to and insertion into the membrane of the ER J. Cell Biol. 113; 35-44

Koivula T, Palva I and Hemila H (1991). Nucleotide sequence of the secY gene from Lactobacillus lactis and identification of conserved regions by comparison of four SecY proteins. FEBS Letts. 288; 114-118

Kyte J and Doolittle RF (1982). A simple method for displaying the hydrophobic character of a protein. J. Mol. Biol. 157, 105-132.

Meyer DI, Krause E and Dobberstein B (1982). Secretory protein translocation across membranes-the role of the docking protein. Nature 297; 647-650

Michalowski CB, Pfanzagl B, Loffelhardt W and Bohnert HJ (1990). The cyanelle S10 spc ribosomal protein gene operon from Cyanophora paradoxa. Mol. Gen. Genet. 224; 222-231

Musch A, Wiedmann M and Rapoport TA (1992). Yeast sec proteins interact with polypeptides traversing the endoplasmic reticulum membrane. Cell 69; 343-352

Nakamura K et al. (1990a) Cloning and characterization of a Bacillus subtilis gene homologous to E coli secY. J. Biochem. (Tokyo) 107; 603-60

Nakamura K et al. (1990b). Complementation of the protein transport defect of an Escherichia coli secY mutant (secY24) by Bacillus subtilis secY homolog. FEBS Lett. 273; 75-7

Ohama T, Muto A and Osawa S (1989). Spectinomycin operon of Micrococcus luteus: evolutionary implications of organization and novel codon usage. J. Mol. Evol. 29; 381-395

Ohkubo S et al. (1987). The ribosomal protein gene cluster of Mycoplasma capricolum. Mol Gen. Genet 210; 314-322

Prehn S et al. (1990). Structure and biosynthesis of the signal sequence receptor. Eur. J. Biochem. 188; 439-445

Rapoport TA (1990). Protein transport across the ER membrane. Trends Biochem Sci 15 (9), 355-358

Rothblatt JA et al (1989). Multiple genes are required for proper insertion of secretory proteins into the endoplasmic reticulum in yeast. J Cell Biol 109 (6 Pt 1); 2641-2652

Rothblatt JA and Meyer DI (1986) Secretion in yeast: translocation and glycosylation of prepro-alpha-factor in vitro can occur via an ATP-dependent post-translational mechanism. EMBO J. 5; 1031-1036

Sadler I et al. (1989). A yeast gene important for protein assembly into the endoplasmic reticulum and the nucleus has homology to DnaJ, and Escherichia coli heat shock protein. J. Cell Biol. 109; 2665-2675

Sanders SL et al. (1992). Sec61p and BiP directly facilitate polypeptide translocation into the ER. Cell 69; 353-365

Stirling CJ and Hewitt EW (1992). The S.cerevisiae SEC65 gene encodes a component of the yeast signal recognition particle with homology to human SRP19. Nature 356; 534-537

Stirling CJ et al. (1992). Protein translocation mutants defective in the insertion of integral membrane proteins into the endoplasmic reticulum. Mol. Biol. Cell 3; 129-142

Stirling CJ and Schekman R (1992). In: Op den Kamp JAF (ed) Dynamics of membrane assembly. NATO ASI Series Vol.H63 277-291, Plenum press.

Suh JW et al. (1990). Isolation of a secY homolog from Bacillus subtilis: evidence for a common protein export pathway in eubacteria. Mol. Microbiol. 4; 305-314

Toyn J et al. (1988). In vivo and in vitro analysis of ptlI, a yeast ts mutant with a membrane-associated defect in protein translocation. EMBO J. 7; 4347-4353

Walter P and Blobel G (1980). Purification of a membrane-associated protein complex required for protein translocation across the endoplasmic reticulum. Proc. Natl. Acad. Sci. USA 77; 7112-7116

Walter P and Blobel G (1981). J. Cell Biol. 91,557-561, Translocation of proteins across the endoplasmic reticulum III. Signal recognition particle (SRP) causes signal-dependent and site-specific arrest of chain elongation that is released by microsomal membranes.

Waters MG and Blobel G (1986a). Secretory protein translocation in a yeast cell-free system can occur post translationally and requires ATP hydrolysis. J. cell Biol. 102; 1543-1550

Weidmann M, Gorlich D, Hartmann E and Rapoport T (1989). Febs Letts. 257; 263-268

Weidmann M, Kurzchalia TV, Hartmann E and Rapoport TA (1987). A signal sequence receptor in the endoplasmic reticulum membrane. Nature 328; 830-833

# ATP transport into yeast ER is a prerequisite for preprotein transfer across the ER membrane

Peter Mayinger and David I. Meyer
Department of Biological Chemistry
UCLA School of Medicine
Los Angeles, CA 90024, USA

The molecular mechanism by which precursors of secretory proteins traverse the membrane of the endoplasmic reticulum (ER) is still poorly understood. Of particular interest is the question how this process is linked energetically to the hydrolysis of ATP. When post-translational modes of translocation in yeast and in mammalian ER were examined, one role for ATP could be attributed to the function of cytosolic stress factors, which maintain nascent polypeptides in a translocation competent state (Meyer, 1991). Recently, an additional ATP requirement of the translocation step at the membrane level has been found for both systems. In yeast it was shown that ATP is required for the membrane transfer of preproteins, bound to receptor sites on the ER membrane (Sanz and Meyer, 1989). With regard to this finding basically two models can be proposed how ATP is used for ER-membrane transfer of preproteins.

(i) Preproteins are transported by a translocation ATPase. Recently a new class of transmembrane transport ATPases has been characterized. This family of so called ATP-binding cassette (ABC) proteins is known to transport small molecules like peptides (Blight and Holland 1990). It has been proposed that a member of this family could facilitate transport of peptides across the ER membrane a process that is likely required for the assembly of cellular antigen complexes (Barinaga, 1990). It is therefore conceivable that an analogous translocation ATPase in the ER membrane drives the translocation of secretory precursors. This ATPase would bind preproteins and subsequently translocate them across the membrane. Such an "ATP-dependent

NATO ASI Series, Vol. H 71
Protein Synthesis and Targeting in Yeast
Edited by A. J. P. Brown, M. F. Tuite, and J. E. G. McCarthy

pump" would directly couple the hydrolysis of ATP to the translocation process.

(ii) Preproteins cross the membrane by passive diffusion through a pore. Translocated proteins differ significantly in their physical properties; they may be extremely hydrophobic or, in contrast, are charged either positively or negatively. Thus the translocation machinery must be rather unspecific regarding the nature of its substrate. Recent evidence suggests that a large aqueous pore is involved in translocation (Simon and Blobel, 1991). After targeting, preproteins may initiate their transport across the membrane by a simple diffusion mechanism through this proteinaceous pore, driven by Brownian motion (Simon et al., 1992). This mechanism alone cannot explain the vectorial movement of preproteins. Luminal factors or protein folding would be required to keep the translocated portions of the protein from diffusing back through the pore. In this model luminal molecular chaperones would interact with the translocating chain in an ATP dependent manner thereby transducing the chemical energy of ATP to the translocation process (Simon et al., 1992).

## Translocation in Yeast ER

To address these questions, experiments were designed to identify microsomal factors linked to the ATP-dependent step in translocation. Genetic studies in yeast have shown that the luminal stress protein Kar2p, a homologue of the mammalian Bip, is involved in translocation (Rose et al., 1989; Normington et al., 1989; Nguyen et al., 1991). Since this factor interacts with polypeptides in an ATP-dependent manner, a system for keeping luminal ATP levels at critical concentrations must exist. Accordingly, we examined yeast rough ER-derived vesicles for a system mediating ATP uptake. We found such activity and determined that the uptake was enzyme-mediated and dependent on intact microsomal membranes. In addition, the stilbene derivative DIDS, a well known inhibitor of numerous anion transport systems and 3'-O-(4-

benzoyl)benzoyl-ATP, ($Bz_2ATP$) was found to effectively block ATP uptake in yeast microsomes.

Specific inhibition of ATP uptake by DIDS and $Bz_2ATP$ enabled us to ask whether the ATP transport system, and hence the maintenance of a luminal ATP pool is required for the translocation process. In yeast, the relationship between ATP and translocation can be investigated both co- and post-translationally (Meyer, 1991). In addition, a cytosol-free assay can be used where chemically-pure radiolabeled preprotein is diluted out of a denaturant, such as 8M urea, into the membrane sample (Sanz and Meyer, 1988). This latter assay has been used to show that translocation in yeast can be dissected into an energy-independent preprotein binding step followed by a subsequent ATP-requiring translocation step (Sanz and Meyer, 1989).

Inhibition of ATP uptake with increasing amounts of DIDS and $Bz_2ATP$ was found to inhibit preprotein translocation in yeast in an identical manner, regardless of the type of assay used (Mayinger and Meyer, submitted). Translocation was found to be somewhat less sensitive to both inhibitors than was ATP uptake. We therefore had to consider the possibility that both processes are affected by DIDS and $Bz_2ATP$ treatment of membranes in a similar yet independent manner. To address this issue, we demonstrated that the DIDS inhibition of translocation could be prevented by coincubation with ATP. The presence of 0.5 mM ATP effectively blocked the effects of 1mM DIDS on translocation. Since the inactivation by DIDS of other stilbene-sensitive anion transporters was prevented in the presence of substrates, it seemed likely that the ATP transporter could be protected from DIDS inactivation when supplemented with ATP. This finding provides evidence for a direct linkage of ATP-transport and preprotein translocation.

To elucidate in more detail how ATP transport is connected with preprotein transport, the DIDS effect on translocation was further quantified and analyzed. In Figure 1, a semi-reciprocal blot of DIDS inactivation of preprotein translocation is shown. Two distinct phases of DIDS action on translocation were obtained. The region of the curve corresponding to low inhibitor concentrations shows only minor effects of DIDS on preprotein translocation. This was followed by a region of rapidly increasing inhibition at higher DIDS concentrations. Such behavior is typical for the inhibition of an enzyme catalyzing a non-rate-limiting step (Rognstad, 1979) and we conclude that ATP transport possesses a considerable excess capacity over the rate of translocation. This observation would rule out the possibility that translocation is regulated by the luminal ATP concentration. To further analyze these data, preprotein translocation as a function of DIDS inhibition of ATP

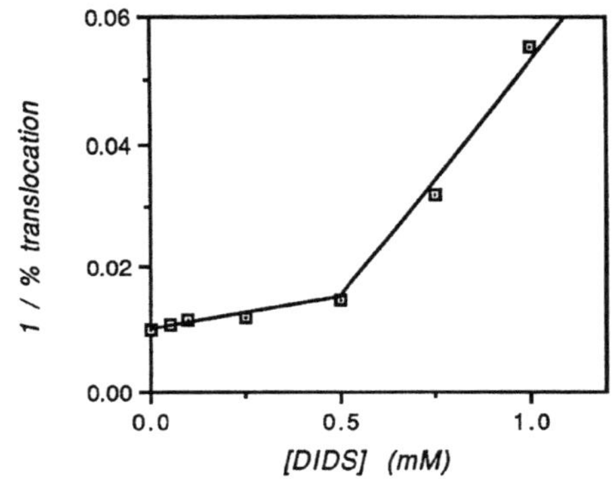


**Figure 1. DIDS does not inhibit a rate-limiting step of translocation in yeast.** Semireciprocal plot of the inhibition of prepro-α-factor translocation by DIDS.

transport is depicted in Figure 2. Again a biphasic curve was obtained. Here it can be seen directly that ATP transport can be reduced over a wide range without a major effect on translocation. However after reaching a critical point, preprotein translocation was inhibited in a linear manner by further reduction of ATP transport. The DIDS concentration from which a rapidly increasing inhibition of preprotein translocation was observed (Fiugre 1) corresponded exactly to the concentration required to obtain a linear relationship between the inhibition of ATP transport and a reduction of translocation (Figure 2).

We postulate from these observations, that ATP transport is indeed required for preprotein translocation in yeast and that a luminal ATP dependent protein is directly connected with the translocation process. Our findings are consistent with current views that the yeast Bip homologue, Kar2p, participates in translocation. Although the mechanism of Kar2p action has not yet been elucidated, it has been shown that that defects in or reduced expression of the *KAR2* gene, affects translocation (Rose et al., 1989; Normington et al., 1989; Nguyen et al., 1991). Recent data indicate that Kar2p is involved in the ATP-dependent association of preproteins with Sec61p (Sanders et. al.,

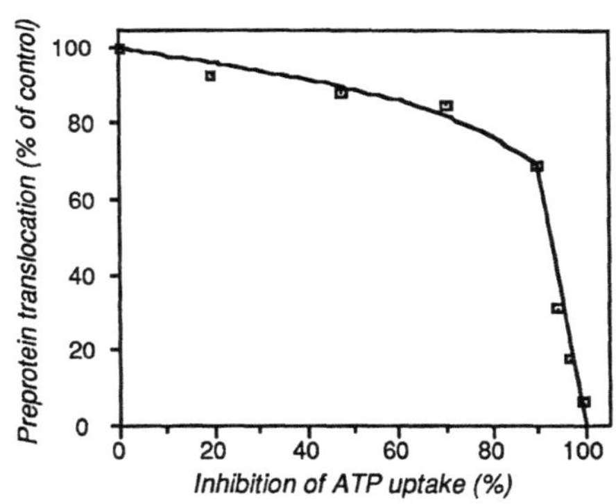


**Figure 2. Preprotein translocation is directly connected to ATP uptake into yeast microsomes.** Preprotein translocation is plotted as a function of DIDS inhibition of ATP transport.

1992). The ATP transporter's role in this process would be to supply this factor with sufficient amounts of ATP. In conclusion, these data seem to favor the diffusion model mentioned above, and it therefore may be questionable whether in yeast a translocation ATPase exists. However, further experiments are required to determine if additional ATP-dependent factors are involved in preprotein translocation in yeast.

**Translocation in Mammalian ER**

Similar to yeast, translocation in mammalian ER has been reported to be ATP-dependent at the membrane level (Klappa et al., 1991; Zimmerman and Walter, 1991). In addition, an ATP uptake system has recently been described for mammalian microsomes (Clairmont et al., 1992). As in yeast, the mammalian ATP transporter is specifically inhibited by DIDS (Clairmont et al., 1992). It would therefore be interesting to investigate if a mechanism, similar to yeast can be found for mammalian translocation.

By examining the effects of DIDS on mammalian ER we found a significant inhibition of preprotein translocation, concommitant with an inhibition of ATP uptake. We conclude that a luminal ATP requirement exists for mammalian translocation. This would be in contradiction to biochemical studies which indicate that translocation in mammalian microsomes may not be dependent on luminal factors. In such studies, either preprotein transport was reconstituted into liposomes from a crude detergent extract of microsomal proteins, in which soluble luminal proteins like BiP were barely detectable (Nicchitta and Blobel, 1990), or luminal contents were removed by high pH treatment of microsomes (Bulleid and Freedman, 1988). One could think of two possible explanations for this discrepancy. A very small subpopulation of BiP or another luminal component, associated with membrane proteins, might still be present in sufficient quantity for activity in the reconstituted system. Since Bip is an abundant ER protein, it is that a minor proportion of this protein participates in the translocation reaction, while the majority is involved in folding of precursor proteins after their translocation has been completed. Alternatively, an ATP-requiring membrane protein, utilizing luminal ATP, may drive preprotein translocation. Accordingly a critical luminal ATP concentration would be required for maintaining the function of this factor, in the absence of any luminal participants in translocation.

In conclusion, there appears to be some similarity between yeast and mammalian translocation regarding ATP dependence. In both systems the presence of ATP on the luminal side of the membrane is required, which in turn implicates that either soluble factors in the lumen or membrane bound components with luminal binding sites for ATP are involved in the translocation of preproteins. In yeast, Kar2p is the most likely candidate for such a factor. However it still remains to be clarified whether in mammalian ER the translocation process is also connected to luminal chaperones. The use of radiolabeled analogues of ATP and improvement in the reconstitution of translocation hold the greatest promise for the identification of those factors that link preprotein transport to ATP-hydrolysis.

## References


Barinage, M. (1990) *Science* **250**, 1657-1658.

Blight, M.A. and Holland, I.B. *Mol. Microbiol* **4**, 873-880.

Bulleid, W. J. and Freedman, R. B. (1988) *Nature* **335**, 649-651.

Clairmont, C. A., De Maio, A. and Hirschberg, C. B. (1992) *J. Biol. Chem.* **267**, 3983-3990.

Klappa, P., Mayinger, P., Pipkorn, R., Zimmermann, M. and Zimmermann R. (1991) *EMBO J.* **10**, 2795-2803.

Meyer, D. I. (1991).*Trends Cell. Biol.* **1**,154-159.

Nicchitta, C.V and Blobel G. (1990) *Cell* **60**, 259-269.

Nguyen, T.H., Law, D.T.S. and Williams, D.B. (1991) *Proc. Natl. Acad. Sci. USA* **88**, 1565-1569.

Normington, K., Kohno, K., Kozutsumi, Y., Gething, M.-J. and Sambrook, J. (1989) *Cell* **57**, 1223-1236.

Rognstad R. (1979) *J. Biol. Chem.* **254**, 1875-1878.

Rose, M. D., Misra, L. M. and Vogel, J. P. (1989) *Cell* **57**, 1211-1221.

Sanders, S. L., Whitfield, K. M., Vogel, J. P., Rose, M. D. and Schekman, R.M. (1992) *Cell* **69**, 353-365.

Sanz, P. and Meyer, D. I. (1989) *J. Cell Biol..* **108**, 2101-2106.

Sanz, P. and Meyer, D.I.(1988) *EMBO J.* **7**, 3553-3557.

Simon, S. M. and Blobel, G. (1991) *Cell* **65**, 371-380.

Simon, S.M., Peskin, C.S. and Oster, G.F. *Proc. Nat. Acad. Sci. USA* **89**, 3370-3774

Zimmerman, D. L. and Walter, P. (1991) *Cell Regulation* **2**, 851-859.

# CROSS-LINKING SIGNAL SEQUENCES TO COMPONENTS OF YEAST MICROSOMES.

Olwyn Westwood and Brian Austen.
Peptide Unit, Dept. of Surgery,
St. George's Hospital Medical School,
Cranmer Terrace, LONDON SW17 ORE

Candidate proteins of the endoplasmic reticulum (ER) membrane functional in the translocation of newly-synthesised proteins from the cytoplasm to the lumen have been identified using cross-linking to photoreactive truncated nascent polypeptides paused in the act of translocation. Proteins include SSRalpha, a glycoprotein of 34kDa $M_r$ (Weidmann et al, 1987), which has one transmembrane region and is part of a heterotetrameric complex, while a glycoprotein of 36kDa, translocating chain-associated membrane protein (TRAM), is cross-linked in reconstituted proteoliposomes to a translocation intermediate with carbodiimide. TRAM is predicted to cross the membrane eight times (Gorlich **et al.**, 1992).

In yeast microsomes, truncated photoreactive pre-pro alpha factor intermediates cross-link to sec61p in the presence of ATP, and to sec62p in the absence of ATP (Musch **et al**, 1992). The same intermediate cross-links to a homologue of sec61p in mammalian microsomes (Rapoport, private communication). These results suggest that during translocation a polypeptide is in continuous attachment to sec61p , while attachment to sec62p is more transient. Pre-pro alpha factor coupled to avidin becomes trapped at translocation sites, and can be cross-linked to sec61p and BiP (Sanders **et al.**, 1992).

ER membranes have been probed with photoreactive synthetic signal peptides, which in mammalian microsomes cross-link to a integral membrane protein of apparent $M_r$ of 43kDa and a 15kDa cytoplasmic domain. Cross-linking is specifically displaced by competing signal peptides, showing that the 43kDa protein has

NATO ASI Series, Vol. H 71
Protein Synthesis and Targeting in Yeast
Edited by A. J. P. Brown, M. F. Tuite, and J. E. G. McCarthy

the properties of a specific receptor with $K_d$ $10^{-7}M$ for signal sequences (Robinson **et al.**, 1987). Because sec61p has a calculated molecular weight of 52kDa, but abnormally fast mobility on SDS-PAGE, it was of interest to discover if the signal peptide cross-linked to sec61p in yeast microsomes.

## Material and Methods

**Yeast strains:** The ABYS-1 (a pra1 prb1 prc1 cps1 ade2) (Achstetter **et al.** 1984) strain of yeast used for lysate and microsome preparation was donated by Dr. M. Tuite at the University of Kent. A strain with a mutation in sec61 (RDM15-5B (sec61, leu2-3,112, ade2, ura3-52, pep4-3, mat alpha) and a strain containing multi copies of SEC61 on YEp351 in W303-1Balpha were kindly provided by Dr R. Schekman of UC Berkeley.

**Spheroplast preparation:** Yeast cells were harvested (see below) and incubated for 15min at 30°C with 100mM Tris-sulphate pH 9.4 containing 10mM dithiothreitol (DTT) [0.25g cells/ml], then washed with 1.2M sorbitol. The cells were suspended in 300ml of media (50% YPD containing 1.2M sorbitol, 5mM DTT, 10mM $K_2HPO_4$ and adjusted to pH 7.4 with ortho-phosphoric acid) at 30°C with gentle agitation, and spheroplasted in the presence of lyticase (Sigma, Poole, UK) at a concentration of 10,000 units per litre of the original culture volume. The extent of spheroplasting was determined by cell lysis with 0.1% w/v SDS (see Tuite & Plesset, 1986).

**Isolation of yeast rough microsomes:** ABYS-1 or the yeast strain bearing multi-copies of SEC61 were grown to stationary phase ($OD_{600nm}$ of 2-3) in YPD media (5 lit) at 30°C. The sec61 mutant was grown overnight at 24°C, then for 2h at 37°C. The cells were spheroplasted as above, and microsomes isolated by the modified method of Rothblatt & Meyer (1986). The spheroplasts were layered on to a 15ml cushion of 0.8M sucrose containing 1.5% (w/v) Ficoll-400, and 20mM Hepes/KOH pH 7.5, and centrifuged at 3000rpm at 4°C for 15min. The pellet was homogenized with 30 strokes in a Dounce glass/glass homogenizer in 0.1M sorbitol containing 40ml of 50mM potassium acetate, 20mM Hepes/KOH, 2mM EDTA, 1mM DTT and 1mM PMSF. The homogenate was layered on a 20ml cushion of 1M sucrose containing 20mM Hepes/KOH pH 7.5 and 2mM EDTA and centrifuged at 6500rpm in a Sorval SS34 rotor for 15min at 4$^{o}$C. The supernatant was ultra-centrifuged at 35000rpm at 4$^{o}$C in a Beckmann SW51 rotor for 40min. The pelleted microsomes were resuspended in 0.3M sorbitol containing 50mM potassium acetate, 20mM Hepes/KOH pH 7.4, 1mM DTT, 1mM PMSF [concentration of 50 at $OD_{260nm}$], flash frozen in liquid nitrogen and stored at -80$^{o}$C.

**SDS-PAGE and Western Blot analysis:** Electrophoresis was performed on gradients (7-16%) of acrylamide. Transfer to nitrocellulose and Western blot analysis was performed as described (Robinson & Austen, 1985). Antisera against a peptide sequence of sec61 was kindly provided by Dr Colin Stirling of Manchester University.

**Photoactivated cross-linking.**

The synthesis and photoactivation of $^{125}$I-($^{12}$Cys-azidophenacyl signal peptide has been described previously, and UV photoactivation was performed on yeast microsomes at a concentration of 10A$^{260}$ units per ml (Robinson et al., 1987).

## Results.

Yeast microsomes prepared by the above procedure from ABYS-1, sec61 and a SEC61 multi-copy strain co-translationally processed bovine preprolactin and glycosylated pre-pro alpha factor newly-synthesised in an ABYS-1 yeast lysate. The synthetic azidophenacyl signal peptide at concentrations between 5 and 50 uM inhibited processing and glycosylation (data not shown), competing with native signal sequences as reported in translation systems from wheat-germ or reticulocytes supplemented with pancreatic microsomes (Robinson et al., 1987). Thus, the signal peptide competes with native signal sequences for binding to a component that is involved in translocation of these newly-synthesised proteins into the lumen of yeast microsomes where proteolysis by signal peptidase and glycosylation takes place.

ABYS-1 microsomes uv-irradiated in the presence of the radioactive photoreactive signal peptide and separated on SDS-PAGE gave one major radioactive 52kDa product (Fig. 1; lane labelled: "WT"). The contribution to the $M_r$ by the photoreactive peptide is 2kDa, so that the initial $M_r$ of the protein cross-linked is 50kDa.

Cross-linking of the 50kDa protein was more intense in microsomes isolated from the temperature-sensitive sec61 mutant (Fig 1, "Sec61"), but less intense in microsomes from a strain containing multi-copies of SEC61 ("MC 61"). Levels of sec61p, which runs with an apparent $M_r$ of 35kDa $M_r$, detected by Western blots in corresponding microsomes from the three strains do not correspond with the intensities of cross-linking 52kDa (Fig 1).

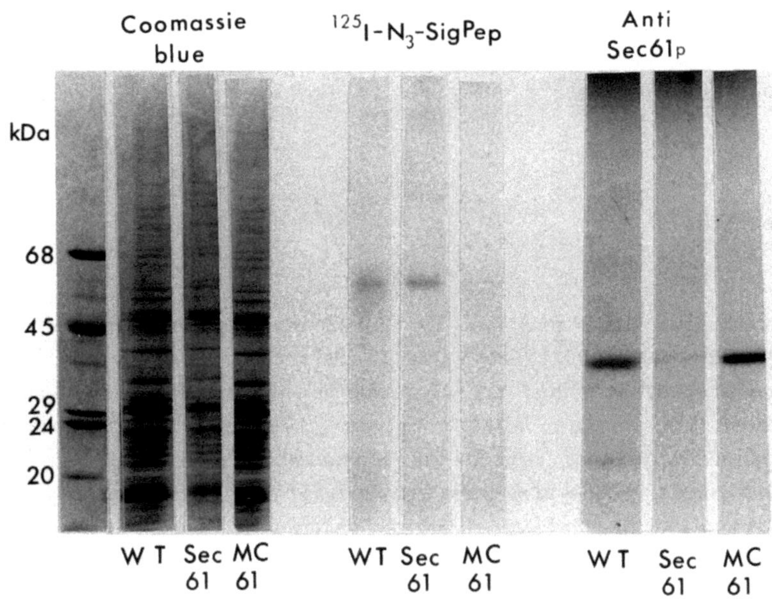


Fig 1. A comparison of SDS-PAGE protein profiles, proteins cross-linked to 125I-labelled signal peptide, and sec61p levels in ABYS-1(WT), sec61 and multi-copy SEC61 strains.

partially to fragments of 37kDa, 24kDa, 14kDa and 10kDa (Fig 2; Tracks 10). When photolabelled microsomes were disrupted by Triton X-100 (1%), trypsin did not yield the 37kDa fragment, suggesting that it is protected by the intact surrounding membrane (Fig 2; Track 11). Cross-linking at 52kDa was not effected by ATP (1mM) (data not shown).

Purification of the 52kDa cross linked species was obtained by ion-exchange in 1% Nikkol, 50 mM KCl, phosphatidyl choline (0.8mg/ml) on DE-cellulose (Fig 3), from which most of the 52kDa

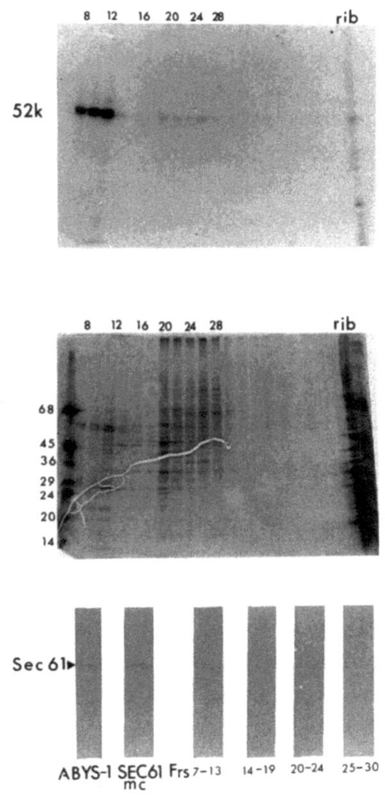


Fig 4. Photolabelled yeast microsomes extracted into 1% Nikkol, 500 mM KOAc, phosphatidyl choline (0.8 mg/ml and protease inhibitors, were first centrifuged to remove the insoluble, ribosomal proteins (rib) and eluted from Sephacryl S-200. Shown are fractions from DE-cellulose eluted with increasing KCl.

was eluted by 50mM KCl. Western blots of TCA-precipitated material from equal volumes of pooled fraction 7-13, 14-19, 20-24 and 25-30 showed that sec61p, like the cross-linked 52kDa protein, was most concentrated in fractions 7-13 (Fig 3). Proteins seen in fractions 7-13 of the DE-Cellulose column, including the labelled 52kDa protein, also eluted without retention from SP-Sephadex, Con A Sepharose and GTP-agarose in 50 mM KCl (data not shown).

## Discussion.

Mammalian signal recognition particle (SRP) functions as an adaptor between secretory protein translation and translocation across the ER membrane. However, in yeast cell-free lysates, pre-pro alpha mating factor translocates into yeast microsomes in the absence of ongoing protein translation (Rothblatt and Meyer 1986b), Disruption of the gene coding for the yeast homologue of the 54kDa subunit of SRP disrupts translocation across the ER only for a subset of proteins, and does not prevent it completely (Hann and Walter, 1991). These findings indicate that alternative mechanisms independant of SRP operate in yeast enabling the ER membrane to distinguish those polypeptides containing signal sequences from those that do not.

UV irradiation of a photoreactive signal sequence cross-links it to a 50kDa integral membrane protein in active yeast microsomes. The yeast signal binding protein (SBP) is not a glycoprotein or GTP-binding protein, and is larger in size than the corresponding 43kDa SBP which is cross-linked in pancreatic microsomes (Robinson **et al.**, 1987). It is of similar size to a protein cross-linked by an identical ligand in liver microsomes (Robinson **et al.**, 1986). We have shown that synthetic signal sequences inhibit access of pre-proteins to the lumenal surface of the yeast microsomal membranes **in vitro**, strongly suggesting that the 50kDa SBP is functional in the translocation of proteins across to the ER lumen. The 50kDa protein may have a targeting role, and provide a point of recognition by the ER

A signal peptide isolated from ovalbumin (Robinson **et al.**, 1986) at 10 uM concentration inhibited photo-cross-linking to the 50kDa protein (Fig 2; compare Tracks 1 & 2), indicating that cross-linking to the photoreactive signal peptide results from specific and saturable binding. The 52kDa adduct remained in the pelleted membrane fraction after extraction of microsomes with 0.5 M KCl (Fig 2; compare Tracks 4 & 5), showing that the adduct is neither soluble nor loosely attached to the membranes. In parallel experiments, Western blots showed that a 58kDa protein that cross-reacted with polyclonal antisera against the 54kDa subunit of SRP is solubilized from the pellet by 0.5 M KCl (data not shown). Thus, the cross-linked protein does not behave in the same way as the yeast homologue of SRP54k.

Extraction of photo-cross-linked microsomes with octyl glucoside at a concentration of 75 mM, which converts the membranes to mixed micelles, released most of the adduct into the supernatant (compare Tracks 6 & 7), while in photo-cross-linked membranes treated with 0.1 M sodium carbonate adjusted to pH 11, the 52kDa protein remained in the pellet, showing that it behaves like an integral membrane protein (Tracks 8 & 9). The 52kDa adduct was fairly resistant to trypsin. Added to intact photo-cross-linked yeast microsomes, trypsin digested the adduct

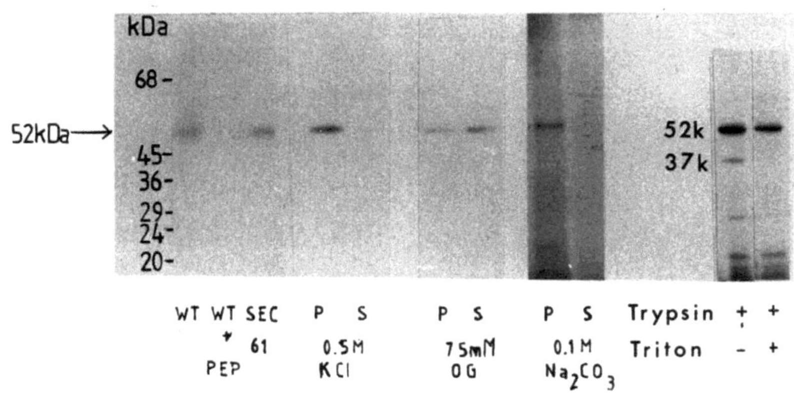


Fig 2. Behaviour of the 52kDa cross-linked protein visualised on SDS-PAGE to post-labelling extraction and proteolysis.

membranes that is alternative to targeting mediated by SRP and docking protein.

Genetic studies of yeast have led to the identification of three membrane proteins (sec61p, sec62p and sec63p) encoded by essential genes, certain mutations in which result in defects in translocation (Rothblatt et al., 1989). Cross-linking from translocation intermediates stuck across the membrane suggest that both sec61p and sec62p are part of the aqueous environment that nascent chains contact during translocation.

ATP is not required for the initial insertion of the nascent chains, wheras translocation of larger domains to the lumen requires ATP (Rothblatt and Meyer, 1986b). Cross-linking of the signal peptide to the 50kDa protein has been shown to be independant of ATP, suggesting that, in contrast to sec61p, it acts as at a relatively early stage of translocation. In keeping with this is the finding that the intact membrane protects a 37kDa domain of the protein against tryptic digestion, so that like its counterpart in mammalian microsomes, the protein may possess a relatively large cytoplasmic domain of about 15kDa. Relatively little of the sec61p sequence is predicted to extend into the cytoplasm.

The 50kDa SBP is clearly differentiated from sec61p by its mobility on SDS-PAGE. Moreover, the extent of cross-linking varies inversely to the levels of sec61p in the microsomes. If overexpression of sec61p were to directly suppress the availability or amount of the signal binding protein (SBP), while a temperature-sensitive mutation in the gene enhances it, it is possible that the two proteins interact. A physical Interaction between sec61p and the 50kDa SBP is indicated by their coexistence in material eluting from ion-exchange and con A columns in low salt, but further purification will be required before conditions that separate the two proteins are established.

A model for the translocation complex put forward by Singer and Yaffe (1990) also implicates a signal binding protein as part of the translocation complex. They envisaged that the signal sequence intercalates into the translocator interface between the hydrophobic surfaces of two of the subunits, while the following stretch of residues is pulled into the interface so that ionic residues face into the central aqueous channel. The 50kDa signal binding protein could be one of the interfacing subunits, as shown schematically in Fig 4, while sec61p and a yeast homologue of TRAM may form other parts of the aqueous channel. After release of the signal sequence from SRP upon making contact with docking protein, or from alternative chaperones such as Hsp70, the signal sequence may then bind the 50kDa SBP. SBP may then bind sec62p and the complex then move to the major part of the translocation complex consisting of sec61p and a yeast homologue of TRAM. SBP and sec62p may move out of the translocation complex after the initial stages of translocation has commenced. Lumenal BiP, with ATP bound, may function in the translocation process by pulling successive segments of polypeptide through the channel, release from which would require ATP hydrolysis.

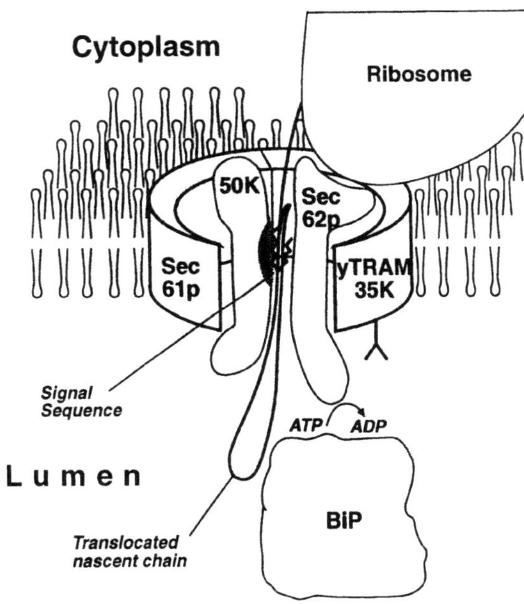


Fig 4. A model for translocation across the ER membrane.

**References.**

Achstetter T, Emter O, Ehmann C & Wolf DH (1984) Proteolysis in eukaryotic cells. Identification of multiple proteolytic enzymes in yeast. J Biol Chem **259**, 13334-43.

Gorlich D, Hartmann E, Prehn S & Rapoport TA (1992) A protein of the endoplasmic reticulum involved early in polypeptide translocation. Nature (Lond.) **357**, 47-52.

Hann BC & Walter P (1991) The signal recognition particle in S. cerevisiae. Cell **67**, 131-144.

Musch A, Wiedmann M, & Rapoport TA (1992) Yeast Sec proteins interact with polypeptides traversing the endoplasmic reticulum membrane. Cell **69**, 1-20.

Robinson A & Austen BM (1985) Fractionation of rat liver proteins involved in translocation of newly-synthesised proteins at the ER membrane. Biochem. Soc. Trans. **13**, 865-866.

Robinson A, Kaderbhai MA and Austen BM (1987) Identification of signal sequence binding proteins integrated into the rough endoplasmic reticulum membrane. Biochem J. **242**, 767-777.

Rothblatt JA, Deschaies RJ, Sanders, SL, Daum G & Schekman R (1989) Multiple genes are required for proper insertion of secretory proteins into the endoplasmic reticulum in yeast. J Cell Biol **109**, 2641-2652.

Rothblatt JA & Meyer DI (1986a) Secretion in yeast: reconstitution and glycosylation of alpha factor and invertase in a homologous cell-free system. Cell **44**, 619-628.

Rothblatt JA and Meyer DI (1986b) Secretion in yeast: translocation and glycosylation of prepro-alpha- factor **in vitro** can occur via an ATP-dependant post-translational mechanism. EMBO J **5**, 1031-1036.

Sanders SL, Whitfield KM, Vogel JP, Rose MD & Schekman RW (1992) Sec61p and BiP directly facilitate polypeptide translocation into the ER. Cell **69**, 353-363.

Singer SJ & Yaffe MP (1990) Embedded or not? hydrophobic sequences and membranes. Trends in Biochem Sci. **15**, 369-373.

Tuite M and Plesset J (1986) mRNA-Dependant yeast cell-free translation system: theory and practice. Yeast **2**, 35-52.

Wiedmann M, Kurzchalia E, Hartmann E & Rapoport TA (1987) A signal sequence receptor in the endoplasmic reticulum membrane. Nature (Lond.) **328**; 830-832.

**Acknowledgements.**

We gratefully acknowledge the SERC and Delta Biotechnology for their financial assistance.

**THE NUCLEOTIDE CYCLE OF *SEC4* IS IMPORTANT FOR ITS FUNCTION IN VESICULAR TRANSPORT**

Peter Novick, Patrick Brennwald, Michelle D. Garrett, Mary Moya, Denise Roberts and Robert Bowser.
Department of Cell Biology,
Yale University School of Medicine,
333 Cedar Street
New Haven, CT 06510

Sec4 is a *ras*-like, GTP-binding protein required for exocytosis in yeast (Novick et al., 1980; Salminen and Novick, 1987). The purified protein displays a high affinity for GTP and a very low intrinsic GTPase activity (Kabcenell et al., 1990). We have found that the addition of yeast lysate greatly stimulates the GTPase activity of the protein, reflecting the presence of a Sec4-GTPase activating protein (GAP) (Walworth et al., 1992). Excess Sec4 bound to GTPγS will inhibit the interaction, but excess Ypt1-GTPγS will not, indicating that the GAP is specific. A mutation ($^{79}$Q->L) in a site of Sec4p which is predicted to interact with the phosphoryl group of GTP, lowers the intrinsic hydrolysis rate to unmeasurable levels. When purified from yeast, the mutant protein is predominantly GTP bound (Walworth et al., 1992), while the wild-type protein is predominantly GDP bound (Kabcenell et al., 1990). This mutant protein can be stimulated by the yeast Sec4-GAP activity, but only to 30% of the GAP-stimulated rate seen with wild type Sec4p. The *sec4-L79* allele can function as the only copy of *sec4* in yeast (Walworth et al., 1992). However, the mutation causes recessive cold sensitivity, slowing of invertase secretion, accumulation of secretory vesicles and displays synthetic lethality with a subset of other secretory mutants, indicating a partial loss of function. This is in contrast to the equivalent mutation in *ras* which is completely insensitive to stimulation by ras-GAP and results in a gain of ras function seen ultimately as oncogenic transformation. Taken together this argues that, unlike the case of *ras*, where the absolute level of the GTP-bound form determines activity, it is the ability of Sec4 to cycle between its GTP- and GDP-bound forms that is critical for Sec4p function in vesicle transport.

The cycle of nucleotide binding and hydrolysis may be coupled to a cycle of Sec4 localization within the cell. Sec4 binds to secretory vesicles which then fuse with the plasma membrane by exocytosis (Goud et al., 1988). Sec4 can recycle from the plasma membrane through a soluble pool to rebind onto a new round of vesicles. We have found a soluble yeast activity which can release Sec4 from membranes, but only when Sec4 is in its GDP-bound form. Pre-binding GTPγS to Sec4 inhibits release. This activity is comparable to that of the GDI protein isolated from bovine brain described in Araki et al. (1990). Association of Sec4 with secretory vesicles requires prior prenylation by Bet2, a component of a geranyl-geranyl transferase (Rossi et al., 1991). Correct targeting of Sec4 may involve the carboxy terminal hypervariable region of Sec4 as well as upstream sequence information since chimeras of Sec4 with the carboxy terminus of the related protein, Ypt1, result in partial mislocalization to the Ypt1 compartment.

*DSS4-1*, a dominant suppressor of the *sec4-8* temperature-sensitive mutation, encodes an exchange protein (Moya et al., submitted). Lysates derived from yeast strains overproducing the suppressor protein or purified Dss4 protein produced in *E. coli* stimulate the dissociation of both GDP and GTP from Sec4. The wild-type Dss4 protein is predominantly insoluble in yeast, while the suppressing allele encodes a partially soluble protein. Disruption of *dss4* is not lethal, yet it can mimic and as well as aggravates a partial loss of *SEC4*, suggesting that Dss4 normally serves to facilitate Sec4 function.

The cycle of Sec4 may function to allow the assembly and subsequent disassembly of a set of proteins necessary for exocytosis. Candidates for members of this set of proteins are encoded by *sec* genes which show strong genetic interactions with *sec4-8*. Two of these, *SEC8* and *SEC15* encode large proteins which are in a 20 *S* complex and which can peripherally associate with the plasma membrane (Salminen et al., 1989; Bowser and Novick, 1991; Bowser et al., 1992). The predicted amino acid sequence of Sec8 includes a 200 amino acid domain that shares 25% sequence identity with the region of yeast adenylate cyclase that is proposed to interact with Ras. By analogy with the Ras/cyclase interaction, the Sec8/Sec15 complex may respond to activated Sec4 to facilitate exocytosis.

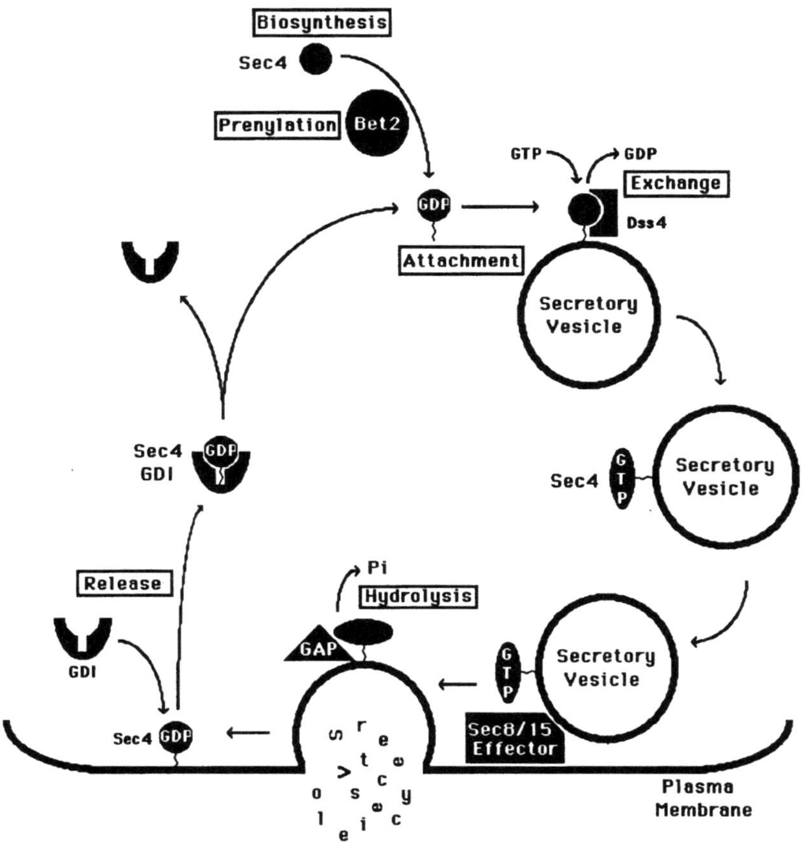


*Figure 1.* As presented in this model, Sec4 functions in a cycle of nucleotide binding and hydrolysis that is coupled to a cycle of subcellular localization. We will consider the start of the cycle to be Sec4 in its GDP-bound state. Since the GDP-bound form can be solubilized by interaction with GDI, the cycle begins with Sec4 as a complex with GDI in the cytosol. Interaction with an exchange protein removes the GDP allowing a GTP to bind. This causes dissociation of GDI allowing Sec4 membrane attachment. The site of membrane association may be dictated by the location of the exchange protein. We propose that membrane association occurs in the trans Golgi at the site of secretory vesicle budding. Sec4 remains on the secretory vesicle in its GTP-bound state until the vesicle interacts with an effector complex, containing Sec8 and Sec15, on the plasma membrane. This interaction triggers a series of events leading to exocytosis. Sec4-GAP also may be a component of the effector complex causing Sec4 to hydrolyse the bound GTP. This leaves Sec4 in its GDP-bound state which allows solubilization by GDI and initiation of a new cycle.

*References*

Araki,S., A. Kikuchi, Y. Hata, M. Isomura, and Y. Takai. 1990. Regulation of reversible binding of *smg* p25A, a *ras* p21-like GTP-binding protein, to synaptic plasma membranes and vesicles by its specific regulatory protein, GDP dissociation inhibitor. J. Biol. Chem. 265:13007-13015.

Bowser, R. and P. Novick. 1991. Sec15 protein, an essential component of the exocytotic apparatus, is associated with the plasma membrane and with a soluble 19.5*S* particle. J. Cell Biol. 112: 1117-1131.

Bowser, R., H. Müller, B. Govindan and P. Novick. 1992. Sec8p and Sec15p are components of a 19.5*S* particle that interacts with Sec4p to control exocytosis. (In press) J. Cell Biol.

Kabcenell, A. K., B. Goud, J. K. Northup and P. Novick. 1990. Binding and hydrolysis of guanine nucleotides by Sec4p, a yeast protein involved in the regulation of vesicular traffic. J. Biol. Chem. 265: 9366-9372.

Goud, B., A. Salminen, N.C. Walworth and P. Novick, 1988. A GTP-binding protein required for secretion rapidly associates with secretory vesicles and the plasma membrane in yeast. Cell 53: 753-768.

Moya, M., D. Roberts and P. Novick. *DSS4-1*, a dominant suppressor of *sec4-8* encodes a nucleotide exchange protein that facilitates Sec4p function. (Submitted).

Novick, P., C. Field and R. Schekman, 1980. Identification of 23 complementation groups required for post-translational events in the yeast secretory pathway. Cell 21: 205.

Rossi, G., Y. Jiang, A. Newman and S. Ferro-Novick. 1991. Dependence of Ypt1 and Sec4 membrane attachment on Bet2. Nature 351: 158-161.

Salminen, A. and P. Novick, 1987. A ras-like protein is required for a post-Golgi event in yeast secretion. Cell 49: 527-538.

Salminen, A. and P. Novick, 1989. The Sec15 protein responds to the function of the GTP binding protein, Sec4, to control vesicular traffic. J. Cell Biol. 109: 1023-1036.

Sasaki ,T., K. Kaibuchi, A.K. Kabcenell, P. Novick, and Y. Takai. 1991. A mammalian inhibitory GDP/GTP exchange protein (GDI) for *smg*p25A is active on the yeast *SEC4* protein. Mol. Cell. Biol. 11: 2909-2912.

Walworth, N. C., P. Brennwald, A. K. Kabcenell, M. D. Garrett and P. Novick. 1992. Hydrolysis of GTP by Sec4 protein plays an important role in vesicular transport and is stimulated by a GTPase activating protein in yeast. Mol. Cell. Biol. 12: 2017-2028.

# VACUOLAR PROTEIN SORTING IN YEAST

Bruce F. Horazdovsky, Todd R. Graham and Scott D. Emr
Division of Cellular and Molecular Medicine
Howard Hughes Medical Institute
UCSD School of Medicine
La Jolla, California 92093-0668

## Introduction.

The compartmentalized nature of the eukaryotic cell requires specialized sorting and localization machinery to correctly deliver specific protein constituents to a wide variety of intracellular organelles. In the yeast *Saccharomyces cerevisiae*, the lysosome-like vacuole serves as one of these specific targets. This organelle is an acidic compartment that houses many of the hydrolytic enzymes responsible for macromolecular degradation. A number of these enzymes including, carboxypeptidase Y (CPY), proteinase A (PrA), proteinase B (PrB) and repressible alkaline phosphatase (ALP) serve as convenient markers to study the vacuolar protein sorting and localization pathway. In a late Golgi compartment, vacuolar proteins are actively sorted away from secretory proteins (Stevens, et al. 1982; Graham and Emr 1991) and are specifically delivered to the vacuole. The analysis of the gene products affected in mutants that specifically disrupt vacuolar protein sorting (*vps* mutants), as well as the role of gene products that function in the secretory pathway (Sec proteins), are beginning to elucidate the molecular basis of this complex localization process.

## Transport of Vacuolar Proteins Through the Secretory Pathway.

Like lysosomal proteins in mammalian cells, vacuolar proteins transit through the early stages of the yeast secretory system together with proteins destined to be secreted from the cell. This was determined by analyzing the transport of vacuolar enzymes in temperature conditional *sec* mutants (Stevens, et al. 1982). The *sec* mutants that exhibit a defect in translocation into the ER, in protein transport from the ER to the Golgi complex, or in protein transport from the Golgi complex, are unable to transport most proteins to the vacuole at the nonpermissive temperature (reviewed in Klionsky, et al. 1990)

The Golgi complex plays an integral role in the sorting of vacuolar proteins from other proteins in the secretory pathway. One goal of our research is to better define the compartmental structure of the yeast Golgi and to potentially localize the vacuolar protein sorting reaction to a specific compartment of this organelle. We have made use of the *sec18* mutant in these studies, based on the finding that the Sec18 protein (Sec18p) is the yeast homologue of the mammalian transport factor NSF (N-ethylmaleimide sensitive factor)(Eakle, et al. 1988, Wilson, et al. 1989). NSF is required for protein transport from the mammalian Golgi to the medial Golgi compartment in vitro, specifically to promote the fusion of transport vesicle intermediates with the target membrane (Malhotra, et al. 1988). The *sec18* mutant had been described to exhibit a block in ER to Golgi protein transport (Novick, et al. 1981). The role of NSF in Golgi transport in vitro suggested that the Sec18p may be required for protein transport through distinct compartments of the yeast Golgi complex in vivo.

We have found that the thermal inactivation of the mutant *sec*18-1 gene product upon temperature shift is very rapid and that multiple steps in α-factor (αf) posttranslational modification could not occur without Sec18p function. Because the Sec18/NSF protein functions to catalyze intercompartmental protein transport, we suggest that the Golgi enzymes required for posttranslational processing of af are sequestered to distinct compartments that contain (from *cis* to *trans*), an α1-6 mannosyltransferase, an α1-3 mannosyltransferase, and Kex2p, the αf processing protease (Graham and Emr 1991).

Consistent with in vitro observations, our data indicates that Sec18p is a general transport factor required at most, if not all, intercompartmental protein transport steps along the secretory pathway in vivo. Surprisingly, we found that the Sec18p is not required for the final intercompartmental transfer of CPY to the vacuole. Our data suggests that the Golgi modified p2CPY precursor is in at least two Golgi subcompartments; an ER proximal compartment that contains the α1-3 mannosyltransferase which is required for the formation of p2CPY, and a more distal, prevacuolar compartment (Graham and Emr 1991).

It was not known if vacuolar proteins are sorted from the secretory pathway before reaching the Kex2p compartment. We have constructed a vacuolar protein that is a substrate for Kex2p by cloning a small fragment of the MFα1 gene into a previously characterized CPY-invertase chimeric gene (Johnson, et al. 1987). The product of this new construct has the CPY vacuolar targeting signal at the amino terminus, separated from enzymatically-active invertase at the carboxyl end by a Kex2p cleavage site (LysArg) encoded within the αf segment. If this trimeric fusion protein was transported through the Kex2p compartment before it was sorted from the secretory pathway, invertase would be cleaved free from the CPY vacuolar targeting signal thus allowing it to be secreted. While the parent CPY-invertase fusion protein is efficiently sorted to the vacuole, most of the invertase activity was secreted from cells expressing the trimeric fusion. This experiment suggests that vacuolar proteins transit through the Kex2p compartment and that sorting of vacuolar proteins from secreted proteins must occur either within the Kex2p compartment or in a more distal compartment. These findings have led us to propose a more detailed pathway for the transport of proteins to the vacuole as depicted in Figure 1.

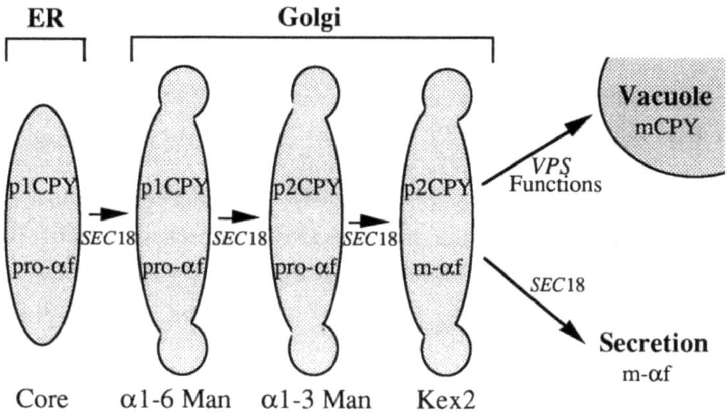


Figure 1. Movement of CPY and α-factor through the secretory pathway.

## Vacuolar Protein Sorting (*vps*) Mutants.

As discussed above, protein localization to the lysosome-like vacuole compartment also involves protein movement to and through the secretory pathway (Stevens, et al. 1982). Upon arrival to a late Golgi compartment, soluble vacuolar proteins are selectively sorted from the secretory protein pool and targeted to the vacuole (Graham and Emr 1991). The targeting information that directs this localization process is contained within specific sequences of the vacuolar proteins themselves, unlike the mammalian system where a specific mannose 6-phosphate modification is involved in the targeting mechanism (Kaplan, et al. 1977; von Figura and Hasilik 1986; Klionsky, et al. 1990). In the case of two soluble vacuolar hydrolases, CPY and PrA, the cis-acting sequence components responsible for vacuole targeting has been mapped to the N-terminal portion of these two proteins (Johnson, et al. 1987; Valls, et al. 1987; Klionsky, et al. 1988). However, there are no obvious primary sequence similarities between the PrA and CPY localization signals, suggesting that the secondary structures of these segments may comprise the sorting signal or protein specific sorting receptors may be required (Johnson, et al. 1987;

Valls, et al. 1987; Klionsky, et al. 1988; Rothman, et al. 1989; Paravicini, et al. 1992).

To identify the cellular components that carry out the localization process, two genetic approaches have been used to identify mutants defective in the vacuolar protein delivery system (Bankaitis, et al. 1986; Robinson, et al. 1988; Rothman, et al. 1989; Rothman and Stevens 1986). These vacuolar protein sorting (*vps*) mutants comprise at least 43 complementation groups whose common phenotype is the secretion of vacuolar protein(s) after they have passed through the Golgi complex (Klionsky, et al. 1990; Robinson, et al. 1988; Rothman, et al. 1989; Raymond, C. and T. Stevens, personal communication). In addition, a number of mutants originally isolated on the basis of a peptidase-deficient phenotype (*pep*) (Jones 1977), also have been shown to mislocalize vacuolar proteins (Rothman, et al. 1989). This large number of mislocalization mutants suggests that the vacuolar protein localization process is complex, presumably including specific steps such as, recognition of vacuolar protein sorting signals, packaging of the proteins into transport vesicles, delivery and fusion of these vesicles with the vacuole, release of vacuole constituents and the subsequent recycling of some transport components.

The *vps* mutants have been categorized into three basic classes based on vacuole morphology studies (Banta, et al. 1988). Class A *vps* mutants exhibit a nearly normal wild-type vacuole morphology, containing several (3-5) clustered vacuolar compartments. Class B mutants lack any of the large vacuolar structures seen in the class A mutants, but rather accumulate many smaller vacuole-like compartments (>20). Class C mutants lack any obvious vacuole-like structures. Instead, the class C mutants accumulate abnormal membrane enclosed compartments, as well as smaller 80nm vesicles. These observations suggest that not only may the protein sorting machinery be affected in these mutants, but functions required for vacuole biogenesis and/or vacuole segregation may also be perturbed (e.g. class B and class C).

## The Role of a Protein Kinase and Lipid Kinase in Vacuolar Protein Sorting.

The *VPS*15 and *VPS*34 gene products serve an essential role in the sorting of soluble proteins to the vacuole. Mutations in these two genes result in a number of similar phenotypes that distinguish them from the other class A *vps* mutants (Herman and Emr 1990; Herman, et al. 1991a). Both *vps*15 and *vps*34 mutant cells show an extreme sorting defect for soluble vacuolar proteins, but the membrane associated vacuolar protein, alkaline phosphatase, is largely unaffected. Although these two genes are not essential for vegetative growth, cells carrying chromosomal deletions of either *VPS*15 or *VPS*34 result in a temperature sensitive growth phenotype, an increased sensitivity to osmotic stress and a defect in vacuole segregation or inheritance. These mutant phenotypes (with the exception of correctly localizing vacuolar membrane proteins) are shared by all of the more severe class C mutants, even though *vps*15 and *vps*34 mutants contain an apparently normal vacuole structure (Banta, et al. 1988; Herman and Emr 1990; Herman, et al. 1991a). These unique phenotypes suggest that these two mutants may disrupt a similar crucial step in the vacuolar protein sorting process.

The *VPS*15 gene encodes a 1455 amino acid phosphoprotein that is peripherally associated with the cytoplasmic face of a late Golgi compartment or membrane vesicle (Herman, et al. 1991a). Amino acid sequence comparisons has revealed that the Vps15 protein (Vps15p) shares striking similarity to the family of serine/threonine-specific protein kinases. This similarity is clustered in the highly conserved kinase catalytic domains of these proteins, indicating that Vps15p is itself a protein kinase and that protein phosphorylation may play an important role in the vacuolar protein sorting process (Herman, et al. 1991a). In support of this conclusion, cells carrying point mutations in the *VPS*15 kinase domain known to disrupt kinase activity in other protein kinases, result in severe vacuolar protein sorting defects.

Examination of the phosphorylation state of Vps15p kinase domain mutants closely correlates with the severity of the vacuolar protein sorting defects (Herman, et al. 1991b). Changes in critical residues for kinase function not only inactivate the *VPS*15 gene product in the sorting process but also result in a marked decrease in the in vivo phosphorylation state of Vps15p. Conversely, mutations of less conserved and critical amino acid residues in the kinase domain have a less profound affect on Vps15p phosphorylation and also exhibit minimal sorting defects. These observations suggest that a functional Vps15p kinase domain is required for Vps15p phosphorylation and that the phosphorylation state of the *VPS*15 gene product is closely coupled to its activity in the vacuolar protein sorting process. Though direct biochemical demonstration of Vps15p kinase activity has not been possible, the genetic analysis of this kinase homologue strongly supports the role of Vps15p kinase activity in this protein localization process.

In addition to the kinase domain point mutants, a small truncation of Vps15p has demonstrated the intimate role of Vps15p function in the sorting process (Herman, et al. 1991b). Deletion of the C-terminal 30 amino acids of Vps15p results in a temperature-conditional defect in the delivery of soluble vacuolar proteins. Within 60 sec of a shift to nonpermissive temperature, cells carrying this *vps*15 mutant allele cease to deliver CPY to the vacuole. In addition, the precursor CPY that is present at the time of this temperature shift is not immediately secreted from the cell, but accumulates as the Golgi modified form. Shortly after the reversal of the temperature block, the retained precursor CPY is rapidly processed to its mature form, indicating delivery to the vacuole. It seems reasonable to suggest that the precursor CPY retained by these mutant cells may define a prevacuolar-endosomal compartment similar to that described in mammalian cells. Unfortunately, the exact nature of this compartment has yet to be defined.

Vps15p is also modified with myristic acid at its N-terminus (Herman, et al. 1991b). This 14 carbon fatty acid has been shown to play an important role in conferring the ability of proteins to associate with cellular membranes. However, mutations that prevent myristoylation of Vps15p do not appear to effect the association of Vps15p with cell membranes. Therefore, the myristic acid modification on Vps15p alone cannot be the only determinant for this protein's association with cellular membranes. Presumably some other component (i.e. a transmembrane protein) helps mediate this association.

The other member of this unique *VPS* gene subgroup, *VPS*34, encodes a protein of 875 amino acids (Herman and Emr 1990). Like Vps15p, the *VPS*34 protein (Vps34p) is a phosphoprotein, a portion of which is associated with a membrane cell fraction (Stack, J. and S. Emr, unpublished observations). Amino acid sequence comparison analysis indicates that the *VPS*34 gene product shares significant similarity (55% similarity, 33% identity) with the catalytic subunit of mammalian phophatidylinositol-3-kinase (PI3-kinase) (Hiles, et al. 1992). Mammalian PI3-kinase activity is thought to be involved in the second messenger cascade responsible for regulation of cell growth (Cantley, et al. 1991). In fact, PI3-kinase activity has been detected in wild-type yeast cells and this activity is markedly reduced in yeast cells carrying a disrupted allele of the *VPS*34 gene (Schu, P., K. Takegawa, J. Stack and S. Emr, unpublished observations). When the *VPS*34 gene product is overproduced, a concomitant increase in the PI3-kinase activity is also observed. These results indicate that the *VPS*34 gene may encode the yeast homologue of mammalian PI3-kinase. This raises the intriguing possibility that lipid phosphorylation may in some way regulate the delivery of proteins to the yeast vacuole.

The unique phenotypes associated with mutations in the *VPS*15 and *VPS*34 genes indicate that these two gene products may functional interact. This appears to be the case. Overexpression of Vps34p in cells carrying *vps*15 kinase domain mutants results in partial or complete suppression of all mutant phenotypes (Herman, et al. 1991b). This suppression is not observed in cells carrying a *vps*15 deletion. Recently, it has been shown that, Vps15p and Vps34p can be coimmunoprecipitated in wild-type yeast cell extracts and the two gene products can be chemically cross-linked (Stack, J., P. Herman and S. Emr. unpublished observations.). Therefore, the Vps15p and Vps34p are part of a protein complex that facilitates a similar critical step(s) in the vacuolar protein localization process.

The studies of *VPS*15 and *VPS*34 genes and gene products indicate that both a protein kinase and a lipid kinase participate in a coordinated fashion to deliver proteins to the yeast vacuole (Figure 2). It is not yet known how the

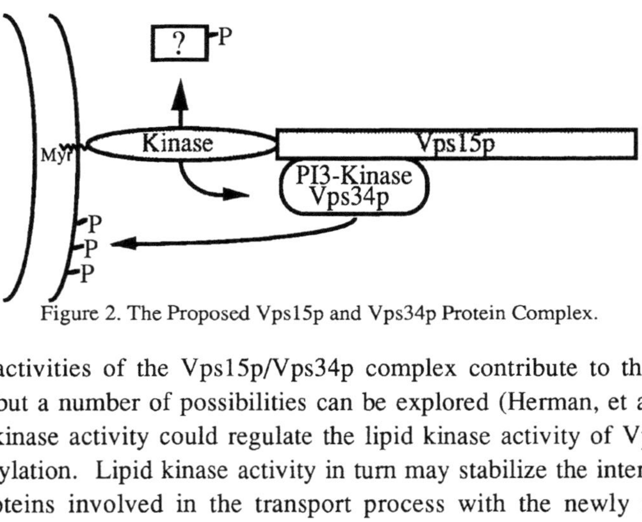


Figure 2. The Proposed Vps15p and Vps34p Protein Complex.

enzyme activities of the Vps15p/Vps34p complex contribute to the sorting process, but a number of possibilities can be explored (Herman, et al. 1992). Vps15p kinase activity could regulate the lipid kinase activity of Vps34p by phosphorylation. Lipid kinase activity in turn may stabilize the interaction of other proteins involved in the transport process with the newly modified lipids. The initiation of this process would be the activation of Vps15p kinase activity. One intriguing possibility is that the activation of the Vps15p could be mediated by an interaction with a transmembrane receptor molecule (i.e. CPY receptor) Therefore, Vps15p interaction with an occupied vacuolar protein sorting receptor could initiate a phosphorylation cascade (including the activation of Vps34p) required in the localization process. We are presently testing this and other possible models.

**Alternative Pathways for Vacuolar Protein Sorting.**

Studying the vacuolar protein sorting properties of *vps*15 and *vps*34 and other *vps* mutants, suggest that the localization of soluble and membrane associated vacuolar proteins involve distinct targeting mechanisms. In fact, the targeting of soluble vacuolar proteins themselves may be mediated by distinct components. Our analysis of the vacuolar protein sorting defects in one of the *vps* mutants, *Δvps35,* indicates that Vps35p function may only be required for the sorting of a small subset of vacuolar proteins (Paravicini, et al. 1992). Even though *Δvps35* cells contain a morphologically wild-type vacuole (class A), CPY sorting is almost completely defective in the *vps35* null mutant; >95% of the protein is secreted as its Golgi modified p2 precursor form. However, the vast majority of two other soluble proteases, PrA and PrB, is retained and matured in *Δvps35* mutant cells. This result indicates that CPY sorting is completely dependent on the presence of Vps35p. Yet, the sorting and processing of other soluble proteases (PrA and PrB), as well as the vacuolar membrane protein, ALP, primarily occurs in a Vps35p-independent manner. Apparently, only a subset of vacuolar proteins, like CPY, depends on Vps35p for efficient sorting to the vacuole.

One possible model we are presently exploring is that, both CPY and PrA use the same transport system (vesicle carriers) to mediate their Golgi to vacuole delivery (disrupted in the *vps*15 and *vps*34 mutants), but may utilize different receptors for their selective packaging into these common carriers. One receptor complex may recognize CPY, whereas a second receptor complex may recognize PrA and PrB. Though Vps35p does not resemble a receptor molecule, it may play a role in the recognition, modification or packaging of the CPY-specific receptor complex. Loss of Vps35p function would inactivate the CPY-specific receptor and result in default CPY secretion. PrA, PrB and ALP localization would largely be unaffected in this model, as their delivery would depend on a different receptor complex(es). Present data is most consistent with this model. Direct proof for the role of multiple receptor complexes in this protein sorting process will require the identification of these complexes. Unfortunately, no good candidate for a receptor molecule has yet been identified among the presently characterized *VPS* gene products. However, further characterization of two other *vps* mutants that share the differential sorting defect observed in *vps*35 mutants is underway. Identification and characterization of a vacuolar protein sorting receptor complex(es) will be essential to test the proposed role of Vps15p and Vps34p, as well as other *VPS* gene product functions in this complex sorting process.

## References.


Bankaitis, V. A., L. M. Johnson and S. D. Emr. (1986). Isolation of yeast mutants defective in protein targeting to the vacuole. Proc. Natl. Acad. Sci. USA. 83: 9075-9079.

Banta, L. M., J. S. Robinson, D. J. Klionsky and S. D. Emr. (1988). Organelle assembly in yeast: characterization of yeast mutants defective in vacuolar biogenesis and protein sorting. J. Cell Biol. 107: 1369-1383.

Cantley, L.C., K.R. Auger, C. Carpenter, B. Duckworth, A. Graziani, R. Kapeller, S. Soltoff. (1991). Oncogenes and signal transduction. Cell 64: 281-302.

Eakle, K. A., M. Bernstein, and S. D. Emr. (1988) Characterization of a component of the yeast secretion machinery; identification of the *SEC18* gene product. Mol. Cell. Biol. 8: 4098-4109

Graham, T. R. and S. D. Emr. (1991). Compartmental organization of Golgi-specific protein modification and vacuolar protein sorting events defined in a yeast *sec*18 (NSF) mutant. J. Cell Biol. 114:207-218.

Herman, P., J. H. Stack and S. Emr. (1992). Essential role for a protein kinase/phosphatidylinositol kinase complex in secretory protein sorting. Trends in Cell Biology. In press.

Herman, P. K. and S. D. Emr. (1990). Characterization of *VPS*34, a gene required for vacuolar protein sorting and vacuole segregation in *Saccharomyces cerevisiae*. Mol. Cell. Biol. 10: 6742-6754.

Herman, P. K., J. H. Stack, J. A. DeModena and S. D. Emr. (1991a). A novel protein kinase homolog essential for protein sorting to the yeast vacuole. Cell. 64: 425-437.

Herman, P. K., J. H. Stack and S. D. Emr. (1991b). A genetic and structural analysis of the yeast Vps15 protein kinase: evidence for a direct role of Vps15p in vacuolar protein delivery. EMBO J. 10: 4049-4060.

Hiles, I.D., M. Otsu, S. Volinia, M. Fry, I. Gout, R. Dhand, G. Panayotou, F. Ruiz-Larrea, A. Thompson, N.F. Totty, J. Hsuan, A.A. Courtneldge, P.J. Paker, M.D. Waterfield. (1992). Phosphatidylinositol 3-kinase: structure and expresion of the 110 kd catalytic subunit. Cell. In press.

Johnson, L. M., V. A. Bankaitis and S. D. Emr. (1987). Distinct sequence determinants direct intracellular sorting and modification of a yeast vacuolar protease. Cell. 48: 875-885.

Jones, E. W. (1977). Proteinase mutants of *Saccharomyces cerevisiae*. Genetics. 85: 23-33.

Kaplan, A., D. T. Achord and W. S. Sly. (1977). Phosphohexosyl components of a lysosomal enzyme are recognized by pinocytosis receptors on human fibroblasts. Proc. Natl. Acad. Sci. USA. 74: 2026-2030.

Klionsky, D. J., L. M. Banta and S. D. Emr. (1988). Intracellular sorting and processing of a yeast vacuolar hydrolase: proteinase A propeptide contains vacuolar targeting information. Mol. Cell. Biol. 8: 2105-2116.

Klionsky, D. J., P. K. Herman and S. D. Emr. (1990). The fungal vacuole: composition, function and biogenesis. Microbiol. Rev. 54: 266-292.

Malhotra, V., L. Orci, B. S. Glick, M. R. Block and J. E. Rothman. (1988). Role of an N-ethylmaleimide-sensitive transport component in promoting fusion of transport vesicles with cisternae of the Golgi stack. Cell. 54: 221-227.

Novick, P., S. Ferro and R. Schekman. (1981). Order of events in the yeast secretory pathway. Cell. 25: 461-469.

Paravicini, G., B. F. Horazdovsky and S. D. Emr. (1992). Alternative pathways for the sorting of soluble vacuolar proteins in yeast: a *vps*35 null mutant missorts and secretes only a subset of vacuolar hydrolases. Molecular Biology of the Cell. 3: 415-427.

Robinson, J. S., D. J. Klionsky, L. M. Banta and S. D. Emr. (1988). Protein sorting in *Saccharomyces cerevisiae*: isolation of mutants defective in the delivery and processing of multiple vacuolar hydrolases. Mol. Cell. Biol. 8: 4936-4948.

Rothman, J. H., I. Howald and T. H. Stevens. (1989). Characterization of genes required for protein sorting and vacuolar function in the yeast *Saccharomyces cerevisiae*. EMBO J. 8: 2057-2065.

Rothman, J. H. and T. H. Stevens. (1986). Protein sorting in yeast: mutants defective in vacuole biogenesis mislocalize vacuolar proteins into the late secretory pathway. Cell. 47: 1041-1051.

Stevens, T., B. Esmon and R. Schekman. (1982). Early stages in the yeast secretory pathway are required for transport of carboxypeptidase Y to the vacuole. Cell. 30: 439-448.

Valls, L. A., C. P. Hunter, J. H. Rothman and T. H. Stevens. (1987). Protein sorting in yeast: the localization determinant of yeast vacuolar carboxypeptidase Y resides in the propeptide. Cell. 48: 887-897.

von Figura, K. and A. Hasilik. (1986). Lysosomal enzymes and their receptors. Ann. Rev. Biochem. 55: 167-193.

Wilson, D. W., C. A. Wilcox, G. C. Flynn, E. Chen, W.-J. Kuang, W. J. Henzel, M. R. Block, A. Ulrich and J. E. Rothman. (1989). A fusion protein required for vesicle-mediated transport in both mammalian cells and yeast. Nature. 339: 355-359.

# EUKARYOTIC MDR1/P-GLYCOPROTEIN HOMOLOGUES: UNCONVENTIONAL SECRETION PROCESSES MEDIATED BY A GROWING FAMILY OF ATP-DEPENDENT MEMBRANE TRANSLOCATORS

Mikael Göransson and Jeremy Thorner
Division of Biochemistry and Molecular Biology
Department of Molecular and Cell Biology
University of California
Berkeley, California 94720
USA

## Introduction

Mdr1/P-glycoprotein, when overexpressed, can confer resistance of human tumor cells to multiple lipophilic cytotoxic drugs and this resistance is a major problem in cancer chemotherapy. When mouse and human *mdr* cDNAs were cloned and sequenced (10,17), one of the intriguing findings was that they showed homology to a family of bacterial active transport systems, including for example the *E. coli hlyB* gene, whose product is a membrane protein involved in the export of a hemolysin. This observation was the start of what has since been a very active and fruitful area of research, and in the past few years many Mdr/P-glycoprotein homologues have been identified in eukaryotes.

The Mdr-like proteins share a proposed secondary structure, composed of two quasi-homologous halves, each comprised of six membrane-spanning domains and a cytoplasmic domain containing the consensus sequence for a nucleotide binding fold (*e.g.* Mdr1, Ste6 in *S. cerevisiae*, and Pgh1 in *P. falciparum*). Some gene products are "half-molecules", and contain only one set of six transmembrane segments and only one nucleotide-binding fold; the nucleotide-binding fold can be situated at the C-terminus, as is seen for Pmp70 in rats, or at the N-terminus, which is the case for the White and Brown proteins of *D. melanogaster*. Sequence homology between the different proteins is largely confined to the nucleotide-binding folds, where, over a stretch of about 200 amino acids, prokaryotic as well as eukaryotic members show a pronounced relatedness. Generally, it is thought that these membrane proteins transport their substrates, either to the exterior, or into membrane-enclosed compartments within the cell, and that the energy for this process is derived from ATP hydrolysis. Hence, these proteins are also known as traffic ATPases or ABC (*A*TP-*B*inding-*C*assette) transporters. Support for this functional model comes mainly from studies in prokaryotes, although findings with proteins such as Ste6 and Mdr1 have contributed significantly.

NATO ASI Series, Vol. H 71
Protein Synthesis and Targeting in Yeast
Edited by A. J. P. Brown, M. F. Tuite, and J. E. G. McCarthy

Many eukaryotic ABC-transporters have been identified either because of their homology to the *mdr*-genes, or because of the phenotype they confer when overexpressed, which typically means increased resistance to one or a number of cytotoxic compounds, often hydrophobic in nature. In most instances, however, little is known about the function of these transport proteins under normal physiological conditions. This situation contrasts to that in prokaryotes, where proteins belonging to this class are known to transport compounds as diverse as inorganic ions, sugars, polysaccharides, amino acids, oligopeptides, and large proteins (20). We will focus the remaining part of this contribution on a few, selected eukaryotic ABC-transporters, with a certain emphasis on *S. cerevisiae* and the *STE6* gene product.

## Functional aspects of the Mdr proteins, the prototypic eukaryotic ABC-transporters

The mammalian *mdr* gene products were the first ABC-transporters to be identified in eukaryotes. They have been studied extensively, and predictions about new gene products belonging to this family are often modelled upon what we know about the topology and the function of the Mdr proteins. The prototypic example, Mdr1, is 1280 amino acids in humans (10). The gene can be found amplified in tumour cells, or *in vitro* in cell lines, after selection with a single lipophilic drug. Even when selected initially by one agent, the overexpressed plasma membrane-associated Mdr1 transporter can pump a variety of other hydrophobic, but otherwise structurally unrelated, compounds (*e.g.* doxorubicin, vinblastine, etoposide, actinomycin D, taxol, colchicine, valinomycin) out of cells. The drug extrusion process is believed to derive energy from ATP hydrolysis. Mdr1 has been shown to bind ATP-analogs *in vitro* (3), and the purified protein has ATPase activity (18). Moreover, Mdr1-dependent transport of vinblastine requires ATP in an *in vitro* vesicle import system (21). Also, the proposed membrane disposition of Mdr1 has been corroborated to some extent (44).

The substrate profile of Mdr1 is very broad. Despite this fact, a close homologue to Mdr1, called Mdr2, has never been seen to mediate multiple drug resistance. The two proteins show about 85 % homology at the amino acid level. The mouse genome carries an additional gene, *mdr3*, which is more closely related (92 % homology) to *mdr1* than is *mdr2*. Mdr3 overexpression will confer a drug-resistant phenotype, with a substrate profile that largely overlaps the one seen for Mdr1, but the two proteins have different relative affinities for the various substrates (12).

Despite this overall picture, much remains to be elucidated about the function of Mdr.

The transport mechanism(s) need to be explained, and the normal physiological substrates of the Mdr-transporters need to be identified. Two recent reports have added intriguing aspects to this difficult question. One report shows that expression of Mdr1 correlates with the activity of ATP-dependent, cell volume-regulated chloride-selective channels, the most straightforward interpretation being that Mdr1 is itself a chloride channel, or at least a component thereof (39). The other report strongly suggests that Mdr1, when overexpressed, can efficiently extrude a short linear peptide, N-acetyl-leucyl-leucyl-norleucinal, thus expanding the possible range of substrates to include peptides, and maybe even proteins (36).

Human Mdr1 shows a relatively close relationship to the *S. cerevisiae* **a**-mating factor transporter, Ste6 (see below). When *mdr1* was expressed in yeast, most of it was correctly localized to the plasma membrane, it could bind an ATP-analogue, and overexpression resulted in increased resistance towards valinomycin, a compound known to interact with Mdr1 in animal cells. All in all, Mdr1 seemed to be properly folded. However, the particular *mdr1* allele used, which carries an amino acid substitution at position 185, could not functionally complement a yeast strain for the loss of Ste6 function (25). A later report demonstrated that expression of the mouse *mdr3* gene in a yeast *ste6* mutant could restore the ability to export **a**-factor, and consequently also the ability to mate. In addition, a mutant form of *mdr3*, impaired in its transport capacity, was unable to complement the *ste6* mutation (33). These results again argue that the function of some Mdr-like proteins in normal mammalian cells could be to conduct the transmembrane export of endogenous proteins and peptides. These results also suggest that *S. cerevisiae* could be a powerful model system for structure-function studies of at least some members of the Mdr-like protein family.

## The cystic fibrosis gene product, CFTR, is a member of the Mdr transporter family

The cloning and sequencing of the cystic fibrosis gene revealed that the predicted gene product, the cystic fibrosis transmembrane conductance regulator (CFTR), is clearly a *mdr*-homologue (34). However, the CFTR protein is larger than human Mdr1, 1480 *vs* 1280 amino acids, and contains in its middle, after the first nucleotide binding fold, the so-called R (regulator) domain. This domain is 241 amino acids long, highly charged, and contains multiple potential phosphorylation sites for protein kinases A and C (34). The protein has since been intensely studied, and so has its involvement in the various pathophysiological aspects of the disease. With respect to the transport functions of CFTR, it has been established that it is a cAMP regulated low-conductance $Cl^-$ channel, a conclusion reached after CFTR had been purified to homogeneity and incorporated into lipid bilayers to enable ion channel activity measurements (5). There are, however, other aspects to CFTR, indicating for instance that CFTR is critical for cAMP-dependent regulation of membrane recycling in epithelial tissues (7), and that CFTR may regulate the activity of another ion channel (14).

ABC-transport proteins involved in intracellular transport processes

Many ABC-transporters have been found to reside in, or are thought to reside in, the cytoplasmic membrane. However, recent evidence suggests that these proteins can also participate in intracellular protein trafficking, and perhaps in the biogenesis of certain organelles.

One of the major integral membrane proteins of rat liver peroxisomes is a 70 kDa protein. cDNA clones for this protein, Pmp70, were isolated and sequenced, showing the deduced polypeptide to be an ABC-transporter with six membrane-spanning domains followed by a nucleotide-binding fold. Limited protease treatment of peroxisomes showed the ATP-binding domain to be exposed to the cytosol. Such protease-treated peroxisomes became incapable of *in vitro* import of acyl-CoA oxidase (22). Although Pmp70 may be part of a peroxisomal protein import machinery, this proposal needs to be further investigated.

The malaria parasite, *P. falciparum*, contains at least two genes, *pfmdr1* and *pfmdr2*, related to the *mdr* transporter genes (42). Pgh1, the product of the *pfmdr1* gene, is implicated in mefloquine-resistance as well as in low-level resistance to chloroquine (4,15), although this conclusion is under some debate (41). It has been suggested that it is a few specific alleles of *pfmdr1* that correlate with the chloroquine resistance phenotype (15), and that overexpression is not involved. High-level chloroquine resistance on the other hand, seems to involve *pfmdr1*-independent mechanisms (4). Pgh1 has been localized to the membrane of the digestive vacuole of the parasite, and again, the normal function of the protein remains unclear (11).

Major histocompatibility class (MHC) I molecules must associate with antigen-derived peptides in the endoplasmic reticulum before being transported through the secretory pathway to the cell surface. But it has long been a puzzle how the antigens, without signal sequence, can move into the ER. Excitingly, two putative peptide transporters, both clearly belonging to the ABC-transporter family, have been identified in the class II region of the MHC (30). Transfection experiments using the cloned genes show that they can restore antigen presentation in cell lines with defined lesions in the MHC II locus, supporting the proposed function. Both proteins are "half-molecules", with the same predicted topology as Pmp70, and they are both needed simultaneously, probably associating to form a heterodimeric complex in the membrane (23,37).

## ABC-transporters in *Saccharomyces cerevisiae*

In the pathway leading to diploid cell formation, haploid yeast cells communicate by virtue of production of peptide hormones, also called mating pheromones. *MATα* cells produce α-factor, and *MAT***a** cells produce a peptide hormone called **a**-factor. α-factor secretion has been well studied, and is known to proceed through the normal secretory pathway (16). Several observations indicated that **a**-factor used a different export route: the **a**-factor precursors are short and lack a typical hydrophobic signal peptide (8); they are devoid of sites for attachment of Asn-linked oligosaccharides (8); protease susceptibility experiments show that **a**-factor is not found in a membrane-bound secretory compartment (38); and, most importantly, secretion-defective (*sec*) strains, unable to secrete α-factor, continue to export **a**-factor (38). Finally, the C-terminal cysteine residue of **a**-factor is both farnesylated, via a thioether linkage, and methyl esterified at the carboxylate group (1). These modifications suggest that **a**-factor might be targeted to the plasma membrane.

How, then, is **a**-factor exported? Mutational analysis has allowed the identification of a number of sterile (*STE*) genes. The phenotype of such mutants is that they are mating-deficient. A certain sub-class causing *MAT***a** sterility was very interesting in this context. These mutants respond normally to α-factor, and transcribe the **a**-factor structural genes *MF***a***1* and *MF***a***2* normally, yet are unable to secrete **a**-factor (8). Among these, the *STE6* gene was chosen for analysis. A *ste6* mutant was shown to produce normal amounts of intracellular **a**-factor, but released little or no extracellular pheromone, indicating a direct role for Ste6 in **a**-factor export. Moreover, experiments involving overexpression of Ste6, and its substrate, **a**-factor, showed the *STE6* gene product to be able to increase both the rate and the extent of **a**-factor export. The ensuing DNA sequence analysis of the *STE6* gene revealed that the predicted polypeptide, 1290 amino acids in length, was remarkably similar to the mammalian *mdr1* gene products. The homology is greatest in the ATP-binding region, where you find an amino acid identity of about 40 %, but readily extends over the entire coding region (24). Taken together, it was concluded that Ste6 is part of a novel export machinery, which secretes the **a**-factor mating pheromone by a mechanism distinct and independent from the classical secretory pathway (24). Thus, not only is Ste6 the first ABC-transporter to be identified in *Saccharomyces cerevisiae*, but it is also the first eukaryotic member in this class for which the normal, physiological substrate has been unambiguously identified, and it is the first eukaryotic ABC-transporter that has been shown to export a polypeptide.

Further biochemical analysis has largely corroborated the predictions made by these findings. The Ste6 protein is found to be 145 kDa, in good agreement with the proposed molecular weight as deduced from the nucleotide sequence. It is found in the plasma membrane, and has the solubilization properties of an integral membrane protein. It can be photoaffinity-labeled with ATP-analogues. Upon treatment of *MAT***a** cells with α-factor, production of Ste6 is induced, and the protein is now found along the pronounced projection ("shmoo tip"), that is formed in pheromone-treated cells, and at which an **a** cell fuses with an α cell partner. This behaviour, seen also for the pheromone receptors, Ste2 and Ste3, as well as for a number of other proteins that perform crucial tasks in the mating process, agrees well with the proposed function for Ste6 as the **a**-factor exporter (27).

As is the case of Mdr1, Ste6 is composed of two homologous halves, each with six membrane-spanning domains and a nucleotide-binding fold. To assess the role of each half in **a**-factor export, the coding sequence was manipulated so that each half could be expressed as a separate polypeptide. When expressed separately in *ste6*Δ cells, each "half-molecule" was unable to support export on its own; in contrast, when both halves were simultaneously expressed in the same cell, **a**-factor export was restored, suggesting that association of the two halves in the membrane reconstituted the functional transporter. This observation was interpreted to mean that both halves of the native Ste6 protein are required for and contribute to its transport function (6). Although the C-terminal half appeared to be relatively stable when expressed alone, the stability of the N-terminal half alone was not examined in this study. Thus, it is possible that the N-terminal half alone could be competent to transport **a**-factor, but that its apparent lack of function is merely due to its rapid degradation. Thus, the transport activity observed when both halves were co-expressed might simply be due to stabilization of the N-terminal half by the C-terminal half, even if the C-terminal half does not play a direct role in the transport mechanism.

With regard to the number of ABC-transporters found in the genome, yeast is now catching up with other organisms. *ADP1* was found as part of the European sequencing project of chromosome III. The deduced protein is basically a "half-molecule", but with a slightly unusual membrane topology. The best homology is seen when compared to the *D. melanogaster* White and Brown proteins. No function has as yet been assigned to it (32). *PDR5* was isolated on the basis that it can cause resistance to cycloheximide when overexpressed. Simultaneous cross-resistance to an unrelated inhibitor, sulfometuron methyl, was observed; *PDR5* thus mediates one of the hallmark features of multiple drug resistance transporters (28). The DNA sequence shows Pdr5 to be a 1511-residue ABC-transporter, again with the highest homology to the White and Brown proteins of *D. melanogaster* (40).

We have looked for additional *mdr*-like genes in *S. cerevisiae*, using the polymerase chain reaction method (PCR) and primers derived from highly conserved regions in the nucleotide-binding folds of the ABC-transporters. We have identified two new genes, tentatively designated *SSH1* and *SSH2*. *SSH1* shows highest homology towards the *mdr*-genes, and *SSH2* is most similar to the Pmp70 cDNA. We are currently characterizing these genes, with the aim of understanding what functions the gene products perform *in vivo*.

## Conclusions and perspectives

ABC-transporters have turned out to be considerably more abundant in eukaryotes than first anticipated. We have here discussed a few well-studied members of this protein superfamily. Some of them are of obvious medical importance, such as: the cystic fibrosis gene product, CFTR; the Pgh1 protein of *P. falciparum*, which is involved in the resistance of parasites to chloroquine, and likely impedes the successful treatment of malaria; and, the Mdr1 protein in humans, which when overexpressed will allow tumor cells to escape from the effects of an array of structurally unrelated cytotoxic drugs. Other examples could be added. *Leishmania* species contain at least three *mdr*-like genes (19,31). Overexpression of two of these correlates with drug resistance, and in one case the pattern seen is virtually the same as for multiple drug resistance caused by *mdr1* overexpression in mammalian cells (19). It is of obvious importance to understand the physiological functions, as well as the mechanism(s) of substrate recognition and transport used by these proteins. This task is intriguing, but complicated by the substantial size of the proteins, as well as by the possibility that the transporters might be bi- or multifunctional. Model systems, such as *S. cerevisiae*, and also *D. melanogaster* where so far five *mdr*-homologues have been found (13,43), should be useful for experimental approaches to these questions. The work described on *STE6*, and the fact that mammalian Mdr-proteins have been functionally expressed in yeast, exemplifies the utility of the yeast system.

Prokaryotic ABC-transporters will transport substrates ranging from ions to large proteins (20), and there is no reason why their eukaryotic counterparts should be less versatile. Ste6 was the first eukaryotic ABC-transporter shown to transport a peptide. However, the secretion of a number of proteins from animal cells does not seem to rely upon the normal secretory pathway, but instead there seem to be striking parallels to **a**-factor export (26). The interleukins IL-1α and IL-1β will serve as an example. Like for yeast **a**-factor, it is the carboxy-terminal half of the precursors that encode the mature protein. No obvious signal sequence is found within the precursors (2,29). The interleukins are not glycosylated, but the precursors are modified by myristylation (9), which could lead to membrane association. In agreement with this, these cytokines are mainly found in the cytoplasm after their synthesis. A detailed study of IL-1β secretion showed that drugs which block transport of proteins along the classical secretory pathway cause, if anything, an increase in secretion of this interleukin (35). The argument could be developed further, and similar cases presented for other proteins, such as the acidic and basic fibroblast growth factors, but our point is that these proteins are candidate substrates for hitherto unidentified ABC-transporters. Whether or not this will turn out to be true, the next couple of years promise to yield fascinating new insights about the role of eukaryotic ABC-transporters, as well as about their mechanism of action.

References


1. Anderegg RJ, Betz R, Carr SA, Crabb JW, Duntze W (1988) Structure of *Saccharomyces cerevisiae* mating hormone **a**-factor. Identification of S-farnesyl cysteine as a structural component. J Biol Chem 263:18236-18240
2. Auron PE, Webb AC, Rosenwasser LJ, Mucci SF, Rich A, Wolff SM, Dinarello CA (1984) Nucleotide sequence of human monocyte interleukin 1 precursor cDNA. Proc Natl Acad Sci USA 81:7907-7911
3. Azzaria M, Schurr E, Gros P (1989) Discrete mutations introduced in the predicted nucleotide-binding sites of the *mdr1* gene abolish its ability to confer multidrug resistance. Mol Cell Biol 9:5289-5297
4. Barnes DA, Foote SJ, Galatis D, Kemp DJ, Cowman AF (1992) Selection for high-level chloroquine resistance results in deamplification of the *pfmdr1* gene and increased sensitivity to mefloquine in *Plasmodium falciparum*. EMBO J 11:3067-3075
5. Bear CE, Li C, Kartner N, Bridges RJ, Jensen TJ, Ramjeesingh M, Riordan JR (1992) Purification and functional reconstitution of the cystic fibrosis transmembrane conductance regulator (CFTR). Cell 68:809-818
6. Berkower C, Michaelis S (1991) Mutational analysis of the yeast a-factor transporter STE6, a member of the ATP binding cassette (ABC) protein superfamily. EMBO J 10:3777-3785
7. Bradbury NA, Jilling T, Berta G, Sorscher EJ, Bridges RJ, Kirk KL (1992) Regulation of plasma membrane recycling by CFTR. Science 256:530-532
8. Brake AJ, Brenner C, Najarian R, Laybourn P, Merryweather J (1985) Structure of

genes encoding precursors of the yeast peptide mating pheromone a-factor. In Gething MJ (ed) *Protein Transport and Secretion*. Cold Spring Harbor Laboratory Press, Cold Spring Harbor, NY, pp. 103-108
9. Bursten SL, Locksley RM, Ryan JL, Lovett DH (1988) Acylation of monocyte and glomerular mesangial cell proteins. Myristyl acylation of the interleukin 1 precursors. J Clin Invest 82:1479-1488
10. Chen C, Chin JE, Ueda K, Clark DP, Pastan I, Gottesman MM, Roninson IB (1986) Internal duplication and homology with bacterial transport proteins in the *mdr1* (P-glycoprotein) gene from multidrug-resistant human cells. Cell 47:381-389
11. Cowman AF, Karcz S, Galatis D, Culvenor JG (1991) A P-glycoprotein homologue of *Plasmodium falciparum* is localized on the digestive vacuole. J Cell Biol 113:1033-1042
12. Devault A, Gros P (1990) Two members of the mouse *mdr* gene family confer multidrug resistance with overlapping but distinct drug specificities. Mol Cell Biol 10:1652-1663
13. Dreesen TM, Johnson DH, Henikoff S (1988) The Brown protein of *Drosophila melanogaster* is similar to the White protein and to components of active transport complexes. Mol Cell Biol 8:5206-5215
14. Egan M, Flotte T, Afione S, Solow R, Zeitlin PL, Carter BJ, Guggino WB (1992) Defective regulation of outwardly rectifying $Cl^-$ channels by protein kinase A corrected by insertion of CFTR. Nature 358:581-584
15. Foote SJ, Kyle DE, Martin RK, Oduola AMJ, Forsyth K, Kemp DJ, Cowman AF (1990) Several alleles of the multidrug-resistance gene are closely linked to chloroquine resistance in *Plasmodium falciparum*. Nature 345:255-258
16. Fuller RS, Sterne RE, Thorner J (1988) Enzymes required for yeast prohormone processing. Ann Rev Physiol 50:345-362
17. Gros P, Croop J, Housman D (1986) Mammalian multidrug resistance gene: complete cDNA sequence indicates strong homology to bacterial transport proteins. Cell 47:371-380
18. Hamada H, Tsuruo T (1988) Purification of the 170- to 180-kilodalton membrane glycoprotein associated with multidrug resistance. 170- to 180-kilodalton membrane glycoprotein is an ATPase. J Biol Chem 263:1454-1458
19. Henderson DM, Sifri CD, Rodgers M, Wirth DF, Hendrickson N, Ullman B (1992) Multidrug resistance in *Leishmania donovani* is conferred by amplification of a gene homologous to the mammalian *mdr1* gene. Mol Cell Biol 12:2855-2865
20. Higgins CF, Hyde SC, Mimmack MM, Gileadi U, Gill DR, Gallagher MP (1990) Binding protein-dependent transport systems. J Bioenerg Biomembr 22:571-592
21. Horio M, Gottesman MM, Pastan I (1988) ATP-dependent transport of vinblastine in vesicles from human multidrug-resistant cells. Proc Natl Acad Sci USA 85:3580-3584
22. Kamijo K, Taketani S, Yokota S, Osumi T, Hashimoto T (1990) The 70-kDa peroxisomal membrane protein is a member of the Mdr (P-glycoprotein)-related ATP-binding protein superfamily. J Biol Chem 265:4534-4540
23. Kelly A, Powis SH, Kerr LA, Mockridge I, Elliott T, Bastin J, Uchanska-Ziegler B, Ziegler A, Trowsdale J, Townsend A (1992) Assembly and function of the two ABC transporter proteins encoded in the human major histocompatibility complex. Nature 355:641-644
24. Kuchler K, Sterne RE, Thorner J (1989) *Saccharomyces cerevisiae STE6* gene product: a novel pathway for protein export in eukaryotic cells. EMBO J 8:3973-3984
25. Kuchler K, Thorner J (1992) Functional expression of human *mdr1* in the yeast *Saccharomyces cerevisiae*. Proc Natl Acad Sci USA 89:2302-2306
26. Kuchler K, Thorner J (1992) Secretion of peptides and proteins lacking hydrophobic signal sequences: The role of adenosine triphosphate-driven membrane translocators. Endocrine Reviews 13: in press
27. Kuchler K, Thorner J (1992) The a-factor transporter (*STE6* gene product) and cell polarity in the yeast *Saccharomyces cerevisiae*. J Cell Biol: in press

28. Leppert G, McDevitt R, Falco SC, Van Dyk TK, Ficke MB, Golin J (1990) Cloning by gene amplification of two loci conferring multiple drug resistance in Saccharomyces. Genetics 125:13-20
29. March CJ, Mosley B, Larsen A, Cerretti DP, Braedt G, Price V, Gillis S, Henney CS, Kronheim SR, Grabstein K, Conlon PJ, Hopp TP, Cosman D (1985) Cloning, sequence, and expression of two distinct human interleukin-1 complementary DNAs. Nature 315:641-647
30. Monaco JJ, Cho S, Attaya M (1990) Transport protein genes in the murine MHC: possible implications for antigen processing. Science 250:1723-1726
31. Ouellette M, Fase-Fowler F, Borst P (1990) The amplified H circle of methotrexate-resistant *Leishmania tarentolae* contains a novel P-glycoprotein gene. EMBO J 9:1027-1033
32. Purnelle B, Skala J, Goffeau A (1991) The product of the YCR105 gene located on the chromosome III from *Saccharomyces cerevisiae* presents homologies to ATP-dependent permeases. Yeast 7:867-872
33. Raymond M, Gros P, Whiteway M, Thomas DY (1992) Functional complementation of yeast *ste6* by a mammalian multidrug resistance *mdr* gene. Science 256:232-234
34. Riordan JR, Rommens JM, Kerem B-S, Alon N, Rozmahel R, Grzelczak Z, Zielenski J, Lok S, Plavsic N, Chou J-L, Drumm ML, Iannuzzi MC, Collins FS, Tsui L-C (1989) Identification of the cystic fibrosis gene: cloning and characterization of complementary DNA. Science 245:1066-1073
35. Rubartelli A, Cozzolino F, Talio M, Sitia R (1990) A novel secretory pathway for interleukin-1ß, a protein lacking a signal sequence. EMBO J 9:1503-1510
36. Sharma RC, Inoue S, Roitelman J, Schimke RT, Simoni RD (1992) Peptide transport by the multidrug resistance pump. J Biol Chem 267:5731-5734
37. Spies T, Cerundolo V, Colonna M, Cresswell P, Townsend A, DeMars R (1992) Presentation of viral antigen by MHC class I molecules is dependent on a putative peptide transporter heterodimer. Nature 355:644-646
38. Sterne RE (1989) A novel pathway for peptide hormone biogenesis: processing and secretion of the mating pheromone, a-factor, by *Saccharomyces cerevisiae*. Ph D Thesis, University of California, Berkeley
39. Valverde MA, Diaz M, Sepúlveda FV, Gill DR, Hyde SC, Higgins CF (1992) Volume-regulated chloride channels associated with the human multidrug-resistance P-glycoprotein. Nature 355:830-833
40. Wang M, Balzi E, Van Dyck L, Golin J, Goffeau A (1992) Sequencing of the yeast multidrug resistance *PDR5* gene encoding a putative pump for drug efflux. Abstract, 16th Int. Conf. on Yeast Genetics and Molecular Biology
41. Wellems TE, Panton LJ, Gluzman IY, do Rosario VE, Gwadz RW, Walker-Jonah A, Krogstad DJ (1990) Chloroquine resistance not linked to *mdr*-like genes in a *Plasmodium falciparum* cross. Nature 345:253-255
42. Wilson CM, Serrano AE, Wasley A, Bogenschutz MP, Shankar AH, Wirth DF (1989) Amplification of a gene related to mammalian mdr genes in drug-resistant *Plasmodium falciparum*. Science 244:1184-1186
43. Wu C-T, Budding M, Griffin MS, Croop JM (1991) Isolation and characterization of *Drosophila* multidrug resistance gene homologs. Mol Cell Biol 11:3940-3948
44. Yoshimura A, Kuwazuru Y, Sumizawa T, Ichikawa M, Ikeda S, Uda T, Akiyama S (1989) Cytoplasmic orientation and two-domain structure of the multidrug transporter, P-glycoprotein, demonstrated with sequence-specific antibodies. J Biol Chem 264:16282-16291

# THE ROLE OF PROTEIN DISULFIDE ISOMERASE IN YEAST

T. Mizunaga, H. Tachikawa and T. Miura
Department of Agricultural Chemistry
Faculty of Agriculture
The University of Tokyo
Bunkyo-ku,
Tokyo 113
Japan

## Introduction

Most disulfide-containing proteins are extracellular and most extracellular proteins contain disulfide bonds. Secretory enzymes, polypeptide hormones, immunoglobulines, and other serum proteins all have disulfide bonds. These extracellular and cell surface proteins are translated on membrane-bound ribosomes and concurrently translocated into the lumen of the endoplasmic reticulum (ER) in a largely unfolded state. Several lines of evidence suggest that disulfide formation is an early event in the processing of such proteins, occurring on nascent chains in some cases, and usually completed on polypeptides located in the rough ER. Very little is known about the formation of these disulfide bonds compared on post-translational modifications such as signal-peptide cleavage and glycosylation.

*In vitro* studies on the refolding of reduced, denatured bovine pancreatic ribonuclease and other proteins demonstrated that the enzyme protein disulfide isomerase (PDI) [EC 5.3.4.1] may be involved in thiol-oxidation and disulfide exchange reactions in the ER of mammalian cells. However, *in vitro* studies have been restricted to relatively small proteins, usually of about 10 to 15 kDa. Moreover, refolding of a

NATO ASI Series, Vol. H 71
Protein Synthesis and Targeting in Yeast
Edited by A. J. P. Brown, M. F. Tuite, and J. E. G. McCarthy

completely denatured molecule *in vitro* cannnot acculately reflect the *in vivo* process because of the serial way in which protein domains are synthesized.

Over the past few years it has become increasingly clear that protein transport from the ER is often critically dependent on protein folding (Pelham, 1989). Genetic and biochemical methods should be useful for analyzing the folding of secretory and membrane proteins *in vivo.* Many secretory mutants of the yeast *Saccharomyces cerevisiae* have been obtained and characterized (Novick et al.,1980) but the gene related to correct disulfide formation has not been isolated yet.

Studies on PDI in the yeast *Saccaromyces cerevisiae* sholud be useful in obtaining information on the function of PDI *in vivo* because yeast is eukaryotic organism that can be handled genetically. In this paper, we describe purification and some properties of yeast PDI, its ability to bind photoreactive peptides, cloning and sequencing of the PDI gene, and show that the enzyme is essential for yeast cell growth.

Purification and characterization of yeast PDI

PDI was purified 1850-fold to apparent homogeneity by five purification steps: 30-70 % ammonium sulfate fractionation, DEAE Toyopearl and Butyl Toyopearl chromatographies, and differential Phenyl-5PW HPLC with or without cystein. The yield was 4 %. From the calculation, the enzyme constituted about 0.05 % of the total cellular protein. In mammalian tissues, PDI is an abundant protein constituting about 0.35-0.79% of the total cellular proteins (Lambert and Freedman, 1983), indicating that its content in yeast cells is about one-tenth of that in mammalian cells. On SDS-PAGE, the purified preparation gave a single band staining with Coomassie Blue of *Mr* 70,000 (Fig 1,lane 2). Gel filtration chromatography on TSKgel G3000SW showed that the *Mr* of the native PDI is 140,000, indicating that yeast PDI is probably composed of two identical subunits. When yeast PDI was treated with Endo H, its *Mr* on SDS-PAGE shifted to 60,000, suggesting that the enzyme is N-glycosylated and that the sugar moiety accounts for about 10,000 for the PDI subunit. The migration of bovine PDI on SDS-

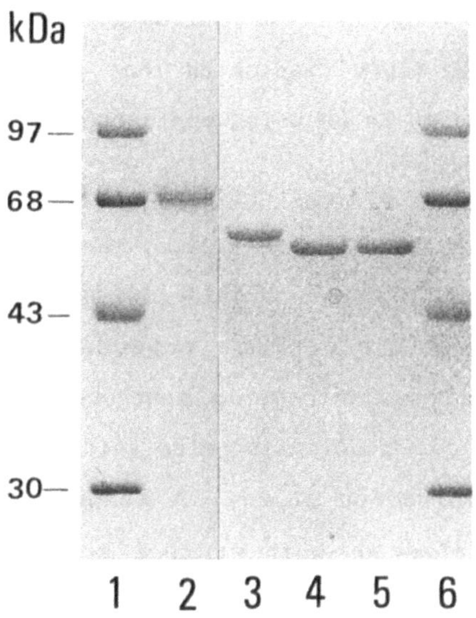


Fig. 1 SDS-PAGE analysis of purified yast and bovine PDIs. Lane 2, yeast PDI; lane 3, yeast PDI treated with Endo H; lane4, bovine PDI; lane 5, bovine PDI treated with Endo H. Lanes 1 and 6, molecular weight markers.

PAGE was not changed on Endo H treatment (lanes 4 and 5), consistent with the fact that less carbohydrate was detected in it. Native yeast and bovine PDIs had a pI of 4.15 and 4.2, respectively, indicating that both are acidic and molecular properties of yeast PDI are very similar to those of mammalian PDIs (Mizunaga et al., 1990).

Location and recycling of yeast PDI

PDI was found to be co-localized with NADPH-cytochrome C reductase known as an ER marker enzyme. The Golgi-specific $\alpha$1-6 mannose modification occurred in the SEC18-dependent manner (data not shown), indicating that PDI was translocated into an early Golgi compartment. These results suggest that yeast PDI, a native protein recycles from the Golgi compartment to the ER. Dean and Pelham (1990) indicated that luminal proteins of ER recycles from Golgi to ER using engineerd proteins.

Cloning and sequencing of yeast *PDI1* gene

For the isolation of yeast PDI gene, the purified yeast PDI was fragmented with cyanogen bromide or digested with lysyl endopeptidase, and the peptide fragments obtained were sequenced in an automated peptide sequencer. We synthesized oligonucleotides corresponding to amino acid sequence, and used them as the hybridization probes. A genomic DNA clone was isolated by hybridization with synthesized oligonucleotide probes. The DNA fragment was sequenced. A single open reading (ORF) that could encode a polypeptide of 522 amino acids was found (Fig. 2). The deduced amino acid sequence contains all of the sequnences of peptide sequences derived from purified PDI, confirming that the ORF encodes PDI. We thus designated the gene *PDI1* (for protein disulfide isomerase). The predicted amino acid sequence is related to those of PDI from mammalian sources (about 30 % identity). Yeast PDI also contains two WCGHCK sequences, presumed to be the PDI active sites (Edman et al., 1985), and the regions containing these sequences are the best conserved. In the N-terminal region, there is a hydrophobic sequence preceded by a positively charged amino acid, which has the feature of a signal peptide. In the C-

```
Yeast  MKFSAGAVLSWSSLLLASSVFAQQEAVAPEDSAVVKLATDSFNEYIQSHDLVLAEFFAPW
                                    *...*. * ...*.* ..*. .*.**.***
Rat         MLSRALLCLALAWAAARVGADALEEEDNVLVLKKSNFAE-PAAHNYLLVEFYAPW

  61'  CGHCKNMAPEYVKAAETL--VEKNITLAQIDCTENQDLCMEHNIPGFPSLKIFKNSDVNN
       *****..****.***..*   ...* **..* **. **   .....*.*..*.***.*...
  55"  CGHCKALAPEYAKAAAKLKAEGSEIRLAKVDATEESDLAQQYGVRGYPTIKFFKNGDTAS

 119'  SIDYEGPRTAEAIVQFMIKQSQPAVAVVADLPAYLANETFVTPVIVQSGKIDADFNATFY
       . .*.. *.*..**... *.. **.....* .*  .  .  . ...    * ... .*. .
 115"  PKEYTAGREADDIVNWLKKRTGPAATTLSDTAAAESLVDSSEVTVIGFFKDAGSDSAKQF

 179'  SMANKHFNDYDFVSAENADD--DFKL---SIYLPSAMDE-PVVYNGKKADIADADVFEKW
        .*.. .* *  ..*.*   ...*   .. * . .**  .  ..*. .. ..   * *
 175"  LLAAEAVDDIPFGITSNSDVFSKYQLDKDGVVLFKKFDEGRNNFEGEITK-EKLLDFIKH

 233'  LQVEALPYFGEIDGS-VFAQYVESGLPLGYLFYNDEEELEEYKPLFTELAKKNRGLMNFV
        *.  .  *.* ... .*.   ... . * .*  .. .. ..  . *.. *.  .* . *.
 234"  NQLPLVIEFTEQTAPKIFGGEIKTHI-LLFL-PKSVSDYDGKLSNFKKAAEGFKGKILFI

 292'  SIDARKFGRHAGNLNMKEQFPLFAIHDMTEDLKYGLPQLSEEAFDELSDKIVLESKAIES
        **.     *..* .. * * *    ..    ...  *  *.**  .  ...  *... *.
 292"  FIDS----DHTDNQRILEFFGL--KKEECPAVR--LITLEEEMTKYKPESDELTAEKITQ

 352'  LVKDFLKGDASPIVKSQEIFEN-QDSSVFQLVGKNHDEIVNDPKKDVLVLYYAPWCGHCK
       . ..**.*. .* ..***. *. .. .*  ***** .*.. * **.*.* .*********
 344"  FCHHFLEGKIKPHLMSQELPEDWDKQPVKVLVGKNFEEVAFDEKKNVFVEFYAPWCGHCK

 411'  RLAPTYQELADTYANATSDVLIAKLDHTENDVRGVVIEGYPTIVLYPGGKKSESVVYQGS
       .***....*..** .. ....***.* *.*.* .* ....**. ..*.. ... . *.*.
 404"  QLAPIWDKLGETY-KDHENIVIAKMDSTANEVEAVKVHSFPTLKFFPASADRTVIDYNGE

 471'  RSLDSLFDFIKENGHFDVDGKALYEEAQEKAAEEADADAELADEEDAIHDEL
       *.**.. .*....*.   ** .  .... . .*. ..* * .*...*..***
 463"  RTLDGFKKFLESGGQ---DGAG--DNDDLDLEEALEPDMEEDDDQKAVKDEL
```

**Fig. 2 Amino acid sequence homology between yeast and rat PDI.** Identical residues and conserved amino acid replacements are indicated by stars and dots, respectively. Dashes represent gaps introduced to obtain maximum matching. Two conserved WCGHCK sequences, presumed to be in the active site of PDI, are boxed.

terminal region of yeast PDI there is a sequence of 4 amino acids (HDEL) that has been proposed to function in yeast as an ER retention signal (Pelham et al., 1988). The deduced polypeptide is acidic and its amino acid composition is consistent with amino acid composition of yeast PDI. Yeast PDI also contains 5 putative glycosylation sites, which is consistent with the fact that yeast PDI contains 10-kDa equivalent of N-linked carbohydrate (Fig. 1)(Tachikawa et al., 1991).

Chromosome Mapping of the *PDI1* Gene

Chromosome mapping of the *PDI1* gene was performed by blotting and hybridization of yeast chromosome separated on CHEF gels. The *PDI1* probe hybridized to the same band as the *LEU2* probe which is known to be on chromosome III (data not shown). Thus the *PDI1* gene was demonstrated to reside on chromosome lll (Tachkawa et al., 1991).

The *PDI1* Gene Is Essential for Cell Growth

The null mutation of *PDI1* was produced by one-step gene disruption, and 87 % of *PDI1* coding sequence was replaced by the selectable marker gene *LEU2*(ppdi1-Δ , Fig. 3A). A fragment containing *PDI1* flanking sequences and the *LEU2* gene was introduced into Leu⁻ (*leu2/leu2*) diploid by transformation. We confirmed that one copy of *PDI1* was replaced by the *pdi1Δ* allele by Southern analysis (Fig. 3B, lane 2). After sporulation of the heterozygous diploid, tetrads were dissected and haploid segregants were scored for viability (Fig. 3C). Of

40 tetrads, 33 gave a viable to nonviable spore ratio of 2:2, while in 7 cases, only a single spore was viable. None of the surviving daughter cells were able to grow on Leu⁻ medium. Southern analysis of DNA from the viable segregants resulted in the hybridization pattern of the wild-type gene (data not shown). These results indicate that the *PDI1* gene is essential for cell viability (Tachikawa et al., 1991).

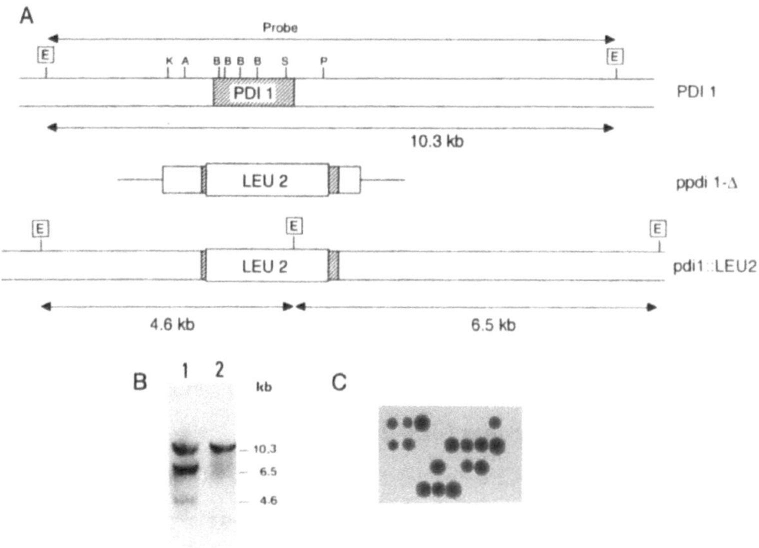


**Fig. 3. Disruption of the PDI1 gene.**
A: The strategy for gene disruption is shown. Hatched boxes represent the regions containing the *PDI1* coding sequence. The *Eco*RI fragments hybridized to the probe are shown by arrows. The abbreviations for restriction enzyme are: K, *Kpn*I; P, *Pst*I; S, *Sal*I; A, *Apa*I; B, *Bal*I. B: Genomic DNAs prepared from the parent strain (*PDI1/PDI1*)(lane 2) and the transformant (*PDI1/pdi1::LEU2*) (lane 1) were digested with *Eco*RI and subjected to Southern hybridization analysis using the probe shown in A. The wild type gene gave a 10.3-kb signal, whereas the disrupted gene gave 6.5- and 4.6-kb signals. C: A transformed diploid strain heterozygous for the *PDI1* gene made by one-step gene disruption was sporulated and tetrads were dissected.

## PDI is Identical to Glycosylation Site Binding Protein in Yeast

PDI is known to be a multifunctional protein resident in the lumen of the ER (Freedman, R.B. 1989), and has been shown to be identical with glycosylation site binding protein (GSBP) in a variety of multicellular organisms (Geetha-Habib et al., 1988). We have determined if GSBP and PDI are identical in yeast. GSBP is an ER protein photolabeled with $^{125}I$-labeled Asn-Lys(N-*p*-azidobenzoyl)Thr-$NH_2$, the acceptor sequence for oligosaccharyltransferase. Yeast PDI could be photolabeled by the glycosylation site photoprobe, and it comigrated with the 70-kDa protein that was specifically photolabeled in yeast microsomes (data not shown). In addition, Endo H treatment followed by photolabeling of both purified yeast PDI and yeast microsomal preparations resulted in the same reduction in apparent molecular mass of the labeled proteins to 60 kDa. Further, an antibody prepared against the 70-kDa photolabeled microsomal protein exhibited crossreactivity with purified yeast PDI. Taken together, these results suggest that yeast microsomal GSBP was identical to yeast PDI. Nucleotide sequence of yeast GSBP gene revealed a polypeptide with sequence identity to yeast PDI (Lamantia et al., 1991, Tachikawa et al., 1991).

## Discussion

We reported purification and some properties of yeast PDI, cloning of the *PDI1* gene, and showed that it is essential for cell viability. Moreover, PDI is identical to GSBP, that is labeled in yeast microsomes by use of an N-glycosylation site peptide photoaffinity probe. These results appeared to suggest involvement of PDI in N-glycosylation. However, recent evidence suggests that PDI is not required for the transfer of oligo-saccharide chains from oligosaccharyl pyrophosphoryldolichol to either nascent polypeptide chains (Bulleid, N. J. and Freedman, R. B. 1990) or to acceptor peptides (Noiva, et al., 1991). In fact, the peptide binding function of PDI is not limited to peptides containing the N-glycosylation site acceptor sequence (Noiva and Lennarz, 1992). Although the biological function of the peptide binding activity is unknown, the location of PDI in the lumen of the ER suggests that peptide binding reflects involvement of this molecule in protein folding.

Gunter et al.(1991) isolated *TRG1* gene, which is probably identical to *PDI1* gene, and observed that when the *TRG1* gene was placed under the control of the glucose-repressible *GAL1* promoter, glucose repression of the *TRG1* gene caused the dissapearance of TRG1 protein and the accumulation of the glycosylated propeptides of vacuolar carboxylase. The observations suggest that *TRG1* -encoded protein is involved in the maturation of carboxypeptidase. Effects of *PDI1* gene on secretory proteins should be clarified further.

## References

Bulleid, N.J., and Freedman, R.B. (1990) Cotranslational glycosylation of proteins in systems depleted of protein disulfide isomerase. EMBO J. 9: 3527-3532

Dean, N., and Pelham, H.B. (1990) Recycling of proteins from the Golgi compartment to the ER in yeast. J. Cell Biol. 111: 369-377

Edman, J.C., Ellis, L., Blacher, R.W., Roth, R.A., and Rutter, W.J. (1985) Sequence of protein disulfide isomerase and implications of its relationship to thioredoxine. Nature 317: 267-270

Freedman, R.B. (1989) Protein disulfide isomerase: multiple

roles in the modification of nascent secretory proteins. Cell 57:1069-1072

Geetha-Habib, M., Noiva, R., Kaplan, H.A., and Lennarz, W.J. (1988) Glycosylation site binding protein, a cmponent of oligosaccharyl transferase, is highly similar to three other 57 kD luminal proteins of the ER. Cell. 54:1053-1060

Gunter, R., Brauer, C., Janetzky, B., Forster, H.H., Ebbrecht, I.M., Lehle, L., and Kuntzel, H. (1991) The *Saccharomyces cerevisiae TRGI* Gene is essential for growth and encodes a lumenal endoplasmic reticulum glycoprotein involved in the maturation of vacuolar carboxypeptidase. J. Biol. Chem. 266: 24557-24563

Lamantia, M., Miura, T., Tachikawa, H., Kaplan, H.A. Lennarz, W.J., and Mizunaga, T. (1991) Glycosylation site binding protein and protein disulfide isomerase are identical and essential for cell viability. Proc. Natl. Acad. Sci. USA. 88: 4453-4457

Lambert, N., and Freedman, R.B. (1983) Structural properties of homogeneous protein disulfide isomerase from bovine liver purified by a rapid high-yielding procedure. Biochem. J. 89: 889-900

Mizunaga, T., Katakura, Y., Miura, T., and Maruyama, Y. (1990) Purification and characterization of yeast protein disulfide isomerase. J. Biochem. 108: 846-851

Noiva, R., Kaplan, H.A. and Lennarz, W.J. (1991) Glycosylation site binding protein is not required for N-linked glycoprotein synthesis. Proc. Natl. Acad. Sci. 88: 1986-1990

Noiva, R., and Lennarz, W.J. (1992) Protein disulfide isomerase: A multifunctional protein resident in the lumen of the endoplasmic reticulum. J. Biol. Chem. 267: 3553-3556

Novick,P., Field, C. and Schekman, R. (1980) Identification of 23 complementation groups required for post-translational events in the yeast secretory pathway. Cell 21: 205-215

Pelham, H.R.B., Hardwick, K.G., and Lewis, M.J. (1988) Sorting of soluble ER proteins in yeast. EMBO J. 7, 1757-1762

Pelham, H.R.B. (1989) Control of protein exit from the endoplasmic reticulum. Annu. Rev, Cell Biol. 5: 1-23

Tachikawa, H., Miura, T., Katakura, Y., and Mizunaga, T. (1991) Molecular structure of a yeast gene, *PDI1*, encoding protein disulfide isomerase that is essential for cell growth. J. Biochem. 110: 306-313

# NOVEL CHAPERONE-LIKE PRO-SEQUENCES ALLOW SECRETION OF RECOMBINANT HUMAN INSULIN-LIKE GROWTH FACTOR-1 FROM YEAST

B. Chaudhuri and A. Hinnen
Department of Biotechnology
Ciba-Geigy Ltd.
K-681. 1. 06
CH 4002
Switzerland

Protein secretion in eukaryotes is defined as a process where polypeptides, newly translated on ribosomes, undergo translocation across the membrane of the rough endoplasmic reticulum (ER) and then, through vesicular packaging, is ferried through the Golgi cisternae on its route to the periplasmic space (Rothman and Orci, 1992). All secretory proteins need transient N-terminal signals for targeting to the ER (Walter and Lingappa, 1986). The signals are embodied by short peptides (15 to 30 amino acid long), the core of which consists of only hydrophobic residues (von Heijne, 1985). In most cases, the enzyme signal peptidase cleaves the signal peptide after it has engaged the nascent protein in the translocation machinery (Walter and Lingappa, 1986). The resultant mature protein is then free to fold into an appropriate conformation, necessary for the activity of the molecule.

The signal sequences are not the only peptides which exist transiently. Some secretory proteins contain specific stretches of amino acids which, too, are ultimately absent in the native proteins. These peptides, widely varying in lengths and in their positions in the precursor proteins, are known as pro-sequences. Till now, no unique structural property has been identified which would define a region of a polypeptide to be a pro-sequence. There is one aspect, however, which does unify all eukaryotic pro-sequences. It is known that they remain connected to proteins till a late stage in the secretory pathway and are most often cleaved in the late Golgi by proteases referred to as pro-protein converting enzymes (Barr, 1991).

It is conceivable that the cellular role of these grossly dissimilar sequences could be manifold. One proposal supports the idea that pro-peptides influence the stability of newly synthesized

NATO ASI Series, Vol. H 71
Protein Synthesis and Targeting in Yeast
Edited by A. J. P. Brown, M. F. Tuite, and J. E. G. McCarthy

proteins during or after translation (Wiren et al., 1988). Suggestions have been made that, in some cases, they might even be involved in protein transport from the ER to the Golgi (Guan and Rose, 1984). A recent finding tends to propound yet another role (Zhu et al., 1989). Related observations (Silen and Agard, 1989; Winther and Sorensen, 1991) have corroborated that the pro-regions of particular proteases can act as molecular chaperones (Ellis and van der Vies, 1991), in their ability to guide the folding of proteins to their correct three-dimensional conformation. In this chapter, we attempt to characterize the pro-sequences which may perform the task of unfolding a nascent protein to allow translocation into the ER.

## THE ROLE OF THE PRO-REGION OF THE YEAST PREPRO-α-FACTOR IN THE UNFOLDING OF IGF-1

The *Saccharomyces cerevisiae* prepro-α-factor, encoded by the *MFα* gene (Kurjan and Herskowitz, 1982), consists of a 19-amino acid pre- or signal sequence (αFss) and a 66-amino acid pro-region. The complete prepro-sequence, usually referred to as the α-factor leader (αFL), is required for the efficient secretion of the 13-amino acid mating pheromone, α-factor (Fuller et al., 1988). The signal sequence part is cleaved during translocation (Waters et al., 1988). The pro-region (i.e. the proαFL) is processed later in the *trans* Golgi by an endoprotease encoded by the *KEX2* gene (Fuller et al., 1988). The processing occurs at a pair of dibasic amino acids, Lys-Arg, which links the proαFL to the rest of the polypeptide.

It appears, however, that αFss alone suffices for the secretion of some heterologous proteins from yeast (Ernst, 1988), implying that the proαFL may be important only in the secretion of the homologous α-factor, an extremely short peptide. Contrary to these observations, we find that mere N-terminal fusion of the αFss does not permit the secretion of recombinant human insulin-like growth factor-1 (IGF-1) from *Saccharomyces cerevisiae* (Chaudhuri et al, 1992a), IGF-1 being a 70-amino acid polypeptide with three *intra*molecular disulfide bonds. Experimental data indicate that not even translocation, the first step in the secretion pathway, is taking place *in vivo*. Similar results have been obtained when the αFss was substituted by the 17- and 19-amino acid signal peptides belonging to the yeast secretory proteins, acid phosphatase (Meyhack et al., 1982) and invertase (Taussig and Carlson, 1983). Surprisingly, secretion is observed again when the proαFL is covalently linked to the C-terminus of any of the above three signal peptides, (Chaudhuri et al, 1992a). We have explained this phenomenon by suggesting that the proαFL confers on nascent IGF-1 a structure which would allow translocation. For translocation to occur, polypeptides need to be in an unfolded conformation (Deshaies et al., 1988). With only a signal peptide at the N-terminus, IGF-1 probably folds too quickly into a configuration which is incompetent for translocation. It is likely that the N-terminal proαFL extension enforces a delay in the early process of folding (Randall and Hardy, 1989), making translocation possible once again.

The fact that the yeast proαFL can participate in the translocation of a heterologous protein is intriguing. Here, we embark on a study which could answer a pertinent question: does the yeast genome bear other sequences which function in a way similar to the proαFL?

## SEARCH FOR PRO-SEQUENCES WHICH MAY ALLOW THE SECRETION OF IGF-1

For this enquiry, we have chosen an IGF-1 expression cassette (Fig. 1) containing the αFL as a secretion signal. The aim was to substitute the proαFL with a random set of sequences, and then attempt to identify, from that pool, specific DNA sequences which might permit secretion of IGF-1.

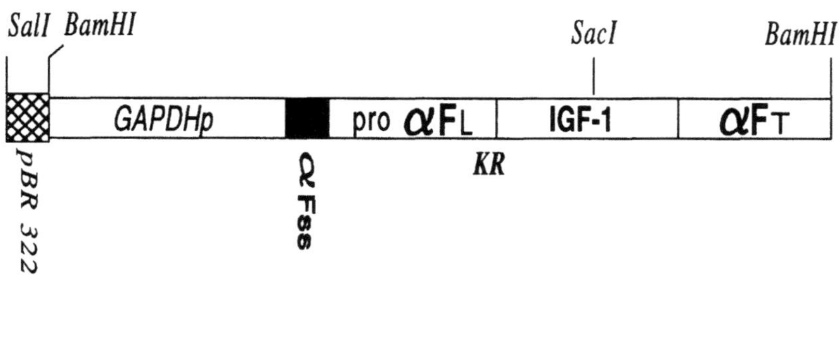


Fig.1. The expression cassette for IGF-1 consists of a 276 bp *SalI-BamHI* fragment from pBR322, a 400 bp promoter fragment from the glyceraldehyde-3-phosphate dehydrogenase gene (*GAPDH*), a 255 bp sequence encoding the 85-amino acid αFL ending with a pair of dibasic amino acids Lys-Arg (KR), a 216 bp fragment encoding the 70-amino acid IGF-1 and two stop codons, and a 275 bp transcription terminator from the MFα gene (Chaudhuri et al., 1992a). The cassette is subcloned in the *SalI/ BamHI* sites of the yeast 2-micron plasmid pDP34 (Steube et al., 1991).

To facilitate the removal of most of the proαFL sequence, two *Bgl*II restriction sites have been introduced (Fig. 2) by site-directed mutagenesis (Zoller and Smith, 1987). Complete digestion of plasmid DNA with the *Bgl*II restriction enzyme, followed by religation, would afford a new plasmid where *Sau*3A-digested yeast genomic DNA could be conveniently subcloned.

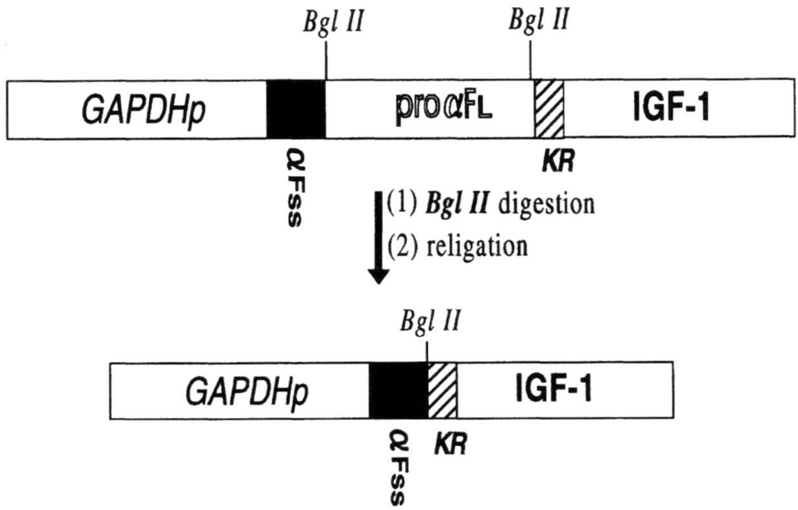


Fig. 2. Introduction of *Bgl*II sites in the proαFL DNA sequence, for subcloning of small *Sau*3A fragments obtained from the yeast genome (the *Bgl*II and *Sau*3A sites have compatible cohesive ends).

The plasmid construction, resulting from the deletion of the DNA encoding proαFL, cannot secrete IGF-1 when expressed in yeast. The missing proαFL could now be replaced by random yeast genomic DNA fragments with the hope that some sequences from this library of DNA fragments may enable again the secretion of IGF-1 from yeast. In order to unravel the nature of these sequences, two different approaches were taken. The secreted IGF-1 is amenable to a direct assay, or else it could be ascertained indirectly as a protein fused to an easily detectable reporter enzyme. The fusion approach, with the secreted yeast invertase (encoded by the *SUC2* gene; Carlson et al., 1983) as a C-terminal tag (Fig. 3B), was chosen at first. The success of a direct screen for the secretion of IGF-1 (Fig. 3A) had initially been thought to be doubtful.

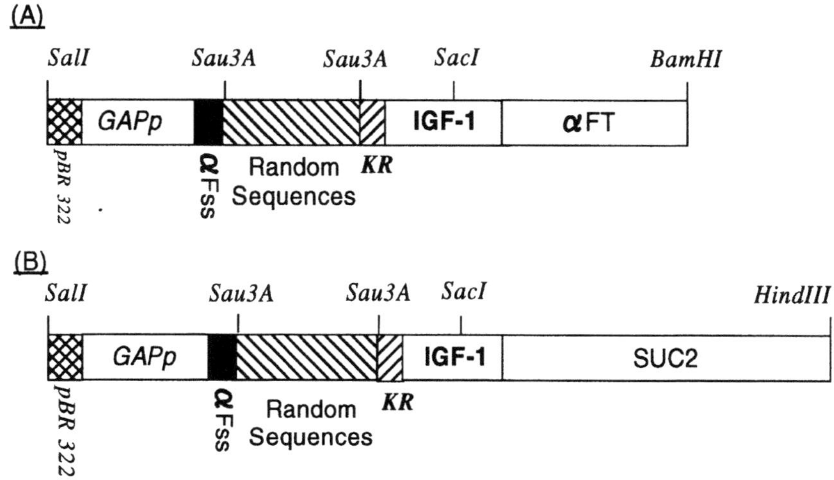


Fig. 3. Expression cassettes, containing *Sau*3A-digested yeast genomic DNA fragments (150-350 bp), to unearth sequences which would allow secretion of IGF-1 (A) and the IGF-1–SUC2 fusion protein (B). (A) The *SalI-Bam*HI cassettes are subcloned in the yeast 2-micron plasmid pDP34 (Steube et al., 1989); (B) the *SalI-Hind*III cassettes are subcloned in the centromere plasmid, pDP83 (bearing the *S. cerevisiae CEN14* sequence; unpublished). Both pDP34 and pDP83 encode the *S. cerevisiae URA3* gene as a selectable marker. Upon transformation of the two pools of DNA in *E. coli*, ~25,000 transformants were obtained.

In the indirect screen (Fig. 3B), ~25,000 yeast transformants in the strain AB110*suc2* were checked for secretion of the fusion protein. All transformants from uracil-selective plates were replica-plated on to full medium plates which contained sucrose as the sole carbon source (Kaiser and Botstein, 1986). Only transformants which secrete active invertase would grow on this medium. Nine colonies grew, implying that nine individual pro-sequences, from the random library of fragments, may have caused the secretion of the IGF-1–SUC2 fusion protein. Plasmid DNA was isolated from these strains and the nine pro-sequences were transposed into a context where solely IGF-1 could be expressed from the plasmid pDP34. We found that only four of the nine sequences permit the secretion of IGF-1 alone.

About 20,000 yeast transformants, in the strain AB110, were screened directly for the secretion of IGF-1 (Fig. 3A). Secreted proteins from the transformants were blotted on to nitrocellulose filters and the IGF-1–like proteins adhering to the filter were detected by a procedure similar to Western blotting (i.e. a colony assay to detect IGF-1 secretion;

Chaudhuri et al., 1992a). A total of 50 filters, containing all the transformants, were probed. Only six individual transformants, each of them on separate filters, clearly showed the secretion of IGF-1.

The few pro-sequences which allow secretion of IGF-1, when covalently linked to IGF-1 at the N-terminus, have been sequenced in both directions (Sanger et al, 1977). The peptide sequences, encoding the DNA, were thereby obtained. Out of the six pro-sequences obtained from the direct approach two pairs were found to be identical. Therefore, merely four distinct sequences have been isolated from this approach. These four peptide sequences (63-, 66-, 69- and 132-amino acid long) appear to be nearly identical, in primary structure, to the four sequences obtained from the indirect approach.

The amounts of active, monomeric IGF-1 secreted from strains, bearing the new pro-sequences, were quantified by reversed-phase HPLC (Steube et al., 1991) and were compared with the values obtained from strains bearing the proαFL sequence. None of the new pro-sequences show any improvement over the proαFL in their ability to secrete IGF-1 (unpublished data).

## CHARACTERISTICS OF THE NEW PRO-SEQUENCES

It appears that the isolated pro-sequences, which permit translocation and secretion of IGF-1, share a common feature. They are distinctly hydrophilic, being rich in acidic, basic, and polar amino acid residues. We believe that the salt-like nature of the polypeptide sequences contributes to the maintenance of an unfolded IGF-1 conformation, which triggers translocation. This suggestion is not improbable in the context of current knowledge of protein-folding *in vitro* (Kim and Baldwin, 1990). The charged residues, positioned in *cis* (i.e. covalently linked to IGF-1), could behave like a detergent *in vivo* preventing the nascent IGF-1 to assume a folded structure.

Three of these polypeptides do not contain any sites for asparagine-linked glycosylation (N-glycosylation). The longest pro-sequence codes for one site which is at the C-terminus. The consensus sequence for sugar-chain attachment is Asn-Xaa-Ser/Thr (Kornfeld and Kornfeld, 1985). The proαFL possesses three such sites and all are utilized for N-glycosylation (Graham and Emr, 1991). It is believed that glycans, attached to Asn, aid in the intracellular transport of proteins (Guan et al., 1985). This is obviously true when the secretion of α-factor is examined (Caplan et al., 1991). Amino acid replacements at the sites for N-glycosylation hinder secretion and cause precursors to accumulate. However, *in vitro* experiments indicate that the proαFL may not have a dramatic influence on the translocation of the prepro-α-factor precursor. Translocation is maintained appreciably, even after complete removal of the proαFL sequence from the cDNA encoding prepro-α-factor (Rothblatt et al., 1987).

Our results indicate that, in the secretion of IGF-1 from yeast, the pro-sequences are essential. This implies that the pro-sequences must be involved in translocation, since signal peptides on their own permit neither translocation nor secretion of IGF-1. An important question to answer would be, whether they are required beyond the translocation step. After fulfilling their role in translocation-competence, the pro-sequences are still conjoined to IGF-1 by the Lys-Arg dipeptide. The Kex2 enzyme separates the pro-sequences from IGF-1 in the late Golgi, only after protein-folding has occurred in the ER. It has been recently reported that the pro-sequence of the yeast carboxypeptidase Y can act as a folding agent (Winther and Sorensen, 1991). Therefore, it could be imagined that the pro-sequences, which allow translocation at first, may have an impact later on the folding of IGF-1. The role of pro-sequences in the folding of IGF-1 can be conveniently determined with the aid of a modified Kex2 endoprotease which functions in the ER (Chaudhuri et al., 1992b). If the pro-sequences are removed early in the ER, then IGF-1 should be able to fold unhindered.

It has been observed that the IGF-1–like molecules which secrete, using the αFL, are preponderantly the inactive dimers (Chaudhuri and Stephan, 1992c and refs. cited therein). Only 10-15% of the total secreted molecules are the correctly-folded, active monomeric IGF-1. When the proαFL is removed in the ER, instead of in the Golgi, with the aid of an ER-mislocated Kex2 enzyme, sKex2pHDEL (Chaudhuri et al., 1992b), the percentage of the active monomers increases dramatically (Chaudhuri and Stephan, 1992c). This indicates that, after translocation, the proαFL influences the formation of disulfide-linked dimeric IGF-1.

Disulfide-bonded multimers of IGF-1 are also noticed when the new pro-sequences are used for secretion. However the molecules which predominate, in these expression systems, are the misfolded positional isomer of IGF-1 (unpublished results), where two of the three *intra*molecular disulfide bonds are misplaced (Steube et al., 1991). Preliminary results indicate that this misfolding of IGF-1 can also be prevented when the ER-retained sKex2pHDEL is co-expressed.

THE PRO-SEQUENCES ALLOW SECRETION OF IGF-1 ALSO IN *TRANS*

We believe that each of the four alternate pro-sequences, which can substitute for the pro$\alpha$FL in the secretion of IGF-1, effectively functions as a cotranslational molecular chaperone (Winther and Sorensen, 1991) by keeping the covalently-attached IGF-1 in an unfolded structure, compatible with translocation. A conventional chaperone (Ellis and van der Vies, 1991), however, would impose its effect even when distally removed (i.e. when in *trans*).

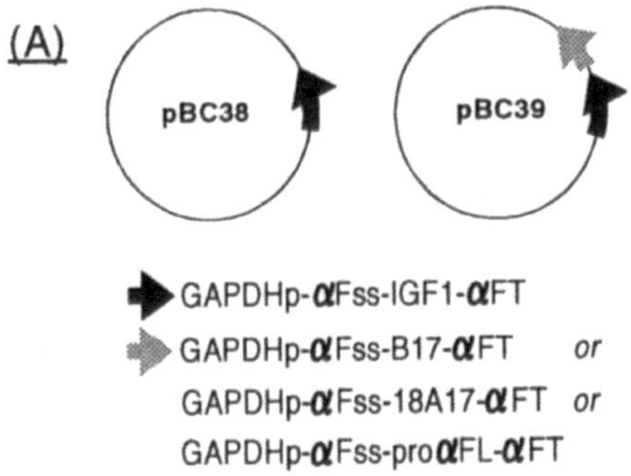


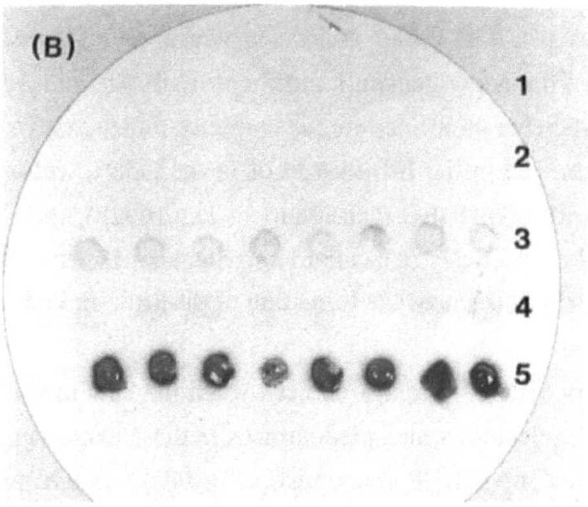


Fig. 4. (A) The plasmids pBC38 and pBC39 are derivatives of pDP34 (see Fig.1). B17 and 18A17 denote the two new pro-sequences isolated from the random screens. (B) Eight transformants in the strain AB110 were analyzed for secretion by the IGF-1 colony assay (see text). 1: pBC38; 2: pBC39 containing pro$\alpha$FL in *trans*; 3: pBC39 containing B17 in *trans*; 4: pBC39 containing 18A17 in *trans*; 5: secreted IGF-1 using the $\alpha$FL as a secretion signal (see Fig.1).

Two of the four pro-sequences have been expressed independently of IGF-1 (Fig. 4A). The individual sequences are under the control of the *GAPDH* constitutive promoter and are subcloned on a plasmid (pBC38) which already encodes the IGF-1 expression cassette containing the αFss. We find that the two sequences clearly do cause secretion when in *trans* (Fig.4B), even though the levels are very poor. When the proαFL sequence was similarly subcloned for expression, no secretion of mature IGF-1 could be detected, suggesting that the proαFL sequence functions only in *cis*, not in *trans*. These results were confirmed by Western blotting (data not shown).

## CONCLUSIONS

A class of hydrophilic sequences have been identified which, like the proαFL, enables the translocation of IGF-1 in yeast. Not only do they act in *cis* but, like true chaperones, they have the propensity to act in *trans*. The proαFL sequence is not merely redundant beyond the translocation step, but has a deleterious effect on the folding of IGF-1 in the ER. The new pro-sequences also have an adverse impact on folding, but the manifestation of the effects appears to be different.

## LITERATURE REFERENCES


Barr PJ (1991) Mammalian subtilisins: the long-sought dibasic processing endoproteases. Cell 66: 1-3

Caplan S, Green R, Rocco, J, Kurjan J,(1991) Glycosylation and structure of the yeast MFα1 α-factor precursor is important for efficient transport through the secretory pathway. J Bacteriol 173: 627-635

Chaudhuri B, Steube K, Stephan C (1992a) The pro-region of the yeast prepro-α-factor is essential for membrane translocation of human insulin-like growth factor 1 *in vivo*. Eur J Biochem 206: 793-800

Chaudhuri B, Latham SE, Helliwell SB, Seeboth P (1992b) A novel Kex2 enzyme can process the proregion of the yeast alpha-factor leader in the endoplasmic reticulum instead of in the Golgi. Biochem Biophys Res Commun 183: 212-219

Chaudhuri B, Stephan C (1992c) A modified Kex2 enzyme retained in the endoplasmic reticulum prevents disulfide-linked dimerisation of recombinant human insulin-like growth factor-1 secreted from yeast. FEBS Lett 304: 41-45

Deshaies RJ, Koch BD, Werner-Washburne M, Craig EA, Schekman R (1988) A subfamily of stress proteins facilitates translocation of secretory and mitochondrial precursor proteins. Nature 322: 800-805

Guan JL, Rose JK (1984) Conversion of a secretory protein into a transmembrane protein results in its transport to the Golgi complex but not to the cell surface. Cell 37: 779-787

Guan JL, Machamer CE, Rose JK (1985) Glycosylation allows cell-surface transport of an anchored secretory protein. Cell 42: 489-496

Ellis RJ, van der Vies SM (1991) Molecular chaperones. Annu Rev Biochem 60: 321-347

Graham TR, Emr SD (1991) Compartmental organization of Golgi-specific protein modification and vacuolar protein sorting events defined in a yeast *sec18* (*NSF*) mutant. J Cell Biol 114: 207-218

Ernst J (1988) Efficient secretion and processing of heterologous proteins in *Saccharomyces cerevisiae* is mediated solely by the pre-segment of α-factor precursor. DNA 7: 355-360

Fuller RS, Sterne RE, Thorner J (1988) Enzymes required for yeast prohormone processing. Annu Rev Physiol 50: 345-362

Kaiser CA, Botstein D (1986) Secretion-defective mutations in the signal sequence for *Saccharomyces cerevisiae* invertase Mol Cell Biol 6: 2382-2391

Kim PS, Baldwin RL (1990) Intermediates in the folding reactions of small proteins. Annu Rev Biochem 59: 631-660

Kornfeld R, Kornfeld S (1985) Assembly of asparagine-linked oligosaccharides. Annu Rev Biochem 54: 631-664

Kurjan J, Herskowitz I (1982) Structure of a yeast pheromone gene (*MF*α): a putative α-factor precursor contains four tandem copies of mature α-factor. Cell 30: 933-943

Meyhack B, Bajwa W, Rudolph H, Hinnen A (1982) Two yeast acid phosphatase structural genes are the result of a tandem duplication and show different degrees of homology in their promoter and coding region. EMBO J 1: 675-680

Rothman JE, Orci L (1992) Molecular dissection of the secretory pathway. Nature 355: 409-415

Randall LL, Hardy SJS (1989) Unity in function in the absence of consensus in sequence: role of leader peptides in export. Science 243: 1156-1159

Rothblatt JA, Webb JR, Ammerer G, Meyer DI Secretion in yeast: structural features influencing the post-translational translocation of prepro-α-factor *in vitro*. EMBO J 6: 3455-3463

Sanger F, Nicklen S, Coulson AR (1977) DNA sequencing with chain-terminating inhibitors. Proc Natl Acad Sci USA 74: 5463-5467

Silen JL, Agard DA (1989) The α-lytic protease pro-region does not require a physical linkage to activate the protease domain *in vivo*. Nature 341: 462-464

Steube K, Chaudhuri B, Märki W, Merrywearther JP, Heim J (1991) α-factor-leader-directed secretion of recombinant human-insulin-like growth factor I from *Saccharomyces cerevisiae*. Eur J Biochem 198: 651-657

Taussig R, Carlson M (1983) Nucleotide sequence of the *SUC2* gene for invertase. Nucl Acids Res 11: 1943-1954

von Heijne G (1985) Signal sequences: the limits of variation. J Mol Biol 184: 99-105

Walter P, Lingappa VR (1986) Mechanism of protein translocation across the endoplasmic reticulum membrane. Annu Rev Cell Biol 2: 499-516

Wiren KM, Potts Jr JT, Kronenberg HM (1988) Importance of the propeptide sequence of human preproparathyroid hormone for signal sequence function. J Biol Chem 263: 19771-19777

Waters MG, Evans EA, Blobel G (1988) Prepro-α-factor has a cleavable sequence. J Biol Chem 263: 6209-6214.

Winther JR, Sorensen P (1991) Propeptide of carboxypeptidase Y provides a chaperone-like function as well as inhibition of the enzymatic activity. Proc Natl Acad Sci USA 88: 9330-9334

Zhu X, Ohta Y, Jordan F, Inouye M (1989) Pro-sequence of subtilisin can guide the refolding of denatured subtilisin in an intermolecular process. Nature 339: 483-484

Zoller MJ, Smith M (1987) Oligonucleotide-directed mutagenesis: a simple method using two oligonucleotide primers and a single-stranded DNA template. Methods Enzymol 154: 329-350

# mRNA TRANSLATION AND PROTEIN FOLDING *IN VIVO*

**Alistair J.P. Brown and Tanya Crombie**
Molecular & Cell Biology
University of Aberdeen,
Marischal College
Aberdeen AB9 1AS, U.K.

## INTRODUCTION

Most current models that describe how polypeptide chains adopt their native three-dimensional structure are based largely upon studies on the refolding of purified denatured proteins *in vitro* (9,10,13,18). Although these models have provided important insights into protein folding pathways, they have focussed attention primarily upon the *spatial* organisation of amino acids within the polypeptide chain. It has now become clear that additional factors promote protein folding inside the cell. These include the chaperonins, protein disulphide isomerases and peptidyl prolyl *cis-trans* isomerases (13-15). In addition, the *temporal* separation of amino acids during the synthesis of the nascent polypeptide chain might be significant for protein folding *in vivo* (4,16,25).

In his review on protein folding, Goldberg (16) suggested that the translation of mRNA into protein is followed by a "second translation of the genetic message": the folding of the polypeptide chain into its native conformation. We have demonstrated that the intracellular folding of a bifunctional protein in yeast is influenced by the way in which its mRNA is translated, and therefore that Goldberg's "two translation events" might be intimately linked. Hence the link between mRNA translation and protein folding might be exploited to increase the efficiency of protein folding *in vivo*, just as the close link between transcription and translation in prokaryotes has been exploited in the control of gene expression by attenuation mechanisms (32).

## THE HYPOTHESIS

The proposal that the intracellular folding of some proteins is promoted by controlled rates of mRNA translation, is based upon a number of observations (20,27).

NATO ASI Series, Vol. H 71
Protein Synthesis and Targeting in Yeast
Edited by A. J. P. Brown, M. F. Tuite, and J. E. G. McCarthy

**FIGURE 1 The distribution of non-preferred codons in some multifunctional yeast genes correlates with their domain organisation.** The length of each contiguous array of non-preferred codons in the *TRP3*, *ARO1* and *FAS1* genes is plotted against the position of the array within the coding region. Arrays that are shorter or equal to the number average mean length of arrays of contiguous non-preferred codons [= 2] are not shown. The domain organisation of each multifunctional protein is shown. The *TRP3* gene encodes anthranilate synthase II [ASII] and indoleglycerol phosphate synthase [IGPS] domains (33). The *ARO1* gene encodes 3-dehydroquinate synthase [B], 5-*enol*pyruvylshikimate-3-phosphate synthase [A], shikimate kinase [L], 3-dehydroquinase [D] and shikimate dehydrogenase [E] domains (12). The *FAS1* gene encodes acetyl transferase [AC], enoyl reductase [ER], dehydratase [DH] and malonyl palmityl transferase [MP] domains (30). Data taken from references 20 and 27.

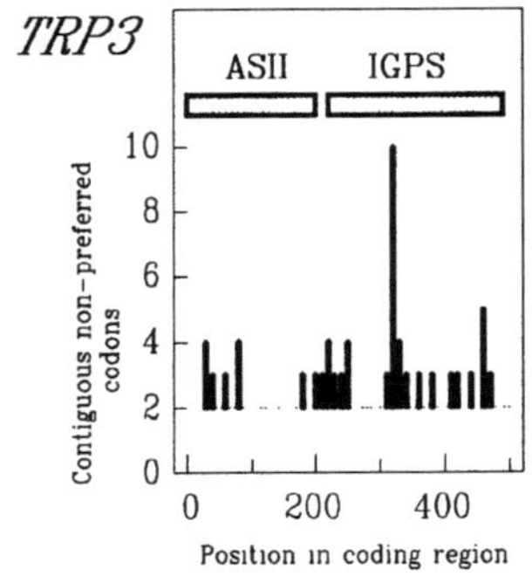


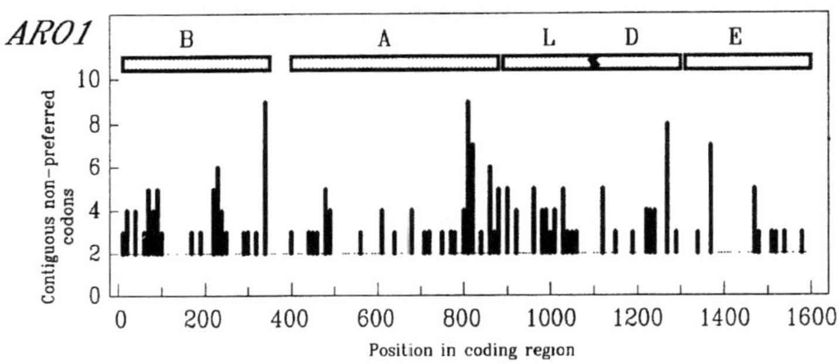


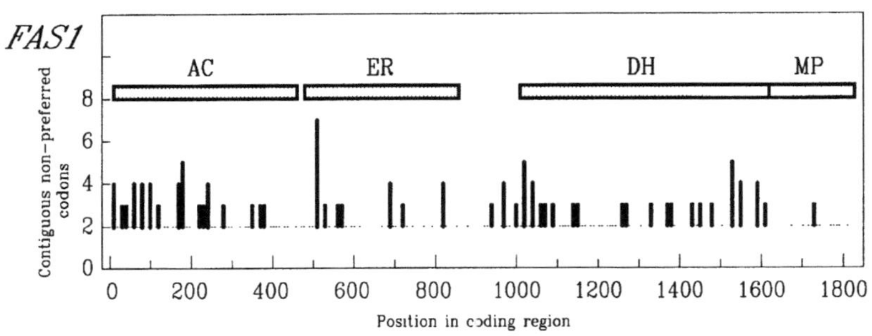

[1] Proteins start to fold as they are being synthesized (5,6,23). This is not surprising since early events in protein folding occur extremely rapidly (probably within milliseconds: 17) relative to the timescale of mRNA translation (takes about two minutes to translate the average mRNA in yeast: 24).

[2] Translational elongation is discontinuous, with distinct translational pauses leading to the accumulation of specific nascent polypeptides (1,7,8,19,26,28,31). Some translational pauses have been attributed to regions of poor codon bias (19,22).

[3] Non-preferred codons are distributed non-randomly in some yeast genes [*Figure 1*]. However, in most of these cases it has not been possible to estimate the possible significance of these data with respect to protein folding, since X-ray crystal structures are available only for a small proportion of the known polypeptide sequences. It has been possible to compare codon distribution with domain organisation for some yeast genes that encode multifunctional proteins which are organised into distinct functional domains. For example, regions of poor codon bias within the yeast *ARO1* and *FAS1* genes correlate with the interdomain regions of the proteins they encode [*Figure 1*].

Therefore, regions of poor codon bias may have evolved in some genes to generate distinct translational pauses which temporally separate the folding of specific regions of their polypeptide chains (27).

Regions of stable mRNA secondary structure formation can temporally slow polypeptide elongation (3). Therefore, translational pauses induced by mRNA secondary structures may also play a role in promoting intracellular protein folding. However, we have limited our analyses to the role of codon bias.

#### THE *TRP3* GENE AS AN EXPERIMENTAL MODEL

The yeast *TRP3* gene was used as a model to test the possible relationship between translational pausing and protein folding *in vivo*. This gene encodes two enzyme functions required for tryptophan biosynthesis: anthranilate synthase II (ASII) and indoleglycerol phosphate synthase (IGPS; 33). The ASII and IGPS activities are separable into distinct structural domains (2). Within the IGPS domain, we identified an array of ten consecutive non-preferred codons which is significantly longer than the other arrays of non-preferred codons within the *TRP3* gene (**Figure**

| | | | | | | | | | | | | | | | |
|---|---|---|---|---|---|---|---|---|---|---|---|---|---|---|---|
| | his | gly | ser | leu | gln | asp | leu | val | asn | val | arg | lys | ile | leu | asp |
| ***TRP3*** | ***CAC*** | ***GGT*** | TCG | TTA | CAG | GAT | TTA | GTA | AAT | GTG | AGG | AAA | ***ATC*** | CTA | GAT |
| | | * | * | | | | | | | | | * | | * | |
| ***TRP3rs*** | ***CAC*** | GGA | ***TCC*** | TTA | CAG | GAT | TTA | GTA | AAT | GTG | AGG | ***AAG*** | ***ATC*** | TTA | GAT |
| | | * | * | * | * | * | * | * | * | * | * | * | | * | |
| ***TRP3pr*** | ***CAC*** | GGA | ***TCC*** | ***TTG*** | ***CAA*** | ***GAC*** | ***TTG*** | ***GTT*** | ***AAC*** | ***GTT*** | ***AGA*** | ***AAG*** | ***ATC*** | TTA | GAT |
| | | *BamHI* | | | | | | | | | | *BglII* | | | |

**FIGURE 2 Removing the translational pause from the *TRP3* gene.** *TRP3*, wild-type gene; *TRP3*rs, Restriction Sites inserted; *TRP3*pr, translational Pause Replaced. *Bam*HI and *Bgl*II sites are underlined, and preferred codons are shown in italics and bold face. The amino acid sequence of the region is given, and each mutation is shown with an asterisk. Adapted from 11.

1). Therefore, our first approach was to establish whether the removal of this potential translational pause from the *TRP3* gene confers any phenotype upon the host yeast cell.

The array of ten non-preferred codons in the IGPS domain was replaced with synonymous preferred codons in two steps. Firstly, *Bam*HI and *Bgl*II restriction sites were introduced at each end of the region by site-directed mutagenesis (to create *TRP3*rs: Restriction Sites inserted). Then a double-stranded oligonucleotide corresponding to the preferred codons was inserted between these restriction sites (to create *TRP3*pr: translational Pause Replaced: **Figure 2**).

#### THE MODIFIED *TRP3* GENES COMPLEMENT A *trp3* NULL MUTATION

The wild-type and modified genes were subcloned into a multicopy, *LEU2*, 2µm-based plasmid (YEp13), and a single-copy centromeric vector (derived from YCp50). Complementation tests were performed using these constructs in the yeast strain OL2 (*trp3::URA3*). The wild-type *TRP3*, the *TRP3*rs and the *TRP3*pr genes all complemented the *trp3* null mutation. Furthermore, no significant differences in the doubling times of the transformants were observed during growth at 25°C, 30°C or 37°C in minimal media lacking tryptophan (11). In addition, the *TRP3*, *TRP3*rs and *TRP3*pr transformants showed the same level of sensitivity to the analogue, 5-methyltryptophan (11). Therefore, the removal of the translational pause does not have a major effect upon the activity of the TRP3 protein.

THE REMOVAL OF THE TRANSLATIONAL PAUSE CAUSES A DECREASE IN THE RELATIVE AMOUNT OF ACTIVE IGPS

More detailed comparisons of the expression of the wild-type and modified *TRP3* genes were carried out in an effort to establish whether the removal of the translational pause has a less obvious effect upon the TRP3 protein.

RNA was extracted from wild-type *TRP3* cells (OL1), the *trp3* null strain (OL2), and from OL2 transformants containing each of the three *TRP3* genes on single-copy plasmids. The RNA was subjected to Northern blotting and the same filter probed sequentially for the *TRP3* and actin mRNAs (**Figure 3**). Using the AMBIS 2D Radioanalytic Imaging System, the level of the *TRP3* mRNA was measured in each strain relative to the actin mRNA which acts as an internal control (21). The *TRP3* mRNA level was elevated 1.7-fold in cells containing the *TRP3*pr gene relative to the control strains (**Table**). This was due to a reproducible increase in the half-life of the *TRP3*pr mRNA of approximately 1.5-fold (**Figure 4**).

Attempts were made to estimate the amount of Trp3 protein in transformed strains using polyacrylamide gel electrophoresis under denaturing conditions. This was not successful because the Trp3 protein is normally present at low

---

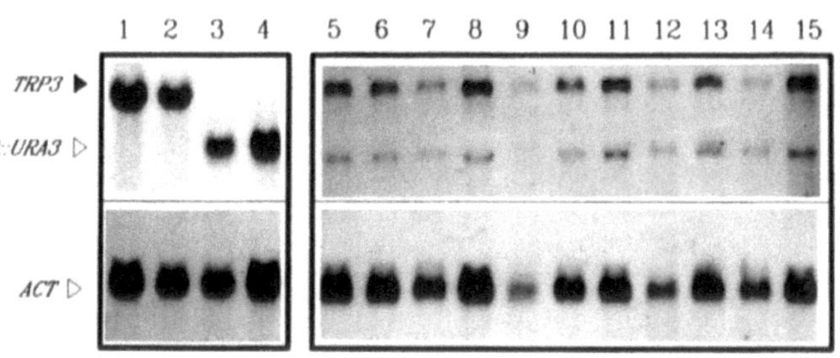


**FIGURE 3** **Northern analysis of the *TRP3* mRNA.**
The abundance of the *TRP3* mRNA was measured relative to the actin mRNA by quantitative Northern analysis of total RNA preparations (21): Lanes 1-2, OL1 (*TRP3*); Lanes 3-4, OL2 (*trp3::URA3*); Lanes 5-7, OL2:YCp*TRP3*; Lanes 8-10, OL2:YCp*TRP3*rs; Lanes 11-15, OL2:YCp*TRP3*pr. Adapted from 11.

**FIGURE 4 The TRP3 protein is over-expressed in yeast strains containing multiple copies of the *TRP3* or the *TRP3*pr genes.** Approximately equal amounts of protein extract from the parental *TRP3* yeast strain (OL1: Lane 1), the *trp3* null strain (OL2: Lane 2), an OL2:YEp13*TRP3* transformant (Lane 3), and an OL2:YEp13*TRP3*pr transformant (Lane 4) were electrophoresed on a 10% SDS-polyacrylamide gel, and the gel stained with Coomassie blue. The positions of molecular weight markers (open arrows) and the TRP3 protein (closed arrow) are shown. Adapted from 11.

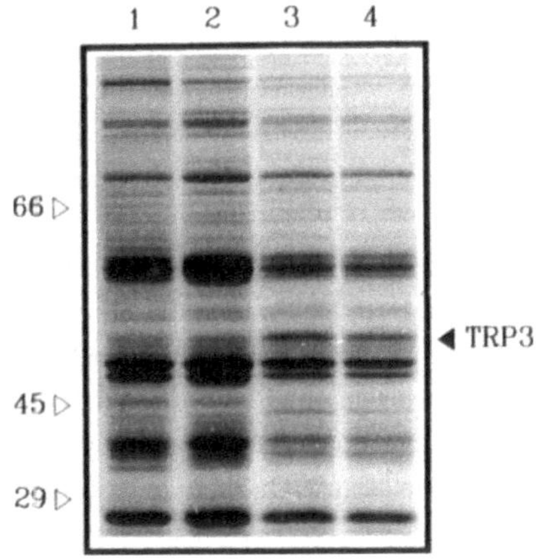


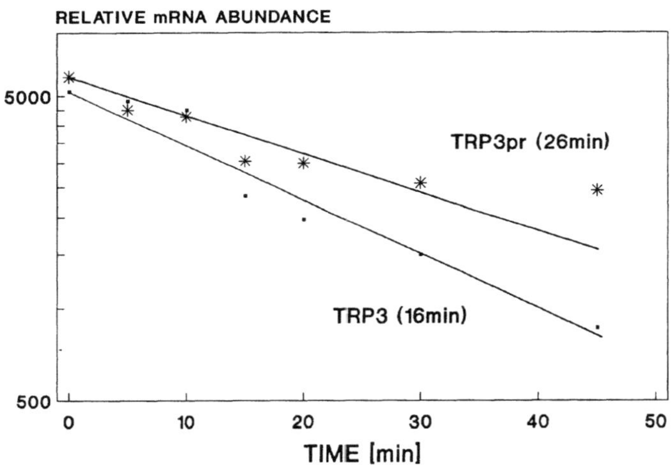


**FIGURE 5 The half-lives of the *TRP3* and *TRP3*pr mRNAs.**
The half-lives of the *TRP3* and *TRP3*pr mRNAs were measured using the transcriptional inhibitor, phenanthroline (29).

**TABLE: Expression of *TRP3* genes**

| Gene | Number of Transformants Assayed | *TRP3* mRNA | IGPS activity | IGPS activity / *TRP3* mRNA |
|---|---|---|---|---|
| *TRP3* | 9 | 0.09 +/- 0.02 | 53 +/- 5 | 589 +/- 58 |
| *TRP3*rs | 3 | 0.10 +/- 0.02 | 67 +/- 5 | 670 +/- 84 |
| *TRP3*pr | 15 | 0.15 +/- 0.04 | 60 +/- 15 | 400 +/- 82 |

*TRP3* mRNA levels are relative to the actin mRNA in each strain (21). IGPS synthase specific activities are given in nmoles of indoleglycerol phosphate released in 20 minutes per mg of total cell protein. Adapted from 11.

concentrations in the cell (**Figure 5**). Furthermore, Western blotting was not possible because no antibodies against yeast Trp3 protein were available. Therefore, the IGPS specific activity was determined for transformants carrying a single copy of the wild-type *TRP3*, the *TRP3*rs, or the *TRP3*pr gene (**Table**). Triplicate assays were performed on individual isolates, and the mean value for each isolate was then divided by its *TRP3* mRNA level to estimate the amount of active IGPS synthesized per *TRP3* mRNA molecule in each individual transformant. A statistically significant decrease (1.5-fold; $P < 0.05$) in enzyme activity per mRNA molecule was observed for the *TRP3*pr transformants relative to the controls. The analysis of a different set of independent transformants confirmed that this observation was reproducible (11).

#### THE DECREASE IN IGPS ACTIVITY IS DUE TO A LOWER EFFICIENCY PROTEIN FOLDING

The mutations in the *TRP3*pr gene improve the codon bias of the region of interest (**Figure 2**). Computer analyses using the FOLD program (34) suggest that the mutations *reduce* the stability of secondary structures formed by the *TRP3* mRNA in this region (11). Therefore, these mutations might be expected to *increase TRP3* mRNA translation. However, a *decrease* in the relative intracellular IGPS activity is observed. Dosage effects upon *TRP3* over-expression in yeast can be excluded, because TRP3 protein is over-expressed from multicopy *TRP3* and *TRP3*pr plasmids

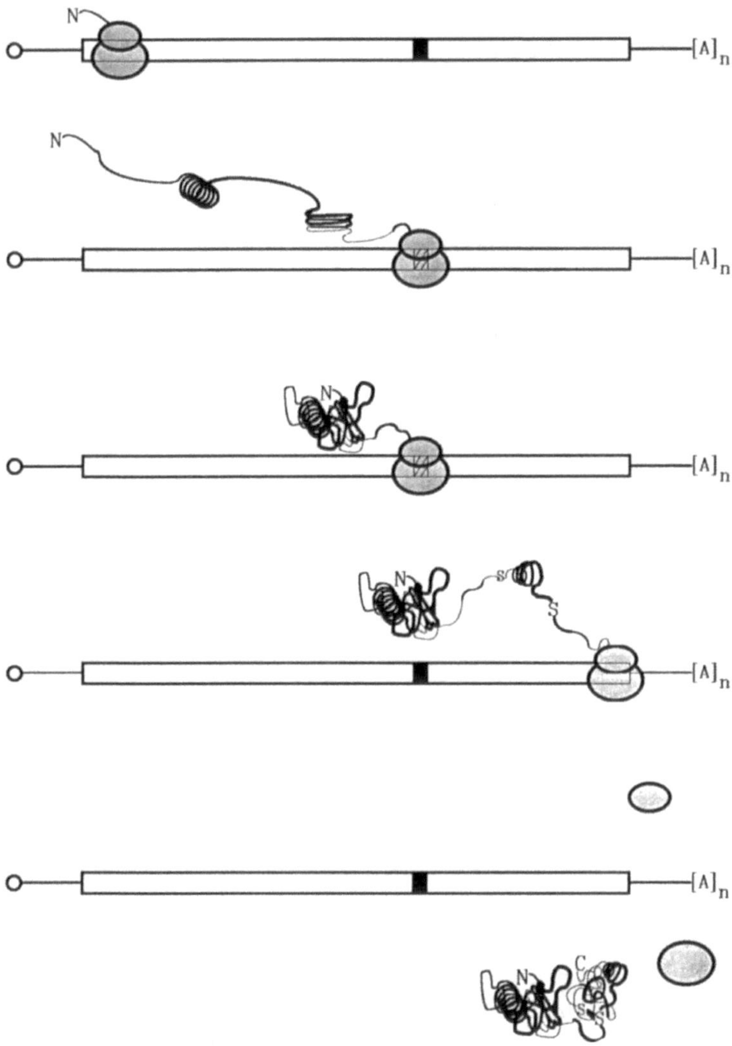


**FIGURE 6** **The model: specific translational pauses temporally separate the folding of regions of some nascent polypeptide chains.**

(**Figure 5**). Therefore, since the amino acid sequence of the protein is unchanged, the relative decrease in IGPS activity is probably mediated at the level of protein folding.

## HOW DO TRANSLATIONAL PAUSES INTERACT WITH PROTEIN FOLDING PATHWAYS?

The folding of a polypeptide chain into its native conformation occurs in stages (for reviews see 10,13,15,18). Even for simple proteins, rapid initial folding events are followed by the formation of distinct intermediates, some of which generate the native conformation of the protein in what is thought to be the rate limiting step in the pathway. For complex proteins, further steps involving the appropriate interaction of domains (or subunits) must take place before the native conformation is achieved. Polypeptide elongation rates are slow compared with the rates at which early folding events take place (27). Therefore, translational pauses are more likely to influence the latter stages of the protein folding pathway. By temporally separating the folding of specific regions of a polypeptide chain (**Figure 6**), they may kinetically bias a protein folding pathway against the formation of some non-productive, but metastable folding intermediates.

We have suggested that translational pauses might temporally separate the folding of structural domains (27). However, the translational pause in *TRP3* occurs within a domain (**Figure 1**). Therefore, these pauses might also promote the folding of individual protein domains.

## CONCLUSION

We have shown that the removal of a specific translational pause from the *TRP3* mRNA adversely affects the intracellular folding of indoleglycerol phosphate synthase. In this case the decrease in enzyme activity is not great, possibly because cytoplasmic chaperonins partially counter the adverse effects of removing the translational pause. In addition to chaperonins, other factors promote protein folding *in vivo*. These include protein disulphide isomerases and peptidyl prolyl *cis-trans* isomerases (13-15). We suggest that controlled rates of polypeptide elongation should be added to this list.

## ACKNOWLEDGEMENTS

We are grateful to Howard Zalkin for the *TRP3* gene, and to Mick Tuite for a yeast strain that overexpresses *TRP3*. This work was supported by The Wellcome Trust.

**REFERENCES**


1. Abraham, A.K. & Pihl, A. (1980). *Eur. J. Biochem.* 106, 257-262.
2. Aebi, M., Furter, R., Prantl, F., Niederberger, P. & Hutter, R. (1984) *Curr. Genet.* **8**, 173-180.
3. Baim, S.B., Pietras, D.F., Eustice, D.C. & Sherman, F. (1985). *Molec. Cell. Biol.* **5**, 1839-1846.
4. Baldwin, R.L. (1975). *Annu. Rev. Biochem.* **44**, 453-475.
5. Bergman, L.W. & Kuehl, W.M. (1979a). *J. Biol. Chem.* **254**, 5690-5694.
6. Bergman, L.W. & Kuehl, W.M. (1979b). *J. Biol. Chem.* **254**, 8869-8876.
7. Candelas, G., Candelas, T., Ortiz, A. & Rodriguez, O. (1983). *Biochem. Biophys. Res. Commun.* **116**, 1033-1038.
8. Chaney, W.G. & Morris, A.J. (1979). *Arch. Biochem. Biophys.* **194**, 283-291.
9. Christensen, H. & Pain, R.H. (1991). *Eur. Biophys. J.* **19**, 221-229.
10. Creighton, T.E. (1990). *Biochem. J.* **270**, 1-16.
11. Crombie, T., Swaffield, J.C. & Brown, A.J.P. (1992) *J. Mol. Biol.*, **228**, 7-12.
12. Duncan, K., Edwards, R.M. & Coggins, J.R. (1987) *Biochem. J.* **246**, 375-386.
13. Fischer, G. & Schmid, F.X. (1990). *Biochemistry* **29**, 2205-2212.
14. Freedman, R.B. (1989). *Cell* **57**, 1069-1072.
15. Gething, M.-J. & Sambrook, J. (1992). *Nature* **355**, 33-45.
16. Goldberg, M.E. (1985). *Trends Biochem. Sci.* **10**, 388-391.
17. Karplus, M. & Weaver, D.L. (1976). *Nature* **260**, 404-406.
18. Kim, P.S. & Baldwin, R.L. (1990). *Ann. Rev. Biochem.* **59**, 631-660.
19. Lizardi, P.M., Mahdavi, V., Shields, D. & Candelas, G. (1979). *Proc. Natl. Acad. Sci. USA* **76**, 6211-6215.
20. McNally, T., Purvis, I.J., Fothergill-Gilmore, L. & Brown, A.J.P. (1989). *FEBS Letters* **247**, 312-316.
21. Moore, P.A., Sagliocco, F.A., Wood, R.M.C. & Brown, A.J.P. (1991). *Molec. Cell. Biol.* **11**, 5330-5337.
22. Morlon, J., Lloubes, R., Varenne, S., Chartier, M. & Lazdunski, C. (1983). *J. Mol. Biol.* **170**, 271-285.
23. Peters, T. & Davidson L.K. (1982). *J. Biol. Chem.* **257**, 8847-8853.
24. Petersen, N.S. & McLaughlin, C.S. (1973). *J. Mol. Biol.* **81**, 33-45.
25. Phillips, D.C. (1966). *Scientific American* **215** (November), 78-90.
26. Protzel, A. & Morris, A.J. (1974). *J. Biol. Chem.* **249**, 4594-4600.
27. Purvis, I.J., Bettany, A.J.E., Santiago, T.C., Coggins, J.R., Duncan, K., Eason, R. & Brown, A.J.P. (1987). *J. Mol. Biol.* **193**, 413-417.
28. Randall, L.L., Josefsson, L.-G. & Hardy, S.J.S. (1980). *Eur. J. Biochem.* **107**, 375-379.
29. Santiago, T.C., Purvis, I.J., Bettany, A.J.E. & Brown, A.J.P. (1986). *Nucleic Acids Res.* **14**, 8347-8360.
30. Schweizer, M., Roberts, L.M., Holte, H.-J., Takabayashi, K., Hollerer, E., Hoffman, B., Muller, G., Kottig, H. & Schweizer, E. (1986) *Molec. Gen. Genet.* **203**, 479-486.
31. Varenne, S., Knibiehler, M., Cavard, D., Morlon, J. & Lazdunski, C. (1982). *J. Mol. Biol.* **159**, 57-70.
32. Yanofsky, C. (1981). *Nature* **289**, 751-758.
33. Zalkin, H., Paluh, J.L., van Cleemput, M., Moye, W.S. & Yanofsky, C. (1984). *J. Biol. Chem.* **259**, 3985-3992.
34. Zucker, M. & Steigler, P. (1981). *Nucleic Acids Res.* **9**, 133-148.

# PROTEIN SPLICING OF YEAST TFP1: EVIDENCE FOR A NEW CLASS OF MOBILE GENETIC ELEMENTS

Antony A. Cooper and Tom H. Stevens
Institute of Molecular Biology
University of Oregon
Eugene, Oregon 97403
U.S.A.

## Summary

The *TFP1* gene of the yeast *S. cerevisiae* has been found to encode two proteins, the 69 kDa subunit of the vacuolar $H^+$-ATPase, and a 50 kDa 'spacer' protein. The 50 kDa spacer protein is encoded by the central 454 codons of *TFP1*, whereas the first 283 and last 334 amino acids are post-translationally connected by a peptide bond to form the 69 kDa polypeptide. Because the reaction involves excision of an intervening protein domain and the religation of the N- and C-terminal regions of the Tfp1 protein, the process was named protein splicing. The Tfp1 precursor has been found to undergo efficient protein splicing even when expressed in *E. coli*, suggesting that this reaction occurs by a self-splicing mechanism. The spacer protein has been found to undergo protein splicing independent of its insertional context, indicating that all the information necessary for protein splicing is contained within the spacer protein. The Tfp1 spacer protein is an endonuclease that specifically cleaves the *TFP1* DNA lacking the intervening sequence, leading to gene conversion of the allele lacking the intervening sequence by the wild-type *TFP1* allele. These data indicate that the *TFP1* spacer DNA defines a new class of mobile genetic elements, which move at the DNA level and excise by protein splicing to remain phenotypically silent.

## *TFP1* Protein Splicing

The *TFP1* gene (also isolated as *VMA1*) is predicted to encode a 1071 amino acid protein (Tfp1p) with a molecular mass of 119 kDa (Shih et al, 1988; Hirata et al, 1990). The first 283 (N-domain) and last 334 (C-domain) residues, which encode the 69 kDa vacuolar $H^+$-ATPase subunit, exhibit about 75% amino acid identity to the 69 kDa vacuolar $H^+$-ATPase subunit from *Neurospora crassa* (Figure 1), whereas the central 454 residue spacer domain shows no homology to any known vacuolar $H^+$-ATPase subunit (Bowman et al., 1988; Hirata et al, 1990; Kane et al, 1990). A single 3.7 kb *TFP1* mRNA was detected in yeast, and this message contained sequences

NATO ASI Series, Vol. H 71
Protein Synthesis and Targeting in Yeast
Edited by A. J. P. Brown, M. F. Tuite, and J. E. G. McCarthy

corresponding to the N-, spacer, and C-domains (Hirata et al, 1990; Kane et al, 1990). No intron consensus sequences exist in the spacer coding region and the production of the 69 kDa subunit continues at the restrictive temperature in yeast strains that are temperature sensitive for nuclear mRNA splicing (Kane et al, 1990). A frameshift mutation (-1) in the spacer domain eliminates production of the 69 kDa subunit, and leads to the synthesis of a truncated Tfp1 protein. The entire open reading frame must be translated since this mutation could be suppressed by a second frameshift mutation (+1) that restored the reading frame before an out-of-frame stop codon was encountered (Kane et al, 1990). The results of such genetic experiments indicated that the spacer sequence of *TFP1* was indeed translated.

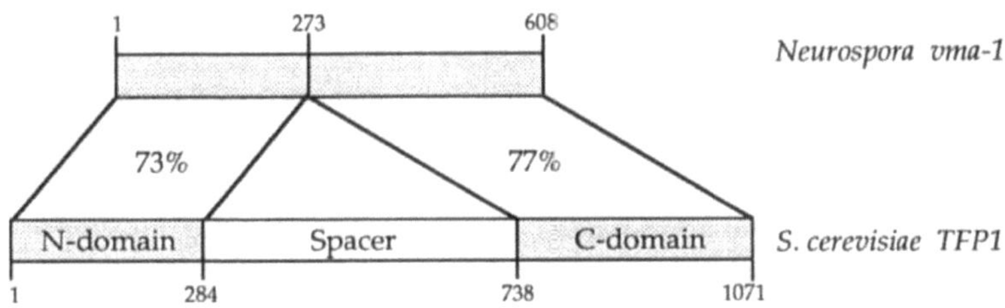


**Figure 1.** The *S. cerevisiae TFP1* N- and C-domain regions are highly homologous to the *N. crassa vma-1* gene. The N- and C-regions of *TFP1* are interrupted by a 1362 bp (454 codon) intervening sequence, called the spacer region.

Further evidence that the *TFP1*-encoded protein undergoes protein splicing comes from biochemical investigations (Kane et al, 1990). Antibodies produced against a predicted peptide of the spacer region identified a 50 kDa protein in yeast, consistent with the predicted size of the translated *TFP1* spacer DNA. Quantitative pulse-chase immunoprecipitation studies indicated that the 69 kDa subunit and the 50 kDa spacer protein were synthesized at the same rate, consistent with the synthesis of one copy of each protein per *TFP1* mRNA translation event. Amino acid sequencing of the tryptic peptide fragment corresponding to the 69 kDa subunit N-/C-domain junction peptide revealed that a peptide bond connects the N- and C-domains in the final 69 kDa spliced protein product (Figure 2). In an attempt to identify and characterize the predicted Tfp1 precursor protein, *TFP1* mRNA was translated *in vitro*. We observed that protein splicing was efficient when *TFP1* was expressed *in vitro*. In addition, when *TFP1* was expressed in *E. coli*, the 69 and 50 kDa protein products were again observed. Based on the very rapid rate of Tfp1p protein splicing and that on the fact that it occurred when Tfp1p was produced in bacteria, we proposed a protein splicing model in which unimolecular processing occurs immediately upon release of the 119 kDa Tfp1p primary translation product from the ribosome (Kane et al, 1990). According to this model, protein splicing is proposed to proceed by a mechanism very similar to the auto-excision of Group I and Group II self-splicing RNA introns (Cech, 1990).

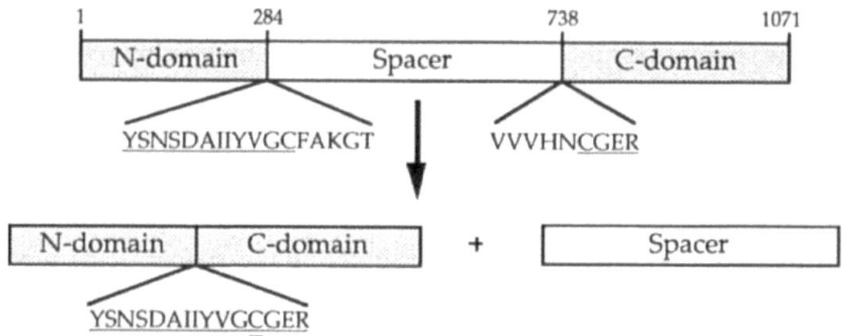


**Figure 2.** The 1071 amino acid *TFP1* open reading frame. The amino acid sequences of the two splice junctions as well as the processed junction peptide are indicated.

Our recent efforts have focused on investigating the mechanism of this extraordinary protein rearrangement reaction. A preliminary genetic analysis has focused on the residues in the vicinity of the two splice junctions of *TFP1*, and on the importance of the N- and C-domains for the protein splicing reaction. Mutational analysis of the splice junctions indicated that residues in these regions are critical for the splicing reaction (Cooper and Stevens, unpublished). Alteration of the $Cys_{284}$ at the N-domain/spacer junction to either Gly, Val, Ser or Arg prevents protein splicing and lead to the accumulation of the 119 kDa Tfp1 precursor protein. Similarly, substitution of either Gly, Ser, or Arg for $Cys_{738}$ at the spacer/C-domain junction also blocked splicing. Whereas the Gly and Arg substitutions at $Cys_{738}$ completely prevented all splicing, Ser in place of Cys at the second splice junction slowed the reaction, such that the conversion of the 119 kDa Tfp1 precursor to the 69 and 50 kDa products could be monitored. This experiment provided conclusive biochemical evidence that splicing of the spacer protein from Tfp1p occurs at the protein level.

To address the importance of the N- and C-domains for the Tfp1p protein splicing reaction, deletions were constructed that removed most of these regions from the *TFP1* gene (Cooper, Chen and Stevens, unpublished). Removal of all but the first 14 amino acid residues of the C-domain (removal of Tfp1 residues 752-1071) did not prevent the splicing process. Deletion of the first 254 residues (leaving 30 N-domain residues) of the N-domain also did not prevent protein splicing. When the N- and C-domain deletions were combined, the spacer protein was still spliced from the precursor protein. In contrast, whereas small in-frame deletions or insertions in the middle of the spacer domain have been found not to interfere with splicing, larger *TFP1* spacer domain deletions result in the synthesis of an unspliced Tfp1 protein. The results from these rather gross deletions indicate that the N- and C-domains are dispensible for protein splicing, and raise the possibility that the structure of the spacer domain is critically important for the splicing reaction.

To test whether the spacer protein contained all of the information required for protein splicing, the spacer protein was engineered into a new context (Cooper and Stevens, 1992b). We have only characterized the site of cleavage to within one amino acid, and thus do not yet know whether the spliced spacer protein has a cysteine at its N- or C-terminus. To be conservative the spacer domain region of *TFP1* was inserted into a position in the open reading frame of the yeast *VAT2* gene (Yamashiro et al, 1990) adjacent to $Cys_{188}$ (*VAT2::spacer*). In this new context, the spacer protein is flanked by Cys residues, but otherwise flanked by unrelated amino acids. Amazingly, the 50 kDa spacer protein was efficiently excised from the Vat2p::spacer fusion protein expressed in yeast, and the spliced 60 kDa Vat2 protein was fully functional (Cooper and Stevens, 1992b). Therefore, given the previous finding that Tfp1 protein splicing occurs in bacteria (Kane et al, 1990), this new result strongly suggests that the Tfp1 spacer protein undergoes a self-excision (or self-splicing) reaction, very similar to self-splicing RNA introns (Cech, 1990).

## Mechanism of Protein Splicing

Our consideration of possible protein splicing models has been influenced by a number observations. The fact that Tfp1p undergoes protein splicing *in vitro* and when expressed in bacteria, indicates that the reaction is likely to proceed by a self-splicing mechanism. The sufficiency of the spacer protein for splicing, together with the mutational analysis of the spacer domain, suggests that the structure of the spacer protein is critical for the reaction. In light of these results, the requirement for a Cys residue at each splice junction suggests that the structure of the spacer protein brings the two junctions close together, allowing some form of interaction between them.

A potentially critical aspect of the protein splicing model was suggested by the results of additional mutational studies of the spacer/C-domain junction. We found that no amino acid residue could substitute for $Asn_{737}$, not even the conservative alteration to Gln. Asn is known to undergo novel rearrangements in peptides and proteins, including deamidation and peptide bond cleavage (Clarke et al, 1992). The spontaneous cleavage of proteins under physiological conditions can begin with an Asn residue forming an intramolecular succinimide ring (Clarke, 1987; Voorter et al, 1988), which results in the cleavage of the peptide bond on the carboxy-terminal side of Asn. Studies with model peptides have indicated that only Asn exhibits this propensity to undergo spontaneous peptide bond cleavage (Stephenson and Clarke, 1989). Interestingly, Clarke et al (1992) speculated that with an optimal fixed alignment of the peptide carbonyl and side chain nitrogen in a 3-dimensional protein structure could allow very rapid succinimide ring formation. Therefore, we propose that the very rapid Tfp1 protein splicing reaction is initiated by spontaneous Asn intramolecular ring formation, thereby breaking the $Asn_{737}$-Cys peptide bond. The free N-terminus of $Cys_{738}$, which is proposed to be in very close proximity to the other splice junction, is now free to attack the Gly-$Cys_{284}$ peptide bond at the first junction. Such a transpeptidation reaction initiated by the Asn rearrangement is diagrammed in Figure 3.

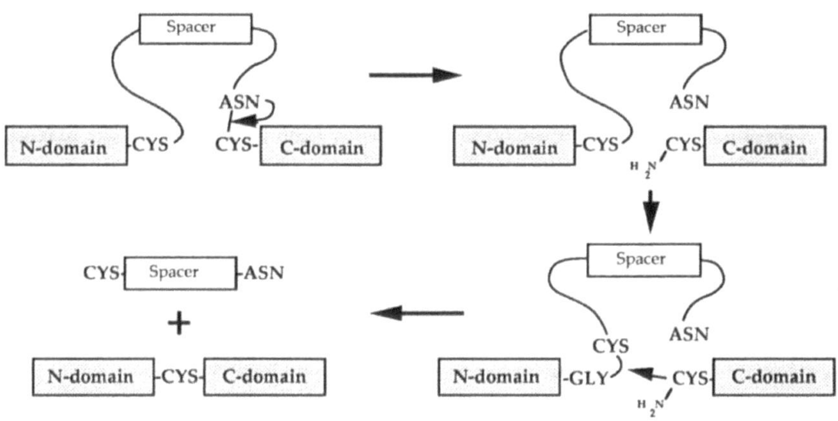


**Figure 3.** Model proposed for Tfp1p protein splicing. Splicing is initiated by Asn ring formation, resulting in cleavage of the Asn-Cys peptide bond. The free N-terminus of $Cys_{738}$ participates in a transpeptidation reaction with the other splice junction.

## Other Examples of Protein Splicing

Concanavalin A (Con A) is a plant lectin that undergoes a post-translational rearrangement that is related to protein splicing (Bowles and Pappin, 1988). Con A isinitially synthesized as a precursor that undergoes processing to a two-chain intermediate (Carrington et al, 1985), and these two protein fragments become ligated through a peptide bond, such that the order of the two fragments is inverted (Bowles and Pappin, 1988). Interestingly, ConA from *Canavalia gladiata* undergoes the same protein rearrangement when expressed in *E. coli* (Yamauchi and Minamikawa, 1990), suggesting that this reaction is self-catalyzed. Whereas the mechanism of Con A rearrangement probably involves a transpeptidation step, this reaction differs significantly from Tfp1 protein splicing in that the Con A precursor does not involve the excision of a large central spacer protein (Cooper and Stevens, 1992a; Hendrix, 1991).

The *recA* gene of *Mycobacterium tuberculosis* is interrupted by a continuous open reading frame intervening sequence that is absent from the *recA* genes of all other species examined (Davis et al, 1991). The sequence of the gene predicts a protein of 85 kDa, of which 45 kDa correspond to the non-homologous spacer protein. Analysis of the protein products of the *M. tuberculosis recA* gene in *E. coli* (Davis et al, 1992) indicated that two proteins of 38 and 47 kDa are produced, and the 38 kDa polypeptide cross-reacted with antibody specific for purified RecA protein. Genetic experiments indicate that the entire *recA* open reading frame is translated, and certain in-frame spacer region deletions have been found to slow the reaction sufficiently to allow a precursor-product analysis to be conducted. Comparison between the *recA* and *TFP1* intervening sequences reveals similarities between the splice junctions (Figure 4). In particular, there is a Cys residue at each splice junction, and the Asn residue is conserved within the VVVHNC motif at the second junction. The sequence similarities at the splice junctions together with the biochemical and genetic results indicate that the *M. tuberculosis* RecA protein undergoes a rapid protein splicing reaction very similar to that followed by yeast Tfp1p.

The archaebacterial *Thermococcus litoralis* DNA polyperase gene was isolated and shown to contain two open reading frame intervening sequences (Perler et al, 1992). The second intervening sequence, IVS2, is predicted to encode a 42 kDa protein that is believed to be removed in a post-translational reaction. Genetic studies indicate that the entire IVS2 region must be translated, and that silent mutations at the IVS2 junctions do not affect the production of the mature DNA polymerase. Perler et al (1992) have found that IVS2 encodes a 42 kDa endonuclease,

I-*Tli* I. As with *M. tuberculosis* RecA, there are conserved residues at the splice junctions, and in particular the Asn is present at the second junction (Figure 4). Interestingly, the N-terminal residue of the 42 kDa I-*Tli* I protein is Ser (Figure 4), consistent with the protein splicing mechanism proposed in Figure 3. Taken together, the data obtained by these investigators make a compelling case that the excision of the 42 kDa I-*Tli* I protein occurs by a protein splicing mechanism very simialr to that proposed for yeast Tfp1 (Figure 3).

```
               N-DOMAIN |------------- SPACER -------------| C-DOMAIN
S. cerevisiae Tfp1        I Y V G C F A K G T N V (454aa) N Q V V V H N C G E R G
                                  |   |   | |                 | | | | | |
M. tuberculosis RecA      V K N K C L A E G T R I (440aa) E G V V V H N C S P P F
                                  :             |                 | | | :
T. litoralis Pol IVS2     L Y A D S V S G E S E I (391aa) N N I L V H N T D G F Y
```

**Figure 4.** Sequence identities at the splice junctions. The numbers in brackets represent the predicted total number of residues comprising each spacer protein. The Ser at the N-domain/spacer junction of *T. litoralis* DNA polymerase is the N-terminal residue of the processed IVS2 protein (I-*Tli* I).

## Invervening Sequences That Excise By Protein Splicing Appear To Be Mobile Genetic Elements

When the *TFP1/VMA1* sequence was compared to others in the protein data base, it was noted that the intervening sequence (spacer protein) exhibits about 34% amino acid identity to the HO endonuclease from *S. cerevisiae* (Hirata et al, 1990; Kane et al, 1990). Gimble and Thorner (1992) have recently reported that the *TFP1*-encoded spacer protein (called VDE, for *VMA1*-Derived Endonuclease) is indeed a double-strand DNA endonuclease. These investigators found that the substrate for the VDE protein was the *TFP1* allele lacking the invervening sequence, and that VDE cleavage occurred at the junction formed by the precise deletion of the intervening sequence. When Gimble and Thorner (1992) generated a heterozygous diploid strain, *TFP1/TFP1Δvde*, they found that cleavage occurred at the *TFP1Δvde* allele, and this cleavage was followed by gene conversion of this allele to the wild-type *TFP1* allele. For reasons that are not understood, this cleavage and resulting gene conversion event only occurred when this heterozygous diploid strain underwent meiosis. The clear message from the studies of Gimble and Thorner is that the *TFP1* intervening sequence is a mobile genetic element, which exhibits 'homing' back to the *TFP1* locus.

The intervening sequence of the *T. litoralis* DNA polymerase gene also encodes an endonuclease (Perler et al, 1992). In fact, this I-*Tli* I endonuclease exhibits 21% amino acid identity to the *S. cerevisiae* HO endonuclease. Removal of IVS2 sequences from the DNA polymerase gene generates a recognition site for the I-*Tli*I endonuclease. Whereas it is unknown whether the I-*Tli* I endonuclease can mediate gene conversion (and thus genetic mobility), the strong prediction is that the *T. litoralis* IVS2 is indeed a mobile genetic element.

The RecA spacer protein from *M. tuberculosis* has not been tested for endonuclease activity, nor has it been determined whether the intervening sequence DNA displays genetic mobility. However, like the spacer proteins from yeast Tfp1p and *T. litoralis* DNA polymerase, the RecA spacer protein exhibits significant amino acid sequence identity (21%) to the yeast HO endonuclease (Davis et al, 1991). This sequence identity is particularly pronounced in the regions displaying the conserved protein motifs, the so-called LAGLI-DADG sequences (in the one-letter amino acid code), which are also found in the open reading frames of mobile Group I introns (Perlman and Butow, 1989).

The genetic mobility of the *S. cerevisiae TFP1* intervening sequence is analogous to the mobility of members of a family of Group I self-splicing introns that encode endonucleases, which endow these introns with the ability to move at the DNA level (Perlman and Butow, 1989). A well characterized member of this family of elements is the ω element in *S. cerevisiae* (Coleaux et al, 1986), which can insert into a recipient allele that lacks the intron in a process called 'intron homing'. Therefore, it is now clear that mobile genetic elements can remain phenotypically silent at their site of insertion by either RNA splicing or protein splicing.

## Conclusions and Perspectives

Recent work from a number of labs has demonstrated that protein splicing is not limited to the *S. cerevisiae TFP1* gene product. There are now three examples of protein splicing that fit the model proposed to explain the Tfp1p splicing reaction. In addition, the post-translational rearrangement of Con A is a form of protein splicing that is likely to differ in detail from the other known examples. In the Tfp1-related protein splicing cases the spacer protein is proposed to function both to bring the splice junctions close together and to increase the reactivity of the Asn residue at the second splice junction to initiate the reaction. All of the available data argue that the mechanism of Tfp1 protein splicing involves the self-excision or self-splicing of the spacer protein. Investigations are currently underway to address in detail the mechanism of the Tfp1 protein splicing reaction.

These investigations have defined a new class of intervening sequence, whose members excise by protein splicing and are active as mobile genetic elements. However, it is too early to predict whether all protein splicing elements will be genetically mobile. Nevertheless, based on the discovery of protein splicing in such diverse organisms as bacteria (and, interestingly, archaebacteria), fungi, and plants, it seems safe to predict that many more examples of protein remain to be discovered.

## References

Bowles DJ and Pappin DJ (1988) Traffic and assembly of concanavalin A. Trends in Bioch Sci 13:60-64

Bowman EJ, Tenney K and Bowman EJ (1988) Isolation of genes encoding the *Neurospora* vacuolar ATPase. J Biol Chem 263:13994-14001

Carrington DM, Auffret A and Hanke DE (1985) Polypeptide ligation occurs during post-translational modification of concanavalin A. Nature 313:64-66

Clarke S (1987) Propensity for spontaneous succinimide formation from aspartyl and asparaginyl residues in cellular proteins. Internat J Peptide Protein Res 30:808-821

Clarke S, Stephenson RC and Lowenson JD (1992) Lability of asparagine and aspartic residues in proteins and peptides: spontaneous deamination and isomerization reactions. Pharmaceutical Biotechnology Manning MC and Berchan R (eds) Plenum Press New York

Coleaux L, d'Auriol L, Betermier M, Cottarel G, Jacquier A, Galibert F and Dujon B (1986) Universal code equivalent of a yeast mitochondrial intron reading frame is expressed into *E coli* as a specific double strand endonuclease. Cell 44:521-533

Cooper AA and Stevens TH (1992a) Protein splicing: excision of intervening sequences at the protein level. BioEssays in press

Cooper AA and Stevens TH (1992b) Protein splicing of the yeast tfp1 intervening sequence occurs by self-excision. submitted to Nature

Davis EO, Jenner PJ, Brooks PC, Colston MJ and Sedgwick SG (1992) Protein splicing in the maturation of *Mycobacterium tuberculosis* RecA protein. submitted to Cell

Davis EO, Sedgwick SG and Colston MJ (1991) Novel structure of the *recA* locus of *Mycobacterium tuberculosis* implies processing of gene product. J Bact 173:5653-5662

Gimble FS and Thorner J (1992) Homing of a DNA endonuclease gene by meiotic conversion in *Saccharomyces cerevisiae.* Nature 357:301-306

Hendrix RW (1991) Protein Splicing. Current Biology 1:71-73

Hirata R, Ohsumi Y, Nakano A, Kawasaki H, Suzuki K and Anraku, Y (1990) Molecular structure of a gene, *VMA1*, encoding the catalytic subunit of $H^+$-translocating adenosine triphosphatase from vacuolar membranes of *Saccharomyces cerevisiae.* J Biol Chem 265:6726-6733

Kane PM, Yamashiro CT, Wolczyk DF, Goebl M, Neff N and Stevens TH (1990) Protein splicing converts the yeast *TFP1* gene product to the 69 kDa subunit of the vacuolar $H^+$-adenosine triphosphatase. Science 250:651-657

Perler FB, Comb DG, Jack WE, Moran LS, Qiang B, Kucera RB, Benner J, Slatko BE, Nwankwo DO, Hempstead SK, Carlow CKS and Jannasch H (1992) Intervening sequences in an Archaea DNA polymerase gene. Proc Natl Acad Sci 89:5577-5581

Perlman PS and Butow RA (1989) Mobile introns and intron-encoded proteins. Science 246:1106-1109

Shih C-K , Wagner R , Feinstein S , Kanik-Ennulat C and Neff N (1988) A dominant trifluorperazine resistance gene from *Saccharomyces cerevisiae* has homology with $F_0F_1$ ATP synthase and confers calcium-sensitive growth. Mol Cell Biol 8:3094-3103

Stephenson RC and Clarke S (1989) Succinimide formation from aspartyl and asparaginyl peptides as a model for the spontaneous degradation of proteins. J Biol Chem 264:6164-6170

Voorter CEM, de Haard-Hoekman WA, van den Oetelaar PJM, Bloemendal H and de Jong WW (1988) Spontaneous peptide bond cleavage in aging α-crystallin through a succinimide intermediate. J Biol Chem 263:19020-19023

Yamashiro CT, Kane PM, Wolczyk DF, Preston RA and Stevens TH (1990) Role of vacuolar acidification in protein sorting and zymogen activation: a genetic analysis of the yeast vacuolar proton-translocating ATPase. Mol Cell Biol 10:3737-3749

Yamauchi D and Minamikawa T (1990) Structure of the gene encoding concanavalin A from *Canavalia gladiata* and its expression in *Escherichia coli* cells. FEBS Lett 260:127-130

# Ubiquitin-dependent protein degradation

*Stefan Jentsch*, Hans-Peter Hauser, Ruth Heinlein, Joern Jungmann, Stephan Schlenker, Wolfgang Seufert, Thomas Sommer, and Sebastian Springer*

Friedrich-Miescher-Laboratorium der Max-Planck-Gesellschaft
Spemannstr. 37-39, 7400 Tübingen, Germany.

*Corresponding author

## Introduction

Intracellular protein levels are controlled by protein synthesis and degradation. One major proteolytic pathway of eukaryotes requires the covalent attachment of ubiquitin, a small (8.5 kDa) and highly conserved protein, to proteolytic substrates prior to degradation (for reviews see Hershko 1991, Finley and Chau 1991, Jentsch et al. 1991, Jentsch 1992a, Jentsch 1992b). This pathway is highly selective and mediates the elimination of abnormal proteins and controls the half-lives of some regulatory proteins. Known targets include transcriptional regulators (Hochstrasser et al. 1991) p53 (Scheffner et al. 1990) and cyclins (Glotzer et al. 1991).

The conjugation of ubiquitin to protein substrates is a multistep process. In an initial activating step, a thioester is formed between the C-terminus of ubiquitin and an internal cysteine residue of a ubiquitin-activating (E1) enzyme. Activated ubiquitin is then transferred to a specific cysteine residue of one of several ubiquitin-conjugating (E2) enzymes. Finally, these E2 enzymes donate ubiquitin to protein substrates. This results in branched protein conjugates in which the C-terminus of ubiquitin in linked by an isopeptide bond to specific internal lysine residues of target proteins. Since ubiquitin itself is a substrate for ubiquitination, certain substrates are modified by the attachment of multiubiquitin chains (Chau et al. 1989). Multiubiquitinated substrates are degraded in vivo by the proteasome, a multisubunit protease complex (Seufert and Jentsch, 1992).

Purified enzymes of the *Saccharomyces cerevisiae* ubiquitin-conjugating system have been used to study the enzymology of the ubiquitin system and allowed the cloning of the corresponding genes. So far, ten genes coding for ubiquitin-conjugating enzymes (*UBC* genes) have been isolated from the yeast *S. cerevisiae* (reviewed in Jentsch et al. 1990). These enzymes are located in the cytosol and the nucleus, indicating that both cellular

NATO ASI Series, Vol. H 71
Protein Synthesis and Targeting in Yeast
Edited by A. J. P. Brown, M. F. Tuite, and J. E. G. McCarthy

compartments have active ubiquitin-conjugating systems. Here we review our data on the genetic analysis of the ubiquitin system (see Table 1).

## Results and Discussion

### *Functions in protein quality control pathways*

Ubiquitin-dependent proteolysis serves essential functions in probably all eukaryotes (McGrath et al. 1991). The yeast proteins UBC1, UBC4, UBC5 and UBC7 have roles in the degradation of abnormal proteins. *UBC1*, *UBC4* and *UBC5* constitute a gene family essential for viability (the triple mutant is inviable) (Seufert and Jentsch 1990, Seufert et al. 1990). The *ubc1* and *ubc4* single mutants and *ubc4 ubc5* double mutants have a slow growth phenotype and are hypersensitive to amino acid analogs. In vivo data indicate that these mutants are largely deficient in the degradation of otherwise short-lived and abnormal proteins. The *ubc1* mutants show specific defects in growth after germination of ascospores, suggesting that UBC1 has a specific function during this transition period in the yeast life cycle. The enzymes UBC4 and UBC5 mediate essential functions of the stress response: the transcription of *UBC4* and *UBC5* genes is induced by heat and *ubc4 ubc5* double mutants are inviable under stress conditions. Transcription of *UBC7* is induced by cadmium but not by heat shock (H.-A. Reins, J. Jungmann, C. Schobert and S. Jentsch, in preparation). Mutations in *UBC7* do not lead to an altered growth rate or a heat shock sensitive phenotype. Yet *ubc7* mutants and also *ubc4 ubc5* double mutants are cadmium hypersensitive, suggesting that cadmium exposure generates abnormal proteins which are normally subject to UBC4, UBC5 and UBC7-mediated proteolysis. A structural close homolog of UBC4 and UBC5 has been cloned from *Drosophila melanogaster* (Treier et al. 1992). This gene, *UbcD1*, has been shown to be functionally equivalent to yeast *UBC4*, suggesting that UBC4-like enzymes mediate similar functions in probably all eukaryotes.

The UBC6 protein has a C-terminal signal-anchor sequence and is an integral membrane protein of the endoplasmic reticulum, with the active site of the enzyme on the cytoplasmic face of the membrane (T. Sommer and S. Jentsch, in preparation). Recent genetic data suggest that the UBC6 enzyme has a function in the secretory pathway, probably by mediating the degradation of mislocalized proteins. Another integral membrane ubiquitin-conjugating enzyme, UbcM1, was recently shown to be expressed in mouse pituitary tumor cells (H.-P. Hauser and S. Jentsch, in preparation). This enzyme is unusually large (500 kDa) and has several potential transmembrane spanning segments.

### *Functions in cellular regulation*

The UBC2 protein (also known as RAD6) plays a prominent role in a central DNA-repair pathway of yeast (Jentsch et al. 1987, Friedberg 1988). Mutants of *UBC2* are highly vulnerable to DNA-damaging agents and deficient in induced mutagenesis. Moreover, *UBC2* is required for sporulation and repression of retrotransposition in yeast (Picologlou et al. 1990). This regulatory role suggests that UBC2 might mediate the degradation of regulatory proteins such as transcriptional repressors for genes encoding proteins for DNA repair, sporulation and transposition functions. Alternatively, since this enzyme conjugates ubiquitin to histones in vitro, UBC2 might modulate chromatin structure (Jentsch et al. 1987).

Only two of the known ten *UBC* genes, *UBC3* and *UBC9*, are essential for viability and both of these encode nuclear enzymes which are required for cell cycle progression (Goebl et al. 1988, W. Seufert and S. Jentsch, unpublished). Whereas the precise function of *UBC9* in the cell cycle is presently unknown, *UBC3* (also known as *CDC34*) is required for the transition from the $G_1$ into the S-phase of the cell cycle (Goebl et al. 1988). Temperature-sensitive *ubc3* mutants arrest at the restrictive temperature in $G_1$ with multiple elongated buds and have unreplicated DNA. The spindle pole bodies duplicate but fail to undergo separation. The in vivo substrates of the UBC3 enzyme are presently unknown but may include the yeast $G_1$-cyclins.

**TABLE 1 THE YEAST UBIQUITIN-CONJUGATING SYSTEM**

| Genes | Function | References |
|---|---|---|
| *UBA1* | ubiquitin-activating enzyme, essential for viability<br>cytoplasmic and nuclear | McGrath et al. 1991 |
| *UBC1* | specifically required after sporulation/germination<br>mediates vital functions together with *UBC4* and *UBC5*<br>$G_0$ induction | Seufert et al. 1990 |
| *UBC2/RAD6* | DNA repair, induced mutagenesis, sporulation,<br>repression of retrotransposition | Jentsch et al. 1987<br>Friedberg, 1988<br>Picologlou et al. 1990 |
| *UBC3/CDC34* | essential for viability, $G_1$-S cell cycle control,<br>DNA replication, spindle pole body separation<br>nuclear protein | Goebl et al. 1988 |
| *UBC4*<br>*UBC5* | bulk degradation of short-lived and abnormal proteins<br>mediate vital functions together with *UBC1*<br>essential under stress conditions<br>heat shock inducible, cadmium inducible | Seufert and Jentsch 1990<br>Seufert et al. 1990<br>Seufert and Jentsch 1990 |
| *UBC6* | integral membrane protein of the endoplasmic reticulum<br>function in the secretory pathway | T. Sommer and S. Jentsch, in prep. |
| *UBC7* | confers cadmium resistance<br>cadmium inducible | H.-A. Reins, J. Jungmann,<br>C. Schobert, S. Jentsch, in prep. |
| *UBC8* | no detectable mutant phenotypes | Quin et al. 1991 |
| *UBC9* | essential for viability, cell cycle progression<br>nuclear protein | W. Seufert and S. Jentsch, unpublished |

**References**


Chau, V., Tobias, J.W., Bachmair, A., Marriott, D., Ecker, D.J., Gonda, D.K., & Varshavsky, A. (1989) A multiubiqquitin chain is confined to specific lysine in a targeted short-lived protein. Science 243:1576-1583

Friedberg, E.C. (1988) Deoxyribonucleic acid repair in the yeast *Saccharomyces cerevisiae*. Microbiol. Rev. 52:70-102

Finley, D. & Chau, V. (1991) Ubiquitination. Annu. Rev. Cell Biol. 7:25-69

Glotzer, M., Murray, A.W., & Kirschner, M.W. (1991) Cyclin is degraded by the ubiquitin pathway. Nature 349:132-138

Goebl, M.G., Yochem, J., Jentsch, S., McGrath, J.P., Varshavsky A., & Byers, B. (1988) The yeast cell cycle gene *CDC34* encodes a ubiquitin-conjugating enzyme. Science 241:1331-1335

Hershko, A. (1991) The ubiquitin pathway for protein degradation. Trends Biochem. Sci. 16:265-268

Hochstrasser, M., Ellison, M.J., Chau, V., & Varshavsky, A. (1991) The short-lived MATα2 transcriptional regulator is ubiquitinated *in vivo*. Proc. Natl. Acad. Sci. USA 88:4606-4610

Jentsch, S. (1992a) Ubiquitin-dependent protein degradation: a cellular perspective. Trends Cell Biol. 2:98-103

Jentsch, S. (1992b) The ubiquitin-conjugation system. Annu. Rev. Genet. 26, in press

Jentsch, S., McGrath, J.P., & Varshavsky, A. (1987) The DNA repair gene *RAD6* encodes a ubiquitin-conjugating enzyme. Nature 329:131-134

Jentsch, S., Seufert, W., & Hauser, H.-P. (1991) Genetic analysis of the ubiquitin system. Biochim. Biophys. Acta 1089:127-139

Jentsch, S., Seufert, W., Sommer, T., & Reins, H.-A. (1990) Ubiquitin-conjugating enzymes: novel regulators of eukaryotic cells. Trends Biochem. Sci. 15:195-198

McGrath, J. P., Jentsch, S., & Varshavsky, A. (1991) *UBA1*: an essential yeast gene encoding ubiquitin-activating enzyme. EMBO J. 10:227-236

Picologlou, S., Brown, N., & Liebman, S.W. (1990). Mutations in *RAD6*, a yeast gene encoding a ubiquitin-conjugating enzyme, stimulate retrotransposition. Mol. Cell. Biol. 10:1017-1022

Qin, S., Nakajima, B., Nomura, M., & Arfin, S.M. (1991) Cloning and Characterization of a *Saccharomyces cerevisiae* gene encoding a new member of the ubiquitin-conjugating protein family. J. Biol. Chem. 266, 15549-15554

Scheffner, M., Werness, B.A., Huibregtse, J.M., Levine, A.J., & Howley, P.M. (1990) The E6 oncoprotein encoded by human papillomavirus types 16 and 18 promotes the degradation of p53. Cell 63:1129-1136

Seufert, W., McGrath, J.P., & Jentsch, S. (1990) *UBC1* encodes a novel member of an essential subfamily of yeast ubiquitin-conjugating enzymes involved in protein degradation. EMBO J. 9:4535-4541

Seufert, W., & Jentsch, S. (1990) Ubiquitin-conjugating enzymes UBC4 and UBC5 mediate selective degradation of short-lived and abnormal proteins. EMBO J. 9:543-550

Seufert, W., & Jentsch, S. (1992) In vivo function of the proteasome in the ubiquitin pathway. EMBO J. 11, 3077-3080

Treier, M., Seufert, W., & Jentsch, S. (1992) *Drosophila UbcD1* encodes a highly conserved ubiquitin-conjugating enzyme involved in selective protein degradation. EMBO J. 11:367-372

An earlier version of this review appeared in Colloquium Mosbach 1992, Springer Verlag Heidelberg, 1992. Research in this laboratory is supported by grants to S. J. from the Deutsche Forschungsgemeinschaft, The German-Israeli Foundation for Scientific Research and Development and The Human Frontier Science Program.

NOVEL CYTOPLASMIC CAP BINDING PROTEINS IN YEAST

Nahum Sonenberg, Antony Blanc, and Charles Goyer
Department of Biochemistry
McGill University
3655 Drummond St.
Montreal, P.Q., Canada, H3G 1Y6

**INTRODUCTION**

The 5' terminal cap structure, $m^7$GpppX (where X is any nucleotide), is a ubiquitous feature of eukaryotic mRNAs that plays an important role in the cytoplasm and the nucleus. In the cytoplasm the cap structure enhances translational efficiency (Shatkin, 1985). In the nucleus, *in vitro* studies have demonstrated that the efficiency of mRNA splicing (Konarska et al,1984;Edery and Sonenberg,1985), and 3' end processing (Georgiev et al, 1984;Hart et al, 1985) is enhanced by the presence of a cap structure. Furthermore, the cap protects the mRNA against 5' exonucleolytic degradation in both the nucleus and the cytoplasm (Furuichi et al,1977;Green et al, 1983) and is implicated in nucleocytoplasmic transport (Hamm and Mataj, 1990). Consistent with the role of the cap in nuclear events, recent data demonstrate that a significant portion of the cap binding protein (eIF-4E) is localized to the nucleus (Lejbkowicz et al,1992).

The best characterized role of the cap structure is its stimulatory effect on ribosome binding (for reviews see, Rhoads 1988 and Sonenberg,1990). The binding of ribosomes to the mRNA is thought to be the rate limiting and a discriminatory step in translation (Ray et al,1983; Sarkar et al,1984). The pivotal role of the cap structure in translation initiation is mediated by a multisubunit complex termed the cap binding protein complex (CBPc or eIF-4F) (Tahara et al,1981;Edery et al,1983;Grifo et al, 1983). eIF-4F, via an associated RNA-dependent helicase activity, is thought to stimulate ribosome binding by unwinding the secondary structure in the mRNA 5' untranslated region (for reviews, see Sonenberg 1988; Hershey 1991).

Several approaches have been used to purify and characterize cellular and viral proteins that interact with the cap structure. Proteins that either specifically cross-link to the cap structure or can be purified by cap-analog affinity chromatography are considered cap binding proteins (Sonenberg et al, 1978;Tahara et al,1983; Grifo et al, 1983). One such CBP, initiation factor eIF-4E, has been isolated and cloned from several species (Altmann et al, 1987;Rychlik et al, 1987;Jaramillo et al, 1991). In all species studied eIF-4E is associated with at least one other subunit ranging in size from 80 to 220 kDa to form eIF-4F. eIF-4E is associated with a 150 kDa subunit in the yeast eIF-4F (Goyer et al, 1989). No specific function has been ascribed to the p150 subunit of

NATO ASI Series, Vol. H 71
Protein Synthesis and Targeting in Yeast
Edited by A. J. P. Brown, M. F. Tuite, and J. E. G. McCarthy

yeast eIF-4F. Recent evidence suggest that an 18 kDa polypeptide is also part of the yeast eIF-4F (Lanker et al, 1992). Thus, the 24 kDa CBP can associate independently with at least two polypeptides of 18- and 150 kDa to form CBP complexes (Lanker et al, 1992).

The p220 component of mammalian eIF-4F is the least characterized of the eIF-4F subunits and its function is moot. The only functional data is derived from translation studies in poliovirus-infected cells. A causal relationship has been demonstrated between the poliovirus-induced cleavage of the p220 and the shut-off of host translation indicating that the p220 is essential for cap-dependent translation (reviewed in Sonenberg, 1987). Moreover, addition of eIF-4F to poliovirus-infected cell extracts rescues cap-dependent translation (Tahara et al, 1981;Lee et al,1985).

In a recent report on yeast CBPs, we described the isolation and characterization of the yeast homolog of eIF-4F (Goyer et al,1989). We also identified a 96 kDa polypeptide capable of cross-linking spontaneously to capped RNA (Goyer et al,1989). Here, we summarize our studies on the characterization of the 96 kDa CBP (referred to as 93 kDa throughout this review) as the coat protein of the yeast dsRNA virus L-A. We also report that the high molecular subunit of yeast eIF-4F is encoded by two homologous genes, *TIF4631* and *TIF4632*.

**THE COAT PROTEIN OF A YEAST VIRUS IS A CAP BINDING PROTEIN**

Incubation of a yeast extract with cap-labeled mRNA results in covalent binding of two polypeptides to the mRNA (Fig. 1, lane 1). Most of the binding occurs to a 93 kDa polypeptide and is completely blocked by the presence of the cap analog $m^7$GDP (compare lane 2 with lane 1). A minor RNA-linked polypeptide migrates as a 75 kDa protein, and its linkage is not affected by the addition of $m^7$GDP. Upon UV irradiation, additional polypeptides cross-link to the mRNA in a cap-specific manner (compare lanes 3 and 4). These polypeptides include subunits of the eIF-4F CBP complex; eIF-4E (molecular weight, 24,000) and p150 (molecular weight, 150,000; Goyer et al, 1989).

To clone the gene coding for the 93 kDa CBP that binds covalently to the mRNA cap structure, we purified the protein and obtained amino acid sequence information. The protein was purified by affinity chromatography on a $m^7$GDP agarose matrix (Edery et al,1988). Surprisingly, the amino acid sequences of several peptides generated from the purified 93 kDa CBP were found to be almost identical in sequence to the coat protein of the L-A dsRNA virus of *Saccharomyces cerevisiae*.

Most laboratory strains of *Saccharomyces cerevisiae* harbor one or more of five major double stranded RNA (dsRNA) virus families (L-A, L-BC, M,T, and W; for a recent review, see Wickner 1992). L-A and L-BC represent distinct groups of closely related variants which are found in intracellular particles composed of a single dsRNA genome encapsidated by

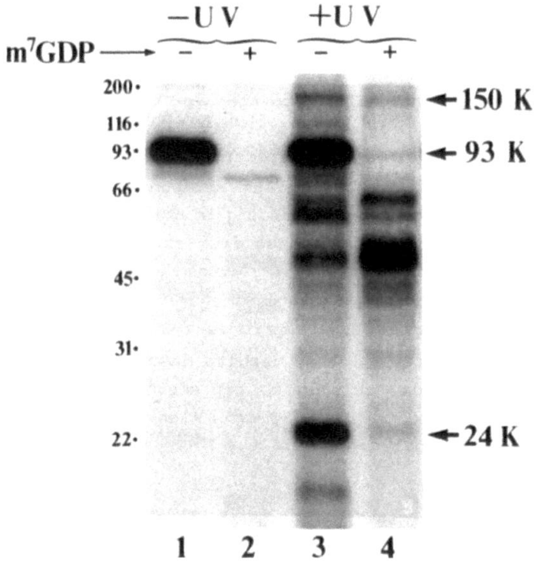


Fig.1. Spontaneous and UV-induced covalent linkage of proteins in yeast S-100 to mRNA cap structure. Cap-labeled RNA (2 x $10^5$ cpm) was incubated with 100 $\mu g$ of an S-100 fraction from *Saccharomyces cerevisiae* lab strain, EJ101, for 10 min at 30°C and then either transferred to ice for 1 h (lanes 1 and 2) or irradiated with UV light for 45 min (lanes 3 and 4), in the presence (lanes 2 and 4) or absence (lanes 1 and 3) of 670 $\mu M$ $m^7GDP$. Samples were processed for SDS-PAGE and autoradiography. Positions of molecular size markers (in kilodaltons) are indicated at the left; positions of the CBPs are indicated by arrows at the right.

60 to 120 coat polypeptide monomers (Bruenn, 1980). The genome of the L-A strain is 4.5 kb in length, and contains two overlapping open reading frames (ORFs). ORF1 encodes the coat protein reported to migrate on SDS polyacryalamide gels as a 81-88 kDa polypeptide (El-Sheirbeini et al, 1984; Sommer and Wickner, 1987). A fusion of ORF1 and ORF2 resulting from a -1 frameshift generates a 180 kDa gag-pol like protein.

To prove that the 93 kDa CBP is indeed the L-A coat protein, yeast strains that were either free of, or that harbored L-A or L-BC dsRNA virus were obtained from Reed Wickner (National Institutes of Health) for use in cross-linking experiments. Cell extracts from these strains were incubated with $^{32}P$-cap-labeled mRNA, and the resulting cap-labeled polypeptides were resolved on SDS-polyacylamide gels (Fig.2). Extract "a" was prepared from a strain commonly used in our laboratory that contains both spontaneously RNA-linked polypeptides (lanes 1 and 2). Extract "b" was prepared from a strain devoid of both L-A and L-BC. No polypeptide in extract "b" became covalently linked to the cap structure (lanes 3 and 4), suggesting that both polypeptides are of viral origin. Only the 93 kDa polypeptide bound to mRNA in a $m^7GDP$-specific manner (compare lane 6 with lane 5), when an extract containing L-A virus was used (extract "c"). Finally, a single polypeptide corresponding to the faster-migrating 75 kDa species was linked in extracts prepared from strains harboring the L-BC virus exclusively (extracts "d" and "e"; compare lanes 7 and 9 with lane 1). As observed above (Fig.1), linkage of the 75 kDa polypeptide was not sensitive to inhibition by $m^7GDP$ (compare lane 8 with lane 7 and lane 10 with lane 9). Taken together, these results demonstrate that the 93 kDa polypeptide, which is linked to the mRNA cap structure in yeast extracts, is the coat protein of the L-A dsRNA virus. Also, these results strongly suggest that the 75 kDa polypeptide is the L-BC coat protein.

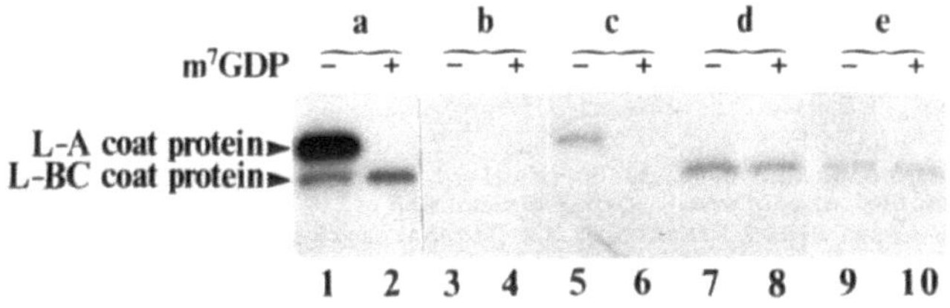


Fig.2. Linkage of CBPs in extracts from virus-harboring yeast strains. Cap-labeled RNA ($10^5$ cpm) was incubated in the absence (odd-numbered lanes) or presence (even-numbered lanes) of 670 $\mu$DM $m^7GDP$. The following strains were used: a, containing both L-A and L-BC; b, free of virus; c, harboring the L-A virus; d and e, harboring the L-BC virus only. Samples were processed, electrophoresed, and analyzed by autoradiography.

We initiated studies to determine the mechanism of covalent bond formation between the coat protein and the mRNA cap structure. Our results suggest that the coat protein engenders the cleavage of the cap structure upon covalent bond formation (Blanc et al, 1992). A model describing this reaction is presented in Fig.3. Tobin et al, (1989) proposed a model involving RNA cleavage for the self-catalyzed covalent attachment of VPg to poliovirus RNA by a transesterification reaction, which, like the covalent linkage between capped RNA and the L-A coat protein, requires $Mg^{2+}$ and is nucleoside triphosphate independent. Similarly, the linkage reaction involving the coat protein could be self-catalyzed, although we have not ruled out catalysis by an enzyme associated with the viral particles.

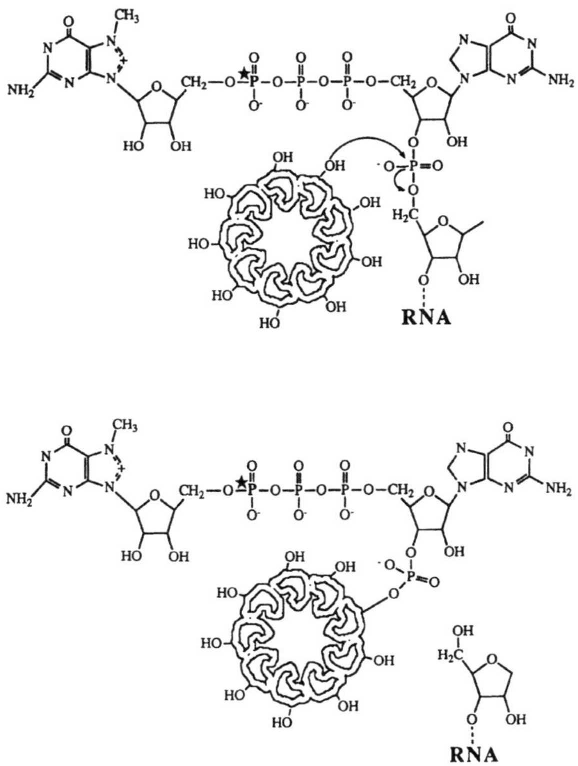


Fig.3. A model for covalent bond formation between L-A coat protein and mRNA 5′ cap structure.

Bruenn and Keitz (1976) and Nemeroff and Bruenn (1987) demonstrated that neither the viral dsRNA nor the *in vitro*-produced positive strand is capped at the 5′ end. It is therefore likely that the cap-binding activity is implicated in some interaction with cellular mRNAs. However, the quantitative binding of L-A virus to the cap column suggests that the coat protein is not covalently attached to cap structures *in vivo* unless only a small fraction of the sites are occupied on each viral particle. One plausible explanation for the apparent absence of covalently attached caps to purified virus is that this interaction is potentially deleterious to the host and therefore is prevented from occurring in the cell. The presence of dsRNA yeast viruses is not deleterious to growth, because the virus copy number (typically 1,000 L-A, 100 L-BC, and 150 $M_1$ molecules per cell) is curtailed by a class of cellular genes termed *SKI* (superkiller [Ridley et al, 1984; Rhee et al, 1989; Wickner, 1992]). An intriguing possibility is that the product of one or more of the *SKI* genes prevents the interaction between the virus coat protein and the cap structure of cellular mRNAs.

**eIF-4F p220 HOMOLOGS IN YEAST**

To clone the yeast homolog of mammalian p220, we have used a polyclonal anti-CBP antibody that recognizes all of the subunits of yeast eIF-4F (Altmann et al,1985), to screen a λgt11 yeast cDNA library. One of the cDNAs clones was shown to encode a 150 kDa polypeptide as determined by the Olmsted method (Olmsted, 1981). Northern blot analysis identified a single mRNA band of approximately 3,800 nt. To isolate the genomic copy of p150, a YEp13 yeast genomic library was screened, and six overlapping clones were obtained. Nucleotide sequencing revealed an open reading frame of 952 amino acids. The AUG at position 1 is most probably the initiator, since it is preceded by two termination codons in the same reading frame (positions -45 and -18). Surprisingly, the molecular weight predicted from this ORF is 107,000, considerably smaller than the estimated 150,000 based on its migration on SDS-polyacrylamide gels. The gene was designated *TIF4631* for Translation Initiation Factor 4 (4 designates RNA binding factors [Safer, 1989]);6 (F, 6th.letter); 3 (γ subunit [Safer, 1989]); 1 (the first of two homologous genes-see below).

A *TIF4631*-related gene termed *TIF4632* was cloned by low stringency hybridization. This gene is predicted to encode a 104 kDa polypeptide with 53% identity to *TIF4631*. Both proteins share common structural domains (Fig.4A). The most striking feature is the presence of a region resembling the RNA recognition motif (RRM;Fig.4B) (Bandziulis et al, 1989;Kenan et al, 1991;Mattaj,1989). The RRM, found in members of a large family of proteins that bind to a wide variety of RNAs, contains a conserved sequence of approximately 80 amino acids that folds into a characteristic tertiary structure consisting of four antiparallel β-strands and two α-helices (for reviews, see Kenan et al, 1991; Haynes 1992). It is also comprised of two conserved RNA-binding protein consensus sequences (RNP-CS): RNP-1, an octamer which is part of the β3 strand, and RNP-2, a hexamer which is part of the β1 strand. Sequences similar to the RNP-CS, and structural motifs similar to the RRM are present in TIF4631 and TIF4632 (Fig.4B). Most RRM proteins contain auxiliary domain(s) that are distinct from the RRM (reviewed in Bandziulis et al, 1989;Keene and Query, 1991 and Haynes 1992). These domains which are usually glycine-, proline- or arginine-serine rich, are believed to contribute to protein-protein and protein-nucleic acid interactions (Bandziulis et al, 1989; Zamore et al, 1990). Present in a number of transcription factors, a glutamine-rich region has been shown to be required for the transcriptional activity of SP1, and is believed to be involved in protein-protein interactions (Courey and Tjian 1988). proline-rich transcription activation domains likely serve similar functions (Mermod et al, 1989). Auxiliary domains are also present in TIF4631 and TIF4632 (Fig.4A). Three findings demonstrate that TIF4631 and possibly TIF4632, are RNA binding proteins. First, TIF4631, and probably TIF4632 can cross-link to RNA. Second, when purified yeast eIF-4F is cross-linked to the cap structure, TIF4631 (p150) cross-linking is not inhibited by cap analogs, suggesting that it interacts with the mRNA independently of eIF-4E. Third, West-Northern analysis reveals that TIF4631 binds to RNA. The RNA binding properties of TIF4631 could be important in the initial interaction of eIF-4F with the mRNA 5' cap structure.

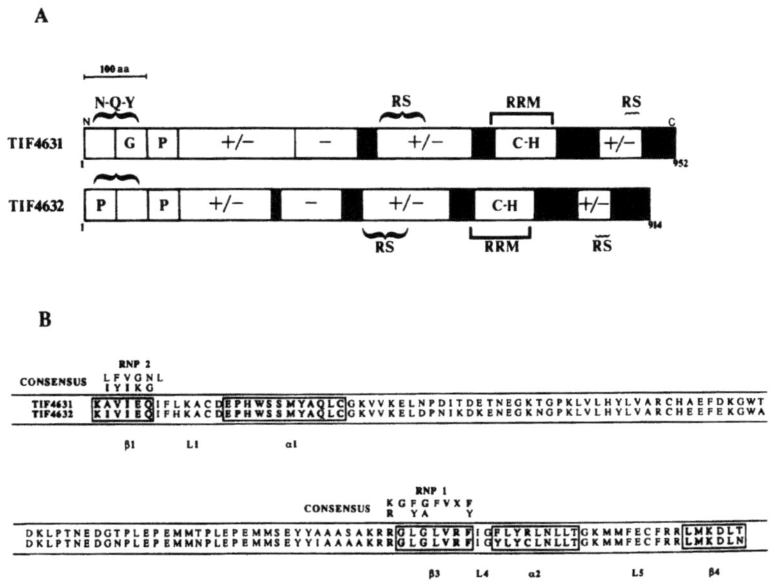


Fig.4. Regions of TIF4631 and TIF4632 predicted to be involved in RNA binding or protein-protein interactions. (A) Schematic representation of TIF4631 and TIF4632. N-Q-Y, G and P indicate regions rich in asparagine-glutamine-tyrosine, glycine and proline, respectively. RS indicate arginine-serine-rich regions. C-H depicts a cysteine-histidine-rich region. RRM denotes the position of the RNA recognition-like motif. The +/- sign depicts a highly charged region. The functional relevance of these domains is discussed in the text. (B) Primary sequence and predicted structure of the RRM-like domain of TIF4631 and TIF4632. The predicted α-helices and β-strands are indicated by shaded boxes.

To examine whether *TIF4631* and *TIF4632* fulfill an essential function in yeast, we performed gene disruption experiments. Gene disruption of one copy of TIF4631 in a diploid strain was performed by replacing 76% of the ORF with *LEU2*. Additionally, *HIS3* was used to replace all but 70 amino acids at the carboxy end of the unique *TIF4631* in a haploid strain. The *LEU2* gene disruption resulted in four viable meiotic products upon sporulation, indicating that the gene is not essential. Leu or His prototrophy, however, co-segregated with a slow growth, cold sensitive phenotype as the growth defect was accentuated at 18°C. The growth defect of the tif4631::*LEU2* haploid cell could be complemented by the introduction of a YCp50-based plasmid harboring *TIF4631*. Gene disruption of *TIF4632* resulted in no detectable phenotype. However, double gene disruption of *TIF4631* and *TIF4632* was lethal. The essential function of *TIF4631*/*TIF4632* is in agreement with the requirement for an intact p220 for efficient cap-dependent translation in mammalian cells (Etchison et al, 1984). We conclude that one of the functions of yeast p150 and its mammalian homolog, p220, is to serve as the RNA-binding component of eIF-4F.

ACKNOWLEDGMENTS:

We thank our collaborators H. Lee, M. Altmann, S. Lanker, P. Müller and H. Trachsel for their invaluable contributions to the work on *TIF4631/4632* cloning. This research was supported by a grant from the Medical Research Council of Canada to N.S. A.B. is the recipient of a predoctoral fellowship from the Medical Research Council of Canada, and C.G. is a recipient of a predoctoral studentship from the Cancer Research Society (Montreal).

REFERENCES

Altmann M, Edery I, Sonenberg N, Trachsel H (1985). Purification and characterization of protein initiation factor eIF-4E from the yeast *Saccharomyces cerevisiae*. Biochemistry 24:6085-6089.

Altmann M, Handschin C, Trachsel H (1987). mRNA cap-binding protein:cloning of the gene encoding protein synthesis initiation factor eIF-4E from *Saccharomyces cerevisiae*. Mol Cell Biol 7:998-1003.

Bandziulis RJ, Swanson MS, Dreyfuss G (1989). RNA-binding proteinsas developmental regulators. Genes & Dev 3:431-437.

Blanc A, Goyer C, Sonenberg N (1992). The coat protein of the yeast double-stranded RNA virus L-A attaches covalently to the cap structure of eukaryotic mRNA. Mol Cell Biol 12:3390-3398.

Bruenn J, Keitz B (1976). The 5' ends of yeast killer factor RNAs are pppGp. Nuc Acids Res 3:2427-2436.

Bruenn JA (1980). Virus-like particles of yeast. Annu Rev Microbiol 34:49-68.

Courey AJ, Tjian R (1988). Analysis of Spl *in vivo* reveals multiple transcriptional domains, including a novel glutamine-rich activation motif. Cell 55:887-898.

Edery I, Hümbelin M, Darveau A, Lee KAW, Milburn S, Hershey JWB,Trachsel H, Sonenberg N (1983). Involvement of eukaryotic initiation factor 4A in the cap recognition process. J Biol Chem 258:11398-11403.

Edery I, Sonenberg N (1985). Cap-dependent RNA splicing in a HeLa nuclear extract. Proc Natl Acad Sci USA 82:7590-7594.

Edery I, Altmann M, Sonenberg N (1988). High level synthesis in *Escherichia coli* of functional eukaryotic initiation factor eIF-4E and affinity purification using a simplified cap-analog resin. Gene (Amst) 74:517-525.
El-Sheirbeini M, Tipper DJ, Mitchell DJ, Bostian KA (1984). Virus-like particle capsid proteins encoded by different L double-stranded RNAs of *Saccharomyces cerevisiae*:their roles in maintenance of M double-stranded killer plasmids. Mol Cell Biol 4:2818-2827.
Etchison D, Hansen J, Ehrenfeld E, Edery I, Sonenberg N, Milburn SC, Hershey JWB (1984). Demonstration *in vitro* that eucaryotic initiation factor 3 is active but that a cap-binding protein complex is inactive in poliovirus infected HeLa cells. J Virol 51:832-837.
Furuichi Y, LaFiandra A, Shatkin AJ (1977). 5' terminal structure and mRNA stability. Nature (London) 266:235-239.
Georgiev O, Mous J, Birnsteil M (1984). Processing and nucleo-cytoplasmic transport of histone gene transcripts. Nucl Acids Res 12:8539-8551.
Goyer C, Altmann M, Trachsel H, Sonenberg N (1989). Identification and characterization of cap-binding proteins from yeast. J Biol Chem 264:7603-7610.
Green MR, Maniatis T, Melton DA (1983). Human $\beta$-globin pre-mRNA synthesized *in vitro* is accurately spliced in *Xenopus* oocyte nuclei. Cell 32:681-694.
Grifo JA, Tahara SM, Morgan MA, Shatkin AJ, Merrick WC (1983). New initiation factor activity required for globin mRNA translation. J Biol Chem 258:5804-5810.
Hamm J, Mattaj IW (1990). Monomethylated cap structures facilitate RNA export from the nucleus. Cell 63:109-118.
Hart RP, McDevitt MA, Nevins JR (1985). Poly(A) site cleavage in HeLa nuclear extract is dependent on downstream sequences. Cell 43:677-683.
Haynes SR (1992). The RNP motif protein family. New Biol 4:421-429.
Hershey JWB (1991) Translational control in mammalian cells. Ann Rev Biochem 60:717-755.
Hoffman DW, Query CC, Golden BL, White SW, Keene JD (1991). RNA-binding domain of the A protein component of the U1 small nuclear ribonucleoprotein analyzed by NMR spectroscopy is structurally similar to ribosomal protein. Proc Natl Acad Sci 88:2495-2499.
Jaramillo M, Pelletier J, Edery I, Nielsen PJ, Sonenberg N (1991). Multiple mRNAs encode the murine translation initiation factor eIF-4E. J Biol Chem 266:10446-10451.
Keene JD, Query CC (1991). Nuclear RNA-binding proteins. Prog Nucl Acid Res Mol Biol 41:179-201.
Kenan DJ, Query CC, Keene JD (1991). RNA recognition: towards identifying determinants of specificity. Trends Bioch Sci 16:214-220.
Konarska MM, Padgett RA, Sharp PA (1984). Recognition of the cap structure in splicing *in vitro* of mRNA precursors. Cell 38:731-736.
Lanker S, Mueller PP, Altmann M, Goyer C, Sonenberg N, Trachsel H (1992). Interaction of the eIF-4F subunits in the yeast *Saccharomyces cerevisiae*. J Biol Chem 267:21167-21171.
Lee KAW, Edery I, Sonenberg N (1985). Isolation and structural characterization of cap-binding proteins from poliovirus-infected HeLa cells. J Virol 54:515-524.
Lejbkowicz F, Goyer C, Darveau A, Neron S, Lemieux R, Sonenberg N (1992). A fraction of the mRNA 5' cap binding protein, eIF-4E, localizes to the nucleus. Proc Natl Acad Sci 89:9612-9616.
Mattaj IW (1989). A binding consensus: RNA-binding-protein interactions in splicing, snRNPs, and sex. Cell 57:1-3.
Mermod M, O'Neill EA, Kelly TJ, Tjian R (1989). The proline-rich transcriptional activator of CTF/NF-I is distinct from the replication and DNA binding domain. Cell 58:741-753.
Nemeroff M, Bruenn J (1987). Initiation by the yeast viral transcriptase *in vitro*. J Biol Chem 262:6785-6787.
Olmsted JB (1981). Affinity purification of antibodies from diazotized paper blots of heterogeneous protein samples. J Biol Chem 256:11955-11957.
Pelletier J, Sonenberg N (1985). Photochemical cross-linking of cap-

binding proteins to eukaryotic mRNAs: Effect of mRNA 5' secondary structure. Mol Cell Biol 5:3222-3230.
Ray BK, Brendler TG, Adhya A, Daniel-McQueen S, Kelvin-Miller J, Hershey JWB, Grifo JA, Merrick WC, Thach RE (1983). Role of mRNA competition in regulating translation:further characterization of mRNA discriminatory initiation factors. Proc Natl Acad Sci USA. 80:663-667.
Rhee SK, Icho T, Wickner RB (1989). Structure and nuclear localization signal of the SK13 antiviral protein of *Saccharomyces cerevisiae*. Yeast 5:149-158.
Rhoads RE (1988). Cap recognition and entry of mRNA into the protein synthesis initiation cycle. Trends Biochem Sci 13:52-56.
Ridley SP, Sommer SS, Wickner RB (1984). Superkiller mutations in *Saccharomyces cerevisiae* suppress exclusion of $M_2$ double-stranded RNA by L-A-NH and confer cold sensitivity in the presence of M and L-A-NH. Mol Cell Biol 4:761-770.
Rychlik W, Domier LL, Gardner PR, Hellmann GM, Rhoads RE (1987). Amino acid sequence of the mRNA cap-binding protein from human tissues. Proc Natl Acid Sci USA 84:945-949.
Safer B (1989). Nomenclature of initiation, elongation and termination factors for translation in eucaryotes. Recommendations 1988. Eur J Biochem 186:3384-3386.
Sarkar G, Edery I, Gallo R, Sonenberg N (1984). Preferential stimulation of rabbit α globin mRNA translation by a cap binding protein complex. Biochem Biophys Acta 783:122-129.
Shatkin AJ (1985). mRNA cap binding proteins:essential factors for initiating translation. Cell 40:223-224.
Sommer S, Wickner RB (1982). Yeast L dsRNA consists of at least three distinct RNAs;evidence that the non-mendelian genes [HOK], [NEX], and [EXL] are on one of these RNAs. Cell 31:429-441.
Sommer S, Wickner RB (1987). Gene disruption indicates that the only essential function of the SK18 chromosomal gene is to protect *Saccharomyces cerevisiae* from viral cytopathology. Virology 157:252-256.
Sonenberg N, Morgan MA, Merrick WC, Shatkin AJ (1978). A polypeptide in eukaryotic initiation factors that cross-links specifically to the 5'-terminal cap in mRNA. Proc Natl Acad Sci USA 74:4288-4292.
Sonenberg N (1987). Translation control by poliovirus. In "Advances in virus research". Academic Press (ed:K.Maramorosch). 33:175-204.
Sonenberg N (1988). Cap-binding proteins of eukaryotic messenger RNA: Functions in initiation of translation. Prog Nucl Acid Res Mol Biol 35:173-207.
Tahara SM, Morgan MA, Shatkin AJ (1981). Two forms of $m^7G$-cap binding protein with different effects on capped mRNA translation in extracts of uninfected and poliovirus-infected HeLa cells. J Biol Chem 256:7691-7694.
Tobin GJ, Young DC, Flanegan JB (1989). Self-catalyzed linkage of poliovirus terminal protein VPg to poliovirus RNA. Cell 59:511-519.
Wickner RB (1992). Double-stranded and single-stranded RNA viruses of *Saccharomyces cerevisiae*. Annu Rev Microbiol 46:347-375.
Zamore PD, Patton JG, Green MR (1990). Cloning and domain structure of the mammalian splicing factor U2AF. Nature 355:609-614.

# INDEX


*A*

*a*-factor export 339
α-factor 359
ABC proteins 307,339
Abf1p (See ABF1)
*ABF1* gene 31,32, 40,41
    Binding site 40,47
    *K. lactis* 47,48
Apocytochrome c 235
Acidic ribosomal proteins 67-77
    *Artemia salina* 71,74
    Cytoplasmic pool 74
    Dephosphorylation 76,77
    Diversity 68
    Phosphorylation 72-74
    *S. pombe* 69
    *Trypanosoma cruzi* 68
3-aminotriazole 5,149
*ANB1* gene 124
Antisuppressors 85, 96-97
*ASU9* gene 96
ATG codon 5
ATPases 132-134
ATP dependence 193
ATP transport 307
    DIDS inhibition 307
    $B2_2$ ATP inhibition 307

*B*

Belly dancer 405
Brome Mosaic Virus
    RNA-4 113,115, 117

*C*

*Candida albicans*
    $tRNA_{CAG}$ 111-119
5' cap 20,131, 395
    Role in translation 20
Cap binding complex 131
Cap binding proteins 395-404
Casein kinase II 73
*Cat* gene
    mRNA abundance 24
    Protein 25
    Reporter gene 21,22
*CDC34* gene 389
*CDC48* gene 13
Cell-free system
    Reticulocyte 26,27
    Yeast 26,27
*Chlamydomonas*
    Chloroplast translation 162
Chloroplast translation 162
Codon bias 369
Codon context 173

Context effects
    Initiation codons 19
    Termination codons 171,172
Cotranslational translocation
    Endoplasmic reticulum 283
*COX2* gene 158,159, 169
*COX3* gene 158,170
Cross-linking 203,315, 395
CUG codon
    Miscoding 118-119
*CYC1* gene 22
Cycloheximide 9,82, 159
Cytochrome $b_2$ 223
Cytochrome *c* 213,235
Cytochrome *c* heme lyase 213,235
Cytochrome oxidase 203

*D*

*DBP2* gene 8
D-E-A-D box 131-134, 136,137
D-E-A-H family 134
Dilsulfide exchange enzyme 349
DnaJ 193,245
Dnak 245
Dosage compensation 14
ds RNA virus 395

*E*

eIF's (See Initiation Factors)
Elongation factors
    EF-1α 84,97, 98
    Mitochondrial 188
Endoplasmic reticulum 283-292, 293-305, 307-313

Endoplasmic reticulum (cont) 315-324
Lumen 349
*ESU1* gene 103,108
[eta] factor 91
Exocytosis 325

*F*
FLAG epitope 7,8
5-fluorotryptophan 145
Frameshift allele 7

*G*
*GAL* promotor 126,127
*GAL4* gene 8,245
*GCD5* gene 143-154
Autoregulation 153,154
Cloning 145,146
*lacZ* fusion 150,151
Mutant allele 146-152
*GCN4* gene 143,144
Derepression 149
*lacZ* fusion 149
GDP dissociation inhibitor 325
Gene disruption 69
Globin mRNA
*In vitro* translation 116,177
Golgi 325-338
GTPase 283
GTPase activating protein 325
GTP-binding protein 325

*H*
Half-mers 127
Heat shock 21,48
Histidinyl-tRNA synthetase 145
Heme 213,235
Holocytochrome c synthase 235
*HSP26* gene 21,22
118
Hsp60p 203,213
Hsp70p 193,203
213
Hypusine 123,126

*I*
*IMP1* gene 213
Import receptor 193
*In vitro* translation 113-118
Indoleglycerol phosphate
Synthase 369
Initiation factor 123
Initiation factors
eIF-2α 19,144
eIF-3B 145
eIF-4A 131-140
eIF-4B 132
eIF-4F 395
eIF-5A 123-128
Initiator tRNA 112
Insertional mutagenesis 187
Insulin-like growth factor-1 359
Internal ribosomal entry site 20
*ISP42* gene 203
*ISU1* gene 103-105
107,108

*K*
*KAR2* gene 307
Kex2p 329,359
*Kluyveromyces lactis*
ABF1p 47,48
Ribosomal protein genes 41-44
*Kluyveromyces marxianus*
Ribosomal protein genes 41-44
*KRS1* gene 146

*L*
*LAC9* gene 45
*lacZ* fusions
*GCD5* 150-151
*GCN4* 149
*PET54* 161
*PET494* 161
Lipid kinase 325
LLrep3 93,94
Lysyl-tRNA synthetase
*gcd5* mutant 146-148
Gene 143-145

*M*
*MBR1* gene 186
*MBR2* gene 186
Mdr homologies 339
Membrane traffic 325
Membrane translocation 213
Methionyl-puromycin formation 123
Assay 125
Synthesis 127
Microsomes 307,315
Mini-Mu transposon 187
Mitochondrial translation 157-161
167-174
177-191
*In vitro* system 167
Initiation 168-171
177
Mutants 168
Termination 171-174

Mitochrondrial protein import 193-202
203-212
213-222
223-233
235-246
Mitochondrion
Inner membrane 203,213
Outer membrane, 193,213
Intermembrane space 213,223
*MNS1* gene 8
Mobile genetic element 379
Modular expression system 21
Molecular chaperone 193,
359-368
*MOM38* gene 193
*MRF1* gene 171-172
mRNA
Half-life 2,3,6
Secondary structure 171
Splicing 3
mRNA decay
Nonsense-mediated 1-10
Model 9,10
mRNA-specific activators
Mitochondrial translation 157-159
160-181
mRNA stability 170
Multicopy suppression 187
*MRS2* gene 185
*MRS3* gene 185
Multifunctional protein 349,369
Multiple drug resistance 339

*N*
*NAM1* gene 185
NLS-binding protein 245
Nonsense-mediated mRNA decay 1-10
Nonsense mutations 1,6
Nonsense suppression 172
*NPL* genes 245
*NSP* genes 269
Nuclear localization sequences 245,251
Nuclear pore complex 257,269
Nuclear suppressors of mitochondrial mutations 814,185
Nuclear transport 245-255
257-267
269-281
Nucleolus 257
Nucleoporins 269
Nucleotide exchange protein 325
Nucleus 245,247
269

*O*
Omnipotent suppression 8
Model 107-108
Omnipotent suppressors 81-88
92-95
101-108

*P*
p220 395
p40 168-171
Binding sites 169
Translational activator 169
Translational repressor 170
p-glycoprotein 339
*PDI1* gene 349
Peptide binding activity 349
Peptidyl transferase centre 181
*PET54* gene 160
*PET111* gene 159
*PET122* gene 160
*PET494* gene 160
*PGK1* gene
Mini-gene 4-6
mRNA decay 2
Nonsense mutations 1-6
Phosphatases 77,97
Phosphoserine 118
*PNM2* gene 86
Poly(A) polymerase 12
Polyadenylation
Role in translation 20
*ssm* mutants 15
Polymerase module 17
Post-translational processing 379
Post-translational translocation
Mitochondrion 193
Endoplasmic reticulum 283
Precursor proteins 213,223
Presequence peptide 203,223
Pro-region 359
Pro-sequence 359
Protein degradation 235
389-393
Protein disulfide isomerase 349-358
Proteins domains 369
Protein folding 93,235
349
359-368
369-378
Protein kinases 77,325
*GCN2* gene 144,145
154

Protein sorting
- Mitochondrion 203,213 223
- Missorting mutants 223
- Vacuole 325

Protein splicing 379-388
[psi] factor 86,87 91,106 108
- Interaction with Omnipotent suppressors 86,87

**R**

rab 325
*RAD6* gene 389
*ram* mutants 93
Rap1p/*RAP1* gene 39
- Ribosomal protein genes 32
- *K. lactis* 46,47

Reb1p
- Binding sites 31

Release factor
- Bacterial 173,174
- Mitochondrial 171-174

Reticulocyte lysate 26,115
Ribonucleoprotein 257
Ribosomal proteins 257
- *B. stearothermophilus* S5 94
- Co-ordinate synthesis 30,39
- Evolution 63-64
- *Kluyveromyces* species 41-46
- Regulation of synthesis 32,33 39-51
- RNA splicing 33
- Transcription of genes 30

Ribosomal proteins, *E. coli*
- L7 60-63 67
- L10 62,63
- L12 60-63 67

Ribosomal proteins, mitochondrial 158,182
Ribosomal proteins, yeast
- Acidic 67-77
- L32 33
- PO 67-68
- S4 93,94
- S13 95

Ribosomal proteins, rat
- L38 15

Ribosomal RNA 257
Ribosomal frameshifting 119
Ribosomal RNA genes
- Enhancer 29,31
- Mini-genes 32
- Mitochondrial mutants 179,181
- Transcription 29,31 32

Ribosome
- Global regulation 33,34
- Regulation of synthesis 29-38

Ribosome assembly 257
Ribosome-associated factor 7
*RNA14* gene 11-17
- Multicopy suppressors 12-15
- Protein 12

*RNA15* gene 11-17
- Protein 12
- Suppressors 13

RNA binding domain 11
RNA helicase 8,131 132
RNA maturase 172
RPG box
- *K. lactis* 44,46
- Ribosomal protein genes 39,94

RNA transport 257
5S RNA 257
7S RNA 257

**S**

*S. pombe*
- eIF-4a gene 139-140

*SAL6* gene 96
Scanning model 19,20
*Sec4* mutant 325
*Sec8* mutant 325
*Sec15* mutant 325
*Sec61* mutant 293
*Sec62* mutant 293
*Sec63* mutant 245,293
*SecY* mutant 293
*Sec* complex 293
*Sec* mutants 245,293 315,325, 329
Second-site mutations 179-181
Secretion 283-292 293-305 307-313 315-324 325-328 339-348 349,359
Signal binding protein 315
Signal recognition particle (SRP) 255,281
Signal sequence 315,359
*SMT1* gene 182,183

Spacer protein 379
SRP receptor 283
*SSA1* gene 193
*SSH* genes 339
*SSM1* gene 14
*SSM2* gene 15
*SSM3* gene 13
*SSM4* gene 12,13
*STE6* gene 339
Stationary phase 75
Stem-loop structures 20-26
Stop codons
Context, mitochondria 172,173
Stop-transfer sequence 203
SUP35p 81-88
92
Interactions 86,87
Functional domains 84.85
Ribosome association 82,83
*SUP44* gene 92-94
SUP45p 81,92
86-88
*SUP46* gene 92,94
95
Interactions 86.87
Ribosome association 82,83
*SUP111* gene 101
Suppressor mutations 137
Synthetic lethality 269

***T***

*TEF1* gene 21,118
*TEF2* gene 102,104
Translation 395-404
*TIF1* gene 132-134
137
*TIF2* gene 132-134
137
*TIF51A* gene 123,124
125
*TIF51B* gene 123,124
125
*TIF4631* gene 395
*TIF4632* gene 395
*TFP1* gene 379
Traffic ATPases 339
Translational arrest 283
Release 283
Translational efficiency 25
Translational elongation 369-378
Translational fidelity 87,88
101,188
Gene dosage effects 97,98
Translational reinitiation 5,8,144
Translocase 293
Translocation complex 313
Translocation mechanism 313
Transport competence 193
tRNA
*C. albicans* 111-119
Maturation 187
Mitochondrial mutant 180-181
Suppressor 96,111
$tRNA_{CAG}$ 112-113
*TRP3* gene 369
*Trypansoma cruzi*
Acidic ribosomal proteins 68
Tryptophan biosynthesis 369
Ty transposition technique 13

***U***

*UBC* genes 389
Ubiquitin fusion 126,127
Universal Code System
for ribosomal proteins 54
5' untranslated regions
Mitochondrial mRNA 169-170
*cat* RNA 21,22
*UPF1* gene 1,3
Protein 7,8
Ubiquitin 389-393
Ubiquitin-activating (E1) enzyme 389
Ubiquitin-conjugating (E2) enzymes 389

***V***

V-ATPase 379
Vacuolar $H^+$-ATPase 379
Vacuole 323-338
*VAR1* gene 179
Vesicular transport 323
Viral coat protein 395
*VMA1* gene 379
*VMAI* derived endonuclease 379
*VPS* genes 325

***W***

Welder 410

***Y***

*YDJI* gene 193
Ypt 325


Printing: Druckhaus Beltz, Hemsbach
Binding: Buchbinderei Schäffer, Grünstadt

# NATO ASI Series H

Vol. 21: **Cellular and Molecular Basis of Synaptic Transmission.** Edited by H. Zimmermann. 547 pages. 1988.

Vol. 22: **Neural Development and Regeneration. Cellular and Molecular Aspects.** Edited by A. Gorio, J. R. Perez-Polo, J. de Vellis, and B. Haber. 711 pages. 1988.

Vol. 23: **The Semiotics of Cellular Communication in the Immune System.** Edited by E.E. Sercarz, F. Celada, N.A. Mitchison, and T. Tada. 326 pages. 1988.

Vol. 24: **Bacteria, Complement and the Phagocytic Cell.** Edited by F. C. Cabello und C. Pruzzo. 372 pages. 1988.

Vol. 25: **Nicotinic Acetylcholine Receptors in the Nervous System.** Edited by F. Clementi, C. Gotti, and E. Sher. 424 pages. 1988.

Vol. 26: **Cell to Cell Signals in Mammalian Development.** Edited by S.W de Laat, J.G. Bluemink, and C.L. Mummery. 322 pages. 1989.

Vol. 27: **Phytotoxins and Plant Pathogenesis.** Edited by A. Graniti, R. D. Durbin, and A. Ballio. 508 pages. 1989.

Vol. 28: **Vascular Wilt Diseases of Plants. Basic Studies and Control.** Edited by E. C. Tjamos and C. H. Beckman. 590 pages. 1989.

Vol. 29: **Receptors, Membrane Transport and Signal Transduction.** Edited by A. E. Evangelopoulos, J. P. Changeux, L. Packer, T. G. Sotiroudis, and K.W.A. Wirtz. 387 pages. 1989.

Vol. 30: **Effects of Mineral Dusts on Cells.** Edited by B.T. Mossman and R.O. Begin. 470 pages. 1989.

Vol. 31: **Neurobiology of the Inner Retina.** Edited by R. Weiler and N.N. Osborne. 529 pages. 1989.

Vol. 32: **Molecular Biology of Neuroreceptors and Ion Channels.** Edited by A. Maelicke. 675 pages. 1989.

Vol. 33: **Regulatory Mechanisms of Neuron to Vessel Communication in Brain.** Edited by F. Battaini, S. Govoni, M.S. Magnoni, and M. Trabucchi. 416 pages. 1989.

Vol. 34: **Vectors asTools for the Study of Normal and Abnormal Growth and Differentiation.** Edited by H. Lother, R. Dernick, and W. Ostertag. 477 pages. 1989.

Vol. 35: **Cell Separation in Plants: Physiology, Biochemistry and Molecular Biology.** Edited by D. J. Osborne and M. B. Jackson. 449 pages. 1989.

Vol. 36: **Signal Molecules in Plants and Plant-Microbe Interactions.** Edited by B.J.J. Lugtenberg. 425 pages. 1989.

Vol. 37: **Tin-Based Antitumour Drugs.** Edited by M. Gielen. 226 pages. 1990.

# NATO ASI Series H

Vol. 38: **The Molecular Biology of Autoimmune Disease.**
Edited by A.G. Demaine, J-P. Banga, and A.M. McGregor. 404 pages. 1990.

Vol. 39: **Chemosensory Information Processing.**
Edited by D. Schild. 403 pages. 1990.

Vol. 40: **Dynamics and Biogenesis of Membranes.**
Edited by J. A. F. Op den Kamp. 367 pages. 1990.

Vol. 41: **Recognition and Response in Plant-Virus Interactions.**
Edited by R. S. S. Fraser. 467 pages. 1990.

Vol. 42: **Biomechanics of Active Movement and Deformation of Cells.**
Edited by N. Akkas. 524 pages. 1990.

Vol. 43: **Cellularand MolecularBiologyof Myelination.**
Edited by G. Jeserich, H. H. Althaus, and T. V. Waehneldt. 565 pages. 1990.

Vol. 44: **Activation and Desensitization of Transducing Pathways.**
Edited by T. M. Konijn, M. D. Houslay, and P. J. M. Van Haastert. 336 pages. 1990.

Vol. 45: **Mechanism of Fertilization: Plants to Humans.**
Edited by B. Dale. 710 pages. 1990.

Vol .46: **Parallels in Cell to Cell Junctions in Plants and Animals.**
Edited by A. W Robards, W. J . Lucas, J . D. Pitts, H . J . Jongsma, and D. C. Spray. 296 pages. 1990.

Vol. 47: **Signal Perception andTransduction in Higher Plants.**
Edited by R. Ranjeva and A. M. Boudet. 357 pages. 1990.

Vol. 48: **Calcium Transport and Intracellular Calcium Homeostasis.**
Edited by D. Pansu and F. Bronner. 456 pages. 1990.

Vol. 49: **Post-Transcriptional Control of Gene Expression.**
Edited by J. E. G. McCarthy and M. F. Tuite. 671 pages. 1990.

Vol. 50: **Phytochrome Properties and Biological Action.**
Edited by B. Thomas and C. B. Johnson. 337 pages. 1991.

Vol. 51: **Cell to Cell Signals in Plants and Animals.**
Edited by V. Neuhoff and J. Friend. 404 pages. 1991.

Vol. 52: **Biological Signal Transduction.**
Edited by E. M . Ross and K . W. A. Wirtz. 560 pages. 1991.

Vol. 53: **Fungal Cell Wall and Immune Response.**
Edited by J. P. Latge and D. Boucias. 472 pages. 1991.

Vol. 54: **The Early Effects of Radiation on DNA.**
Edited by E. M. Fielden and P. O'Neill. 448 pages. 1991.

Vol. 55: **The Translational Apparatus of Photosynthetic Organelles.**
Edited by R. Mache, E. Stutz, and A. R. Subramanian. 260 pages. 1991.

# NATO ASI Series H

Vol. 56: **Cellular Regulation by Protein Phosphorylation.**
Edited by L. M. G. Heilmeyer, Jr. 520 pages. 1991.

Vol. 57: **MolecularTechniques in Taxonomy.**
Edited by G . M . Hewitt, A. W. B. Johnston, and J. P. W. Young .
420 pages. 1991.

Vol. 58: **Neurocytochemical Methods.**
Edited by A. Calas and D. Eugene. 352 pages. 1991.

Vol. 59: **Molecular Evolution of the Major Histocompatibility Complex.**
Edited by J. Klein and D. Klein. 522 pages. 1991.

Vol. 60: **Intracellular Regulation of Ion Channels.**
Edited by M. Morad and Z. Agus. 261 pages. 1992.

Vol. 61: **Prader-Willi Syndrome and Other Chromosome 15q Deletion Disorders.**
Edited by S. B. Cassidy. 277 pages. 1992.

Vol. 62: **Endocytosis. From Cell Biology to Health, Disease and Therapie.**
Edited by P. J. Courtoy. 547 pages. 1992.

Vol. 63: **Dynamics of Membrane Assembly.**
Edited by J. A. F. Op den Kamp. 402 pages. 1992.

Vol. 64: **Mechanics of Swelling. From Clays to Living Cells and Tissues.**
Edited by T. K. Karalis. 802 pages. 1992.

Vol. 65: **Bacteriocins, Microcins and Lantibiotics.**
Edited by R. James, C. Lazdunski, and F. Pattus. 530 pages. 1992.

Vol. 66: **Theoretical and Experimental Insights into Immunology.**
Edited by A. S. Perelson and G. Weisbuch. 497 pages. 1992.

Vol. 67: **Flow Cytometry. New Developments.**
Edited by A. Jacquemin-Sablon. 1993.

Vol. 68: **Biomarkers. Research and Application in the Assessment of Environmental Health.** Edited by D. B. Peakall and L. R. Shugart.
138 pages. 1993.

Vol. 69: **Molecular Biology and its Application to Medical Mycology.**
Edited by B. Maresca, G. S. Kobayashi, and H. Yamaguchi. 271 pages.
1993.

Vol. 70: **Phospholipids and Signal Transmission.**
Edited by R. Massarelli, L. A. Horrocks, J. N. Kanfer, and K. Löffelholz.
448 pages. 1993.

Vol. 71: **Protein Synthesis and Targeting in Yeast.**
Edited by A. J. P. Brown, M. F. Tuite, and J. E. G. McCarthy. 425 pages.
1993.

Vol. 72: **Chromosome Segregation and Aneuploidy.**
Edited by B. K. Vig. 425 pages. 1993.

# NATO ASI Series H

Vol. 73: Human Apolipoprotein Mutants III. In Diagnosis and Treatment.
Edited by C. R. Sirtori. 302 pages. 1993.

Vol. 74: Molecular Mechanisms of Membrane Traffic.
Edited by D. J. Morré, K. E. Howell, and J. J. M. Bergeron. 429 pages. 1993.

Vol. 75: Cancer Therapy. Differentiation, Immunomodulation and Angiogenesis.
Edited by N. D'Alessandro, E. Mihich, L. Rausa, H. Tapiero, and T. R. Tritton.
299 pages. 1993.

MIX
Papier aus verantwortungsvollen Quellen
Paper from responsible sources
FSC® C105338

If you have any concerns about our products,
you can contact us on
**ProductSafety@springernature.com**

In case Publisher is established outside the EU,
the EU authorized representative is:
**Springer Nature Customer Service Center GmbH**
**Europaplatz 3, 69115 Heidelberg, Germany**

Printed by Libri Plureos GmbH
in Hamburg, Germany